Spectroscopy

Spectroscopy

edited by

Preeti Gupta | S. S. Das | N. B. Singh

Published by

Jenny Stanford Publishing Pte. Ltd.
101 Thomson Road
#06-01, United Square
Singapore 307591

Email: editorial@jennystanford.com
Web: www.jennystanford.com

British Library Cataloguing-in-Publication Data
A catalogue record for this book is available from the British Library.

Spectroscopy

ISBN 978-981-4968-32-4 (Hardcover)
ISBN 978-1-003-41258-8 (eBook)

Contents

Preface xv

1. Introduction to Spectroscopy **1**

Preeti Gupta, S. S. Das, and N. B. Singh

1.1 Introduction 1
1.2 Interaction of Electromagnetic Radiation with Matter 2
1.2.1 Wave Properties of Electromagnetic Radiation 2
1.3 Electromagnetic Spectral Range 5
1.3.1 Radiofrequency Region 7
1.3.1.1 Microwave region 7
1.3.1.2 Infrared region 7
1.3.1.3 Visible and ultraviolet region 8
1.3.1.4 X-ray region 8
1.3.1.5 γ-ray region 8
1.4 Born–Oppenheimer Approximation 8
1.5 Energy Levels in Molecules and Quantization of Energy 9
1.6 Mechanism of Absorption and Emission 12
1.7 Transition Probability and Selection Rule 12
1.7.1 Transition Moment and Transition Moment Integral 13
1.7.2 Electronic Selection Rules 15
1.7.3 Vibrational Selection Rules 16
1.7.4 Rotational Selection Rules 16
1.8 Line Width and Line Shapes 17
1.8.1 Spectral Line Broadening and Line Shapes 17
1.8.2 Local Effects Broadening 18
1.8.3 Natural Line Broadening 18
1.8.4 Doppler Broadening 18
1.8.5 Pressure (Collision Broadening) 18

1.8.6 Broadening due to Inhomogeneous Local Effects 18
1.8.7 Broadening due to Non-local Effects 19
1.9 Fourier Transform 19

2. Nuclear Magnetic Resonance Spectroscopy 23

Preeti Gupta, S. S. Das, and N. B. Singh

2.1 Introduction 23
2.2 Nature of Spinning Nucleus 24
2.3 Effect on the Nucleus of an External Magnetic Field 24
2.4 Precessional Motion of Nucleus 25
2.5 Precessional Frequency: Larmor Frequency 26
2.6 Energy Transitions and the Origin of NMR 27
2.7 Theory of NMR 27
2.7.1 Nuclear Spin and Magnetic Moment and Interaction Between Nuclear Spin and Magnetic Field 27
2.7.2 Population of Nuclear Energy Levels 32
2.7.3 Saturation and Relaxation 33
2.8 Concept of Chemical Shift 35
2.9 Types of Shielding 40
2.10 Factors Affecting Chemical Shift 41
2.10.1 Effect of Electronegativity (Inductive Effect) 41
2.10.2 Van der Waal's Deshielding 43
2.10.3 Magnetic Anisotropy 43
2.10.3.1 Alkenes 44
2.10.3.2 Alkynes 45
2.10.3.3 Carbonyl compounds 45
2.10.3.4 Aromatic compounds 46
2.10.3.5 Effect of hydrogen bonding 47
2.10.3.6 Effect of solvents (types, concentration, temperature, and hydrogen bonding) 48
2.10.3.7 Solvents used in NMR 49
2.11 Spin–Spin Coupling 50
2.11.1 The Splitting of Signals in Proton NMR Spectra 50

2.11.2 Theory of Spin–Spin Coupling 52
2.12 The General Rule of Spin–Spin Splitting: The $(n + 1)$ Splitting Rule 56
2.13 Complex NMR Spectra: Breakdown of $(n + 1)$ Splitting Rule 58
2.13.1 Multiplicative Splitting 59
2.13.2 Non-first-order Spectra: Breakdown of the $(n + 1)$ Rule 61
2.14 The Coupling Constant (J) 64
2.14.1 Factors Influencing Coupling Constant, J, and Various Types of Couplings 67
2.14.1.1 Geminal coupling 67
2.14.1.2 Vicinal coupling 68
2.14.1.3 Long-range proton–proton coupling 71
2.14.1.4 Heteronuclear coupling 71
2.14.1.5 ^{1}H–^{13}C spin–spin coupling 72
2.14.1.6 Deuterium coupling 72
2.15 Proton Exchange Reaction in NMR Spectrum of Alcohol 73
2.16 Restricted Rotation 75
2.17 Deuterium Exchange in NMR Spectroscopy 77
2.18 Simplification of Complexities of Proton NMR Spectra 78
2.18.1 By Increase in the Field Strength 78
2.18.2 By Spin Decoupling Method 78
2.18.3 By Using Chemical Shift Reagents 80
2.18.4 NMR of Nuclei Other Than Protons 82
2.18.4.1 ^{13}C NMR 82
2.18.4.2 ^{19}F NMR spectroscopy 83
2.18.4.3 ^{31}P NMR 86
2.19 Instrumentation of NMR Spectroscopy 86
2.20 Nuclear Overhauser Effect 89
2.21 FT-NMR 90

3. Carbon-13 NMR **93**

Tamanna Yakub, Bhawana Jain, Anupama Asthana, Ajaya K. Singh, and Md. Abu Bin Hasan Susan

3.1 Introduction 93

3.2 Comparison with ^{1}H NMR Spectroscopy 96
3.2.1 Characteristic Features of ^{13}C NMR Spectra 97
3.2.2 Referencing ^{13}C NMR Spectra 97
3.3 Instrumentation 99
3.4 Interpretation of ^{13}C Spectra 100
3.4.1 Broadband Decoupling 101
3.4.2 Off-Resonance Decoupling 101
3.5 Chemical Shift 103
3.5.1 Origin 103
3.5.2 Scale 104
3.5.3 Effects on sp^3 Carbons 104
3.5.4 α-Substituent Effects 105
3.5.5 Heavy-Atom α-Effect 106
3.5.6 α-Effect of Triple Bonds 106
3.5.7 α-Effect of Double Bonds 107
3.5.8 β-Substituent Effects 107
3.5.9 γ-Substituent Effects 107
3.5.10 δ-Substituent Effects 109
3.5.11 Membered Rings 109
3.5.12 Effects on sp^2 and sp Carbons 110
3.5.13 Conjugation with π-Acceptors and π-Donors 110
3.5.14 Strongly Charged Systems 110
3.5.15 Carbonyl Groups 112
3.5.16 Conjugation Effects 113
3.5.17 Hydrogen Bond Effects 113
3.6 Application of ^{13}C NMR 116
3.6.1 Identification of Molecules 117
3.6.2 Soil Organic Matter Studies 118
3.6.3 Physical Separations and Structural Characterization of Organic C in Bulk Soils 118

4. Electron Spin Resonance Spectroscopy **123**

Preeti Gupta, S. S. Das, and N. B. Singh

4.1 Introduction 123
4.2 Basic Principles and Origin of ESR Spectra 124

4.3 The Electron g-Factor 127
4.3.1 Factor Affecting g-Factor 128
4.3.2 Determination of g-Values 129
4.4 Saturation and Relaxation in ESR Spectroscopy 130
4.4.1 Spin–Lattice Relaxation Process 130
4.4.2 Spin–Spin Relaxation Process 130
4.5 Splitting of ESR Signals: Hyperfine Interaction 130
4.6 Zero-Field Splitting: Kramer Degeneracy 134
4.7 Isotropic Coupling Constant 137
4.7.1 Isotropic Hyperfine Splitting from Single Set of Equivalent Protons 137
4.7.2 Isotropic Hyperfine Splitting from Multiple Set of Equivalent Protons 138
4.8 Anisotropy and Hyperfine Splitting 140
4.8.1 Anisotropy in Crystals 141
4.8.1.1 Cubic system 141
4.8.1.2 Uniaxial symmetry 142
4.8.1.3 Rhombic symmetry 143
4.9 Instrumentation 144
4.9.1 Source 144
4.9.1.1 Microwave oscillator 145
4.9.1.2 Wave guide and wave meter 145
4.9.1.3 Attenuator 145
4.9.1.4 Isolator 146
4.9.2 Sample Cell/Cavity 146
4.9.3 A Magnet System 146
4.9.4 Detector 146
4.9.5 Recorder 146
4.9.5.1 Working of instrument 146
4.10 Application of ESR Spectroscopy 148

5. Infrared Spectroscopy **151**
Preeti Gupta, S. S. Das, and N. B. Singh
5.1 Introduction 151
5.2 Principle of Infrared Spectroscopy 152
5.3 The Vibration in a Diatomic Molecule 153
5.3.1 Zero-Point Energy 155

5.3.2 Infrared Spectra and Selection Rule 157
5.3.3 Limitations and Drawbacks of Simple Harmonic Oscillator Model 158
5.4 The Diatomic Molecules as Anharmonic Oscillator 159
5.4.1 Zero-Point Energy for Anharmonic Oscillator 161
5.4.2 Infrared Spectra and the Selection Rule 161
5.4.3 Hot Bands 164
5.5 The Vibrating Rotating Diatomic Molecule (Vibrational–Rotational Spectra) 165
5.5.1 Energy Levels of Rotating Vibrators 165
5.5.1.1 Case I: Diatomic molecules performing harmonic oscillations and behaving as a rigid rotator 166
5.5.1.2 Case II: Diatomic molecules performing anharmonic oscillation and behaving as non-rigid rotators 169
5.6 Vibration and Rotation Spectra of Polyatomic Molecule 173
5.6.1 Number of Fundamental Vibrations and Their Symmetry in a Polyatomic Molecule 174
5.6.1.1 Normal modes of vibrations in polyatomic molecules 175
5.6.2 The Influence of Rotation on the Spectra of Polyatomic Molecules 176
5.6.2.1 Linear molecules with parallel vibrations 177
5.6.2.2 Linear molecules with perpendicular vibrations 178
5.6.2.3 Symmetric top molecules with parallel vibrations 179
5.6.2.4 Symmetric top molecules with perpendicular vibrations 180
5.7 Factor Affecting the Infrared Absorption 181
5.7.1 Effect of Force Constant, Bond Strength, and Isotopic Substitution 181

5.7.2 Electronic Effects 183
5.7.2.1 Mesomeric effect 183
5.7.2.2 Inductive effect 185
5.7.3 Bond Angles Effect 187
5.7.4 Effect of Hydrogen Bonding 187
5.7.4.1 Intramolecular hydrogen bonding 190
5.8 Application and Analysis by Infrared Spectra 191
5.8.1 Structural Analysis 191
5.9 The Infrared Spectrometer: Instrumentation and Technique 204
5.9.1 Infrared Light Sources 204
5.9.2 Monochromator 204
5.9.3 Detectors 205
5.9.4 Dispersive IR Spectrometer 207
5.10 Drawback of Double Beam Dispersive Spectrometer 207
5.11 Fourier Transform Infrared Spectrometer 207
5.11.1 Basic Principle and Operation of FTIR Fourier Transform Infrared Spectroscopy 208
5.11.2 Advantages of FTIR 210

6. Microwave Spectroscopy **213**

Preeti Gupta, S. S. Das, and N. B. Singh

6.1 Introduction 213
6.2 Rotational Spectra of Linear Diatomic Molecule 215
6.2.1 The Rigid Linear Diatomic Molecule 215
6.2.2 Description of Rotational Spectra of Rigid Rotating Diatomic Molecule 218
6.3 Various Transitions in the Rotational Spectrum 220
6.4 The Selection Rule 220
6.5 Intensities of Rotational Spectral Lines 221
6.6 Isotopic Substitution 224
6.7 Techniques and Instrumentation 226
6.7.1 The Source and Monochromator 226
6.7.2 Waveguide 226

6.7.3 Sample and Sample Space 227
6.7.4 Detector 227
6.7.5 Spectrum Recorder 227
6.8 Applications of Microwave Spectroscopy 227

7. Raman Spectroscopy **229**
Preeti Gupta, S. S. Das, and N. B. Singh
7.1 Introduction 229
7.2 Quantum Mechanical Concept of Raman Effect 231
7.3 Molecular Polarizability and the Raman Effect 233
7.4 Rotational Raman Spectra 235
7.5 Vibrational Raman Spectra 237
7.6 Rotational–Vibrational Raman Spectra 238
7.7 Instrumentation 240
7.7.1 Light (Radiation Source) 241
7.7.2 A Sample Illumination and Light Collection Optic System 241
7.7.3 Detectors and Data Recording System 241
7.8 Comparison Between IR and Raman Spectroscopic Techniques 242

8. Electronic Spectroscopy **245**
Vinay K. Verma
8.1 Introduction 245
8.2 The Nature of Electronic Excitations 246
8.2.1 $\sigma \rightarrow \sigma^*$ Electronic Transition 248
8.2.2 $n \rightarrow \sigma^*$ Electronic Transition 248
8.2.3 $\pi \rightarrow \pi^*$ Transition 249
8.2.4 $n \rightarrow \pi^*$ Transition 250
8.3 Selection Rule for the Electronic Transition 250
8.4 Beer–Lambert Law 252
8.5 Chromophore and Auxochrome 255
8.6 Effect of Solvent on Electronic Transition (Solvent Effect) 257
8.7 Effect of Conjugation 260

8.8 Woodward–Fisher Rule 264
8.8.1 For Dienes 264
8.8.2 For Carbonyl Compounds: Enones 267
8.8.3 Woodward Rule for Enone System 269
8.8.4 Woodward's Rule for α, β-Unsaturated Aldehydes, Acids, and Esters 271
8.9 Absorption in Aromatic Compounds 272
8.10 Color in Transition Metal Complex 275
8.11 Charge Transfer Transition 276
8.11.1 Ligand to Metal Charge Transfer 276
8.11.2 Metal to Ligand Charge Transfer 277

9. Fluorescence and Phosphorescence Spectroscopy 279

M. Swaminathan

9.1 Introduction 279
9.2 Principle of Fluorescence and Phosphorescence 280
9.2.1 Rules Governing Fluorescence 281
9.2.1.1 Kasha's rule 281
9.2.1.2 Frank–Condon (FC) principle 282
9.3 Instrumentation 282
9.4 General Properties of Fluorescence and Phosphorescence 284
9.4.1 Fluorescence 284
9.4.2 Phosphorescence 285
9.5 Fluorescence Parameters 285
9.5.1 Fluorescence Emission Spectra (λ_{ex}: Fixed) 285
9.5.1.1 Stokes's shift 286
9.6 Fluorescence Excitation Spectra [Wavelength of Emission (λ_{em}): Fixed] 286
9.7 Fluorescence Quantum Yield 287
9.8 Fluorescence Lifetime 288
9.9 Synchronous Fluorescence Spectroscopy 288
9.10 Fluorescence and Structure 290
9.11 Fluorescence and Temperature 291

9.12 Intrinsic and Extrinsic Fluorophores 291
9.13 Applications 292
9.14 Fluorophores: Intrinsic 292
9.14.1 Concentration Determination 292
9.14.2 Study of Excited State Properties 293
9.14.3 Excited State Acidity Constants 293
9.14.3.1 Forster cycle method: Theoretical 294
9.14.3.2 Fluorimetric titration method (pH dependence of fluorescence) 294
9.15 Solvotochromic Shifts 297
9.16 Fluorescence Quenching 298
9.17 Synchronous Fluorescence Spectroscopy Applications 301
9.18 Phosphorescence Spectroscopy 302
9.19 Conclusions 303

10. X-ray Fluorescence and Its Applications for Materials Characterization **307**

Nand Lal Mishra

10.1 Introduction 307
10.2 Production of X-rays 308
10.3 Interaction of X-rays with Matter 310
10.4 Characteristic X-rays 311
10.5 X-ray Fluorescence Principle 312
10.5.1 X-ray Fluorescence Analysis 313
10.6 Instrumentation and Geometrical Consideration 314
10.7 Application of XRF in Materials Characterization 319
10.7.1 Some Examples of XRF Application in Different Areas of Science and Technology 320
10.8 Advancements in XRF Analysis 321

Index 325

Preface

Spectroscopy deals with interactions between matter and electromagnetic radiation and is therefore an important branch of science. This book is intended to help undergraduate and postgraduate students of chemistry, physics, and materials science who are studying in various institutions across the globe. The book will be a reference source for those who are involved in the research and study of spectroscopic techniques. For this purpose, the book is written in such a way that it provides a straightforward introduction to spectroscopy in simple language. The book describes the basic principles and applications of different spectroscopic techniques, such as nuclear magnetic resonance spectroscopy, carbon-13 NMR, electron spin resonance spectroscopy, infrared spectroscopy, microwave spectroscopy, Raman spectroscopy, electronic spectroscopy, fluorescence and phosphorescence spectroscopy, and X-ray fluorescence, and their applications in material characterizations. Explanations and examples that were found to be effective in our courses have been incorporated in this book. The book is user-friendly and easy to read and understand. We highly appreciate the efforts made by the contributors in writing the chapters and also thank the team of Jenny Stanford Publishing for the timely publication of the book. Comments and criticism by readers will highly be appreciated.

Preeti Gupta
S. S. Das
N. B. Singh
January 2023

Chapter 1

Introduction to Spectroscopy

Preeti Gupta,[a] S. S. Das,[b] and N. B. Singh[c]
[a]*Department of Chemistry, DDU Gorakhpur University, Gorakhpur, India*
[b]*Department of Chemistry (ex-professor), DDU Gorakhpur University, Gorakhpur, India*
[c]*Department of Chemistry and Biochemistry, SBSR and RDC, Sharda University, Greater Noida, India*
preeti17_nov@yahoo.com

1.1 Introduction

Spectroscopy may be defined as the study of the interaction of electromagnetic radiation with matter [1–6]. During interaction absorption, emission or scattering of radiation may take place. The structure and chemical properties of the system can easily be understood and studied with the help of atomic and molecular spectroscopic techniques [7–11]. With the help of these techniques, molecular geometries (such as bond distances, bond angles, torsion angles, bond strength, etc.) and molecular symmetry, electronic

Spectroscopy
Edited by Preeti Gupta, S. S. Das, and N. B. Singh

ISBN 978-981-4968-32-4 (Hardcover), 978-1-003-41258-8 (eBook)
www.jennystanford.com

distributions, electron densities, and electrical and magnetic properties all can be studied in detail because there exists a fundamental relationship between the properties of the substance and the interaction of radiation with that substance [12–15]. But before discussing the experimental methods of spectroscopy, it is important to understand the nature of electromagnetic radiation and various interactions.

1.2 Interaction of Electromagnetic Radiation with Matter

The study of the exchange of energy between electromagnetic radiations with matter forms the basis of spectroscopy. There exists a fundamental relationship between the properties of a substance and the interaction of radiation with the substance. In other words, one can say that there is a relationship between the basic chemical structure and electronic configuration of the atom or molecule and the specific absorption or emission of radiation. The mechanism by which the exchange of energy between radiation and matter takes place involves the interaction between the oscillating electric or magnetic field of the radiation with the suitable dipole moment present in the molecule. This interaction will be strong when the dipole moment is oscillating at a similar frequency as that of the radiation [1, 4, 5, 7].

1.2.1 Wave Properties of Electromagnetic Radiation

Electromagnetic radiation may be considered as a simple harmonic wave propagating from a source and traveling in straight lines except when refracted or reflected. The visible light forms a very small part of electromagnetic radiation and its complete spectrum includes microwave, infrared, X-rays, and γ-rays [1]. Electromagnetic radiation is a form of energy that consists of electric and magnetic fields. These electrical and magnetic components oscillate in space at right angles to the direction of wave motion. For example, for linearly polarized light, the oscillating electric field might point in the x direction, the oscillating magnetic field in the y direction and z is the direction of propagation (Figure 1.1).

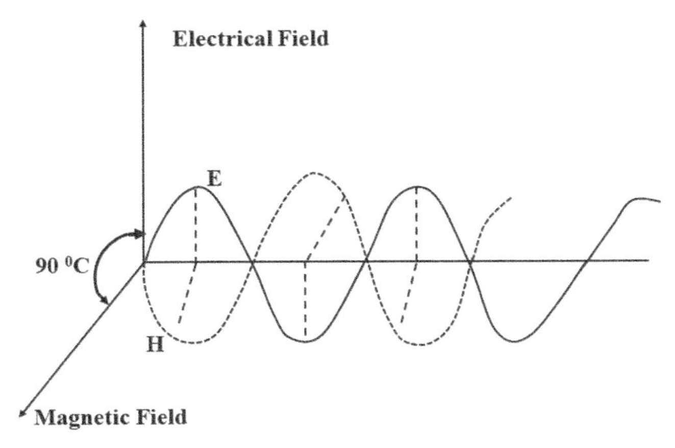

Figure 1.1 The interconnected electric and magnetic fields of electromagnetic radiation.

Any simple harmonic wave has the properties of the sine wave and could be defined by an equation

$$Y = A \sin\theta \tag{1.1}$$

where Y is the displacement and A is its maximum value. θ is the angle that ranges between 0 and 360^0 (0 to 2π radians) (Figure 1.2).

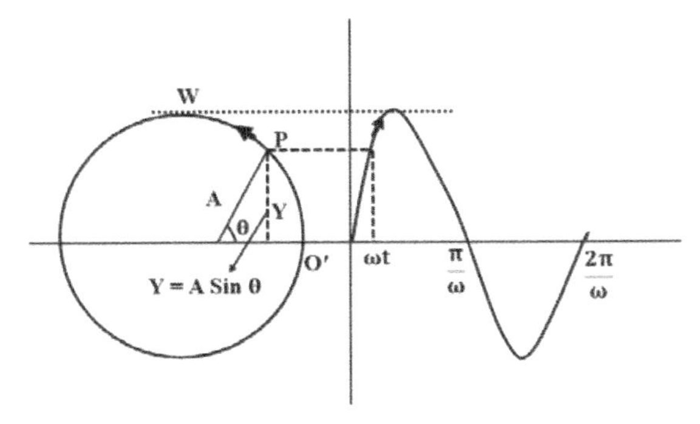

Figure 1.2 The circular motion of a point P in a sine curve with a uniform angular velocity.

As shown in Figure 1.2, if any point P is traveling with a uniform angular velocity ω rad/s in a circular path of radius A and takes t seconds in reaching from the point O′ to point P then it means the point P has covered an angle $\theta = \omega t$ radians. Thus, the sine wave Eq. (1.1) can be written as

$$Y = A \sin\theta = A \sin\omega t \tag{1.2}$$

This displacement Y has been plotted against time t on the right-hand side in Figure 1.2. It is clear from the figure that in time $2\pi/\omega$, the point P returns to the point O′, after completing a full cycle. As P makes further cycles, one may get similar patterns.

In one second the pattern will repeat itself $\omega/2\pi$ times and this is defined as the frequency of the wave, 'υ' (in units of S^{-1} or Hz). Thus, one can write

$$\upsilon = \omega/2\pi \text{ or } \omega = 2\pi\upsilon \tag{1.3}$$

Thus, Eq. (1.3) can be written as

$$Y = A \sin\omega t = A \sin 2\pi\upsilon t \tag{1.4}$$

This is the basic equation of wave motion, which gives the variation of displacement with time. If the wave travels a distance x in time t seconds, then the fundamental relationship between displacement and time is

$$x = ct \tag{1.5}$$

where c is the velocity of propagation of the wave in m/s (In a vacuum, c is equal to the speed of light (c_0), but in general $c = c_0/n$ where n is the index of refraction of the propagation medium). On combining Eqs. (1.4) and (1.2), the final wave equation becomes

$$Y = A \sin 2\pi\upsilon t = A \sin 2\pi x/c \tag{1.6}$$

Further, the wavelength of the radiation, λ, is defined as the distance traveled by the wave during one complete cycle and it is the distance between two consecutive peaks or crests of a traveling wave. Since the frequency is defined as the number of crests that passed by a given point per second or number of cycles per second and frequency both are related to each other by a simple relation.

$$\lambda\upsilon = c \qquad \text{or} \qquad \lambda = c/\upsilon \tag{1.7}$$

Using Eq. (1.7), Eq. (1.6) may finally be simplified as

$$Y = A \sin 2\pi x/\lambda \tag{1.8}$$

Wavelengths, in spectroscopy, are expressed in second units. In the microwave region, λ is measured in centimeters or millimeters, while in IR (infrared) region, it is generally expressed in micrometers, μm ($1\ \mu m = 10^{-6}$ m). In the visible and ultraviolet region, λ is expressed in SI units in nanometer (10^{-9} m) but sometimes it is also expressed in Angstrom units (Å) where 1 Å $= 10^{-10}$ m. But, the proper SI unit of wavelength is nanometer (1 nm $= 10^{-9}$ m $= 10$ Å).

Electromagnetic radiation may also be characterized in terms of wave number, $\bar{\upsilon}$. This is defined as the reciprocal of the wavelength expressed in centimeters.

$$\bar{\upsilon} = 1/\lambda \tag{1.9}$$

Using, wavenumber in Eq. (1.8) we get the following expression:

$$Y = A\sin^2 \pi\bar{\upsilon}.x \tag{1.10}$$

The wavenumber and frequency are related to each other by the following example:

$$\upsilon = c\bar{\upsilon} \tag{1.11}$$

where the proportionality constant c is the velocity of radiation in cm/s (3×10^{10} cm/s). However, the SI unit of velocity is 3×10^{8} m/s.

It would be suitable here to mention the equation for the electrical and magnetic components of electromagnetic radiation which could be written as

$$E = E_0\sin[2\pi\left(\frac{x}{\lambda} - \upsilon t\right) \tag{1.12}$$

$$H = H_0\sin[2\pi\left(\frac{x}{\lambda} - \upsilon t\right) \tag{1.13}$$

where E_0 and H_0 are the maximum value of the electrical and magnetic fields, respectively.

1.3 Electromagnetic Spectral Range

According to the frequency or wavelength of an electromagnetic wave, there are several regions into which the electromagnetic radiation can be divided. This has been represented in Table 1.1.

Table 1.1 The different spectral regions are mentioned in increasing order of frequency, wave number, wavelength, and energy

Change of Spin		Change of Orientation	Change of Configuration	Change of Electron Distribution	Change of Nuclear Configuration	
NMR	ESR	Microwave	Infra-red	Visible and UV	X-ray	γ-ray

Wave number (cm^{-1})		10^{-2}		100	10^{4}	10^{6}	10^{8}
Wave length	10 m	100 cm	100 cm	100 µm	1 µm	10 nm	100 pm
Frequency (Hz)	$3x\ 10^{6}$	$3x\ 10^{8}$	$3x\ 10^{10}$	$3x\ 10^{12}$	$3x\ 10^{14}$	$3x\ 10^{16}$	$3x\ 10^{18}$
Energy (Joules mol^{-1})	10^{-3}	10^{-1}	10	10^{3}	10^{5}	10^{7}	10^{9}

This electromagnetic spectrum (as shown in Table 1.1) spreads over a broad range of radiation frequencies in which the visible region is limited to only a small portion (3×10^{14} to 7.5×10^{14} per second) after then the UV-region spreads over in the frequency range of 7.5×10^{14} to 3×10^{16} per second. The entire regions of this spectrum have been given different names such as radiofrequency/communication to the gamma (γ)-ray region (based on their technological applications or certain spectroscopic methods). Various spectroscopic methods depend on the different energy levels in which the atoms or molecules when exposed to a particular radiation then that radiation causes a transition to occur in the exposed atoms or molecules. The energy associated with each region (in the units of eV) has also been mentioned in Table 1.1. The different regions associated with EMR are discussed below in the order of increasing frequency.

1.3.1 Radiofrequency Region

This region gives rise to nuclear magnetic resonance (NMR) and electron spin resonance (ESR) and the wavelength lies between 3×10^{6} and 3×10^{10} Hz, or 10 m and 1 cm. The energy change which is involved in these arises from the reversal of the spin of a nucleus or electron. The order of energy change is of the order of 0.001 to 10 J/mole.

1.3.1.1 *Microwave region*

The region is associated with rotational spectroscopy and wavelength ranges from 3×10^{10} to 3×10^{12} Hz or 1 cm to 100 μm. The condition for a molecule to be microwave active is that it should have a permanent dipole moment. The energy difference between two rotational levels of a molecule is of the order of hundreds of J/mol.

1.3.1.2 *Infrared region*

This region gives rise to vibrational spectroscopy and the wavelength corresponds to 3×10^{12} to 3×10^{14} Hz or 100 μm to 1 μm. In this region, it is the vibration rather than rotation that gives rise to change in the dipole moment making a molecule IR active. The energy difference between two vibrational levels is of the order of 10^4 J/mol.

Raman discovered another type of spectroscopy known as Raman spectroscopy which gives information similar to that obtained in the microwave and infrared regions. However, the experimental method is such that observations are taken in the visible region. The condition of a molecule to be Raman active is that the polarizability of a molecule must change during the motion.

1.3.1.3 *Visible and ultraviolet region*

The region is associated with electronic spectroscopy and corresponds to wavelength 3×10^{14} to 3×10^{16} Hz or 1 μm to 10 nm. The absorption of light by a molecule in the visible and UV region is due to the transfer of energy from light to the electrons of the molecules which causes the excitation of valence electrons. The separation between energies of the valence electrons is nearly hundreds of kJ/mol.

1.3.1.4 *X-ray region*

This region corresponds to wavelength from 3×10^{16} to 3×10^{18} Hz, or 10 nm to 100 pm and is associated with the excitation of an electron from an atom or molecule. The energy changes involved in this process are of the order of 10,000 kJ.

1.3.1.5 *γ-ray region*

This region involves the rearrangement of nuclear particles and wavelength ranges from 3×10^{18} to 3×10^{20} Hz, or 100 pm to 1 pm. The energies required for the process are of the order of 10^9–10^{11} J/mol.

1.4 Born–Oppenheimer Approximation

In the years 1927, Max Born and J. R. Oppenheimer proposed an assumption that the motion of atomic nuclei and electrons in a molecule can be quantum mechanically studied separately. This approach is called the Born–Oppenheimer (BO) approximation. The BO approximation/separation is of great help in analyzing/studying the molecule's energy levels and the quantum states of a molecule.

The fundamental concept of BO approximation lies in the fact that the electrons and nuclei of a particular molecule experience nuclear motion of equal magnitude. It also suggests that the wave function of an electron is independent of the motion of nuclei and can thus be separated. This is because the mass of an electron is very less compared to the mass of atomic nuclei. Accordingly, the motion of an electron is much more than those of nuclei. In other words, one can say that the motion of the nuclei can be treated as fixed in comparison to the motion of electrons. Therefore, it is possible to write an equation for the total molecular wave function in terms of electron position and nuclei position. It can be represented as [7]

$$\psi_{\text{molecule}} = \psi_{\text{electronic}} \cdot \psi_{\text{nuclear}} \tag{1.14}$$

or

$$\psi_{\text{total}} = \psi_{\text{electronic}}(q_{\text{el}}) \cdot \psi_{\text{nuclear}}(q_{\text{nucl}}) \tag{1.15}$$

Further, the nuclear motion can be split into two parts, i.e., vibrational and rotational. The vibrational motion for a diatomic is caused due to internuclear separation and the rotational motion takes place due to the change in orientation of the molecule. The vibrational frequencies are greater than the rotational frequencies because of the larger separation between vibrational levels in comparison to those between rotational levels. Thus, Eq. (1.14) can be written as

$$\psi_{\text{total}} \approx \psi_{\text{electronic}}(q_{\text{el}}) \cdot \psi_{\text{vibration}}(q_{\text{vib}}) \cdot \psi_{\text{rotation}}(q_{\text{rot}}) \tag{1.16}$$

1.5 Energy Levels in Molecules and Quantization of Energy

The transfer of energy from matter to radiation and vice versa is supposed to take place according to the theory of radiation postulated by Max Planck in 1900 and extended by Einstein in 1905. According to this theory,

1. When a body emits or absorbs energy in the form of radiation, it cannot do so in a continuous manner.
2. The energy can only be emitted or absorbed in integrals multiples of certain units of energy known as *quantum*.

3. The energy E associated with each quantum for a particular radiation of frequency 'υ' is given by the relation $E = h\upsilon$, where h is the universal constant known as Planck's constant.

Thus, any change in energy can only take place by a 'jump' between two distinct energy states.

A molecule in space can have several types of energy. For example, it may contain rotation energy due to rotation of the body about its center of gravity; it may be vibrational energy due to periodic displacement of its atoms from their equilibrium positions; and since the electrons (associated with each atom or bond) are in continuous motion, so it also possesses the electronic energy. Therefore, the total internal energy of a molecule is defined as the sum of electronic, vibrational, and rotational energies, i.e.,

$$E_{\text{internal}} = E_{\text{electronic}} + E_{\text{vibration}} + E_{\text{rotation}} \quad (1.17)$$

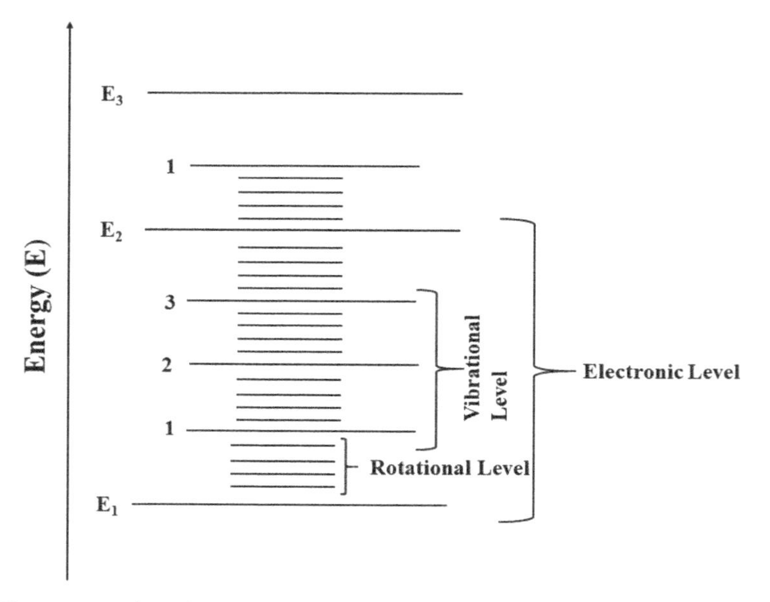

Figure 1.3 The schematic diagram of electronic, vibrational, and rotational energy states of a molecule.

Every electronic level within a molecule is associated with several vibrational levels, and every vibrational level has a set of rotational levels. The electronic energy level has an energy difference nearly equal to 100 kcals/mol whereas the difference in energy between the two vibrational levels is one-tenth to the one-hundredth of the energy between two electronic states. The rotational states have an energy separation of nearly one-hundredth of the energy between two vibrational states. These molecular energy levels are shown in Figure 1.3.

From Figure 1.3, it is very clear that

$$\Delta E_{\text{ele}} \gg \Delta E_{\text{vib}} \gg \Delta E_{\text{rot}}$$

All the electronic, vibrational and rotational energy levels of a molecule are quantized. A particular molecule can exist in a variety of these energy levels and can go from one level to another only by means of a sudden jump involving a finite amount of energy.

To make it more clear, let us consider two possible energy states (E_1 and E_2) of a system as shown in Figure 1.3 where the suffixes 1 and 2 denote the quantum numbers. The transition may occur between the two energy levels E_1 and E_2 either by the absorption or by emission of an appropriate amount of energy, ΔE (where $\Delta E = E_2 - E_1$) (Figure 1.4). These absorbed or emitted energies can take the form of electromagnetic radiation and the frequencies of such radiations are given by:

$$\nu = \Delta E/h \text{ (in Hz)} \tag{1.18}$$

where h is Planck's constant ($h = 6.63 \times 10^{-23}$ J/s). (If one wants to know the total energy change when a gram-molecule of substance changes its energy state, then h is multiplied by Avagadro's number, $N = 6.02 \times 10^{23}$).

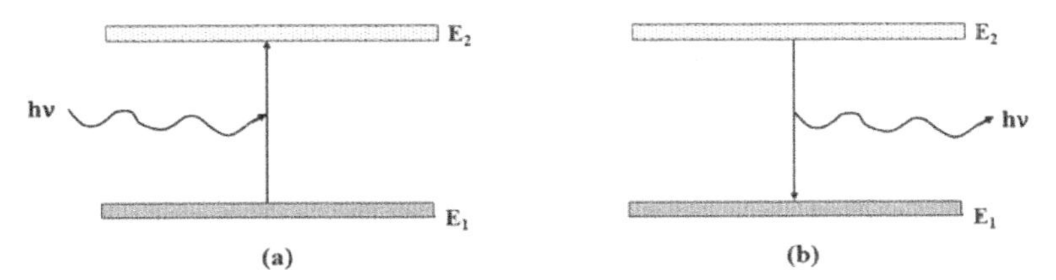

Figure 1.4 Interaction of electromagnetic radiation with matter (a) absorption and (b) emission.

Now, let us consider a molecule in state 1 that interacts with a beam of radiation of a single frequency υ (i.e., by monochromatic light) and the molecule absorbs energy '$h\upsilon$' from the radiation beam and will jump into state 2. Thus, the molecule is said to be in the excited state (Figure 1.4a). Absorption of more energy will further excite the molecule into higher energy levels. Therefore, if a detector is placed to collect the radiation after its interaction with the molecule, then it will show that the intensity of radiation has decreased. This decrease is due to the fact that the molecule has absorbed a certain amount of energy during excitation. The detector will yield a spectrum which is called an absorption spectrum. On the other hand, when the molecule has already reached state 2, then it may return to state 1 with the emission of radiations (Figure 1.4b) having frequency υ. Consequently, an emission spectrum will be obtained. The study of this emitted radiation forms the basis of *emission spectroscopy,* whereas the study of absorbed radiation forms the basis of *absorption spectroscopy.*

1.6 Mechanism of Absorption and Emission

Absorption and Emission spectroscopy are the study of the exchange of energy between radiation and matter. The mechanism of this is simply described as the interaction between the oscillating electric/ magnetic field of the radiation with the suitable dipole moment present in the molecule. The interaction will be strong if the dipole moment of the oscillating molecule (either by rotation or vibration) has the same frequency as that of the radiation field [3, 7]. The exchange of the energy of the oscillating molecule and the electromagnetic radiation is the root cause of the spectroscopic process in which a photon gives its energy ($\Delta E = h\upsilon$) to the molecule and raises it from a lower energy level to a higher energy level. The reverse process i.e. an emission can also take place and energy of absorption and emission has been discussed and described in Sections 1.2 and 1.6 of this chapter.

1.7 Transition Probability and Selection Rule

The process in which a molecule goes from a quantum state (from one eigenstate to another eigenstate) to another is called spectroscopic transition. In this process, the required energy is provided by electromagnetic radiation. Further, the probability of occurrence of a particular spectroscopic transition is defined as transition probability. In fact, when an atom or molecule absorbs a photon (hυ), the probability of transition of an atom or molecule from one energy state to another energy state depends upon two factors: (a) the nature of the initial and final state wave function (eigenstate) and (b) the strength of the interaction of photons with an eigenstate. These transition strengths generally describe the transition probability. Also whether a particular transition is possible (allowed) or not (forbidden) is determined with the help of selection rules. The selection rules can be described into three categories, i.e., electronic selection rule, vibration selection rule, and rotational selection.

1.7.1 Transition Moment and Transition Moment Integral

The electromagnetic radiation induces an oscillating electric or magnetic moment in an atom or molecule upon interaction. When the frequency of this induced electric or magnetic moment becomes equal to the energy difference between two energy states [i.e., one eigenstate (ψ_1) and another eigenstate (ψ_2)], a resonance occurs. This means that the frequency of the induced electric or magnetic moment of an atom or molecule is equal to the frequency of the electric or magnetic field of the radiation. This resonance condition (two motions are in-phase) must be satisfied for spectroscopic transition to occur. The amplitude of this resonating wave (electric or magnetic) is called the *transition moment*. If these two motions are in-phase, the exchange of energy takes place ($h\upsilon = E_2 - E_1$) as described earlier in Section 1.7 of this chapter. One should note that the electric moment transitions are most common in comparison to the magnetic ones in molecular spectroscopy.

The transition moment or transition dipole moment ($\overline{\mathrm{M}}$) between two eigenstates (ψ_1) and (ψ_2) can be expressed as:

$$\bar{M} = \int \psi_1 \, \mu_Z \, \psi_2 \, d\tau \tag{1.19}$$

whereas, the dipole moment of a given molecule is given by

$$\mu_z = \sum_{\alpha = x,y,z} \lambda_{z\alpha} \mu_\alpha \tag{1.20}$$

the quantity $\lambda_{z\alpha}$ mentioned above is a function of the rotational coordinate and μ_α represents the function of electronic and nuclear coordinates.

The intensity of absorption and emission of spectral lines depends upon the transition moment integral and can be represented by Eq. (1.21):

$$\bar{M} = \int \psi_2 \left(\mu^0 + \left(\frac{d\mu}{dr} \right)_{q=0} q \right) \psi_1 d\tau \tag{1.21}$$

According to the Born–Oppenheimer approximation, since the nuclear motion is much slower than the electrons, the total wave function can be separated into electronic, vibrational, and rotational wave functions as shown in the following equations.

$$\psi_{\text{tot}} = \psi_{\text{ele}} * \psi_{\text{vib}} * \psi_{\text{rot}} \tag{1.22}$$

Assuming that the wave functions are real, the transition moment integral Eq. (1.19) can be written as:

$$M = \iiint \psi_1^{\text{ele}} . \psi_1^{\text{vib}} . \psi_1^{\text{rot}} . \mu_Z . \psi_2^{\text{rot}} . \psi_2^{\text{vib}} . \psi_2^{\text{ele}} . d\tau_{\text{rot}} . d\tau_{\text{vib}} . d\tau_{\text{ele}} \tag{1.23}$$

where $d\tau_{\text{rot}}$, $d\tau_{\text{vib}}$, and $d\tau_{\text{ele}}$ are the volume elements for the rotational, vibrational, and electronic coordinates respectively and subscript 1 and 2 denotes the lower and upper energy stage.

Since, the rotational wave function provides information only for the rotational selection rule, therefore at this stage for simplicity, it is not considered and removed. On removing the rotational part of Eq. (1.23), the transition moment integral equation is expressed as

$$M = \iint \psi_1^{\text{ele}} . \psi_1^{\text{vib}} . \mu_\alpha . \psi_2^{\text{vib}} . \psi_2^{\text{ele}} . d\tau_{\text{vib}} . d\tau_{\text{ele}} \tag{1.24}$$

Since μ_α (the dipole moment component) is a function of electronic and nuclear coordinates so one can write

$$\mu_\alpha = \mu_{\text{e}} + \mu_{\text{n}} \tag{1.25}$$

As such the transition moment integral is represented as

$$M = \iint \psi_1^{\text{ele}} . \psi_1^{\text{vib}} . (\mu_{\text{e}} + \mu_{\text{n}}) . \psi_2^{\text{vib}} . \psi_2^{\text{ele}} . d\tau_{\text{vib}} . d\tau_{\text{ele}} \tag{1.26}$$

Eq. (1.23) can be integrated into two parts μ_{e} and μ_{n}, respectively, and in doing so we obtain the following equation:

$$\begin{aligned} M = & \int \psi_1^{\text{ele}} . \mu_{\text{e}} . \psi_2^{\text{ele}} . d\tau_{\text{ele}} \int \psi_1^{\text{vib}} . \psi_2^{\text{vib}} . d\tau_{\text{vib}} \\ & + \int \psi_1^{\text{vib}} . \mu_{\text{n}} . \psi_2^{\text{vib}} . d\tau_{\text{vib}} \int \psi_1^{\text{ele}} . \psi_2^{\text{ele}} . d\tau_{\text{ele}} \end{aligned} \tag{1.27}$$

Hence, the different electronic wave functions must be orthogonal to each other. On applying the conditions of orthogonality

$$\int \psi_1^{ele} . \psi_2^{ele} . d\tau_{ele} = 0 \tag{1.28}$$

$$\int \psi_1^{vib} . \mu_n . \psi_2^{vib} . d\tau_{vib} \int \psi_1^{ele} . \psi_2^{ele} . d\tau_{ele} = 0 \tag{1.29}$$

Therefore, the second part of the integral moment Eq. (1.27) will be equal to zero. As such finally, the simplified form of the transition moment integral equation is written as

$$M = \int \psi_1^{ele} . \mu_e . \psi_2^{ele} . d\tau_{ele} \int \psi_1^{vib} . \psi_2^{vib} . d\tau_{vib} \tag{1.30}$$

This Eq. (1.30) is very important whose defining the electronic and vibrational selection rules. The first integral describes the electronic selection rule whereas the second integral forms the basis of the vibration selection rule.

1.7.2 Electronic Selection Rules

Atoms are known by their primary quantum number (n), angular momentum quantum number (L), spin quantum number (S), and total angular momentum quantum number (J). The selection rules for electronic transition in atoms can be summarized as:

1. The transition between two electronic energy levels is governed by the selection rules:
   a. Spin rule, which suggests that the total spin (S) cannot change, i.e., $\Delta S = 0$. According to this rule for an allowed transition, the promotion of electrons without a change in their spin takes place.
   b. Orbital (Laporte) rule, which suggests that change in total orbital Angular momentum can be $\Delta L = 0$ and $\Delta L = \pm 1$ but $L = 0 \rightarrow L = 0$ transition, is not allowed.
   c. The third rule suggests that $\Delta J = 0$ or ± 1, when ΔJ is the change in the total angular momentum J, however, $J = 0 \rightarrow J = 0$ is forbidden (not allowed).
2. The initial and final wavefunctions must change in parity. The parity selection rule for electronic transition suggests that only even to odd transitions are allowed and even-to-even, or

odd-to-odd transitions are forbidden. Therefore, the parity selection rule for *electric dipole transition* is

$+ \leftrightarrow -$ allowed, $+ \leftrightarrow +$, and $- \leftrightarrow -$ forbidden

However, the parity selection rule for *magnetic dipole transitions* is different

$+ \leftrightarrow +$ and $- \leftrightarrow -$ allowed and $+ \leftrightarrow -$ forbidden

The parity selection rule is based on the symmetry of molecular waves.

1.7.3 Vibrational Selection Rules

The geometry of vibrational wave functions is very important in describing vibrational selection rules. However, the general vibrational selection rules are mentioned below:

1. Transitions with $\Delta v = \pm 1, \pm 2, \ldots$ are all allowed for anharmonic variations, but the intensity of the peaks becomes weaker as the values of Δv increase.
2. $v = 0$ to $v = 1$ transition is normally called the *fundamental vibration*, while those with larger Δv are called *overtones*.
3. $\Delta v = 0$ transition is allowed between the lower and upper electronic states with energy E_1 and E_2 are involved, i.e., $(E_1, v'') \rightarrow (E_2, v')$, where the double prime and prime indicate the lower and upper quantum state respectively.

1.7.4 Rotational Selection Rules

When a molecule absorbs or emits a photon, the total angular momentum must be consumed. It is a fact that a photon contains one unit of quantized angular momentum, therefore, in a rotational spectroscopic transition, during the emission or absorption process of a photon, the value of angular momentum J will vary only be

1. For a diatomic molecule when $\Delta J = 1$, the transition is defined as the R-branch transition. However, when the value of $\Delta J = -1$, the transition is known as the P branch.
2. When transition takes place between two different electronic or two different vibrational states then $\Delta J = 0$ is allowed for transitions.

1.8 Line Width and Line Shapes

Spectral lines are dark and bright lines that appear in a uniform and continuous spectrum obtained as a result of emission or absorption of light in a narrow frequency range (as compared with the nearby frequencies). These spectral lines are characteristic lines for different atoms or molecules and can be used as a fingerprint to identify the atoms and molecules from the previously collected characteristic spectral lines. Generally, these lines are very useful in identifying the atomic and molecular components of stars and planets which are very difficult to predict otherwise. Spectral lines are obtained as a result of interaction between a quantum system (atom and sometimes a molecule or atomic nuclei) and a single photon. What happens actually is that when the photon possesses a sufficient amount of energy to produce a change in the energy state of the system, it gets absorbed. When this photon is spontaneously reemitted with the same original frequency, then spectral lines are observed either as an emission or absorption line.

Spectral lines depend on the type of material and its temperature relative to another emission source. When photons from a hot, broad spectrum source pass through a cold material absorption lines are obtained. On the other hand, bright emission lines are produced when photons from a hot material are detected from a cold source.

Spectral lines are specific to atoms, and thus can be used to identify the chemical composition of any substance which allows light to pass through it (typically gas is used). Spectral lines are widely used to determine the chemical composition and physical conditions of stars and other celestial bodies, which are rather impossible to be analyzed that cannot be analyzed by other means.

1.8.1 Spectral Line Broadening and Line Shapes

Due to the transition between different energy levels, absorption or emission of specific/particular frequencies of the electromagnetic spectrum takes place and one obtains line spectra. But experimentally instead of sharp spectral lines, one obtains broad ones.

1.8.2 Local Effects Broadening

Some local effects take place in the small region around the emitting element and as such the broadening takes place. These effects ensure local thermodynamic equilibrium around the emitting element. The local effects may be homogenous in nature. Broadening due to homogenous local effects are mentioned below.

1.8.3 Natural Line Broadening

It is characteristic of a particular system that is under investigation and does not depend on an experimental setup. It occurs due to the uncertainty principle which states that two conjugated variables can not be determined exactly at the same time. The lifetime Δt of the excited state and ΔE, the uncertainty principle

$$\Delta E \cdot \Delta T \geq \frac{h}{2\pi} \tag{1.31}$$

Thus, an excited state having a shorter lifetime will possess large energy uncertainty, and hence, one obtains broad spectral lines.

1.8.4 Doppler Broadening

The Doppler effect broadening of spectral lines is caused by the thermal motion of atoms and molecules emitting radiations. The higher the temperature of the system, the greater will be the thermal motion of atoms/molecules and the broader will be the spectral lines emitted from that system.

1.8.5 Pressure (Collision Broadening)

The collision of the other particles can occur with the light emitting particles which interrupt the emission process. It shortens the characteristic time for the emission process and increases the uncertainty in the emitted energy. This broadening effect depends on the pressure and the density of the system.

1.8.6 Broadening due to Inhomogeneous Local Effects

When some emitting particles are situated in a different local environment than the other emitting particles, inhomogeneous broadening takes place. Due to this, particles situated in different environments emit at different frequencies. This type of broadening is also known as interaction and proximity broadening. Generally, such broadening occurs in solids where surface grain boundaries and stoichiometric changes lead to different types of local environments.

1.8.7 Broadening due to Non-local Effects

Broadening of spectral lines also takes place due to various changing conditions in the largest area of space around the emitting substance. These broadening do not depend upon the conditions which occur locally around the emitting particles. Hence, they are termed as non-local effect broadening. They may arise due to the changes in the spectral distribution of radiation while traversing their path up to the observer. Non-local effects depend upon the changing condition over a large region of space. Broadening due to non-local effects includes (i) opacity broadening, (ii) macroscopic Doppler broadening, and (iii) combined effects broadening. The details of such types of broadening effects are not being discussed here.

1.9 Fourier Transform

It is a powerful mathematical technique in spectroscopy that is generally used to transform FT a function from one real variable to another. In spectroscopy, one studies electromagnetic radiations having a wide range of frequencies. Due to this fact, the spectral peaks can be processed in the time domain instead of the conventional domain of the frequency.

In the Fourier transform method, a function from the domain of one variable is converted to the domain of another variable, and then it is reconstructed. Fourier transform has been successfully applied in spectroscopic techniques and very good results were obtained. It becomes easier to resolve spectral data with the help of the FT method. When there were difficulties in resolving the data, the FT method found importance in a variety of spectroscopic techniques, and various FT spectroscopic instruments were developed in recent years.

FT spectroscopy is used in all the research areas of science where high accuracy, sensitivity, and resolution are required. In this technique, interferograms are collected by the measurement of the coherence of electromagnetic radiation source either in the time domain or in the space domain and then transferred into the frequency domain with the help of the Fourier transform method. Fourier transform measurements can be done with the help of two types of spectra (i) continuous wave mechanics FT spectrometer and (ii) pulsed FT spectrograph technique. The pulsed FT technique is more suitable and takes less time in comparison to conventional spectroscopic techniques.

References

1. Pérez-Juste, I. and Nieto Faza O., Chapter 1: Interaction of radiation with matter. In: Cid M. M. and Bravo J. (eds.), *Structure Elucidation in Organic Chemistry: The Search for the Right Tools* (Wiley-VCH, Germany), 2015, 1–25.
2. Struve, W. S., *Fundamentals of Molecular Spectroscopy* (Wiley-Interscience, New York), 1989.
3. Herzberg, G., *Molecular Spectra and Molecular Structure, I. Spectra of Diatomic Molecules* (D. Van Nostrand, New York), 1950.
4. Levine, I. N., *Molecular Spectroscopy* (John Wiley, New York), 1975.
5. Banwell, C. N., *Fundamentals of Molecular Spectroscopy*, 4th Ed. (McGraw-Hill, London), 1994.
6. Walker, S. and Straughan, B. P., *Spectroscopy* (Chapman and Hall, New York), 1976.
7. Brown, J. M., *Molecular Spectroscopy* (Oxford University Press, Oxford), 2003.
8. Barrow, G. M., *Introduction to Molecular Spectroscopy* (McGraw-Hill, New York), 1962.
9. Bernath, P. F., *Spectra of Atoms and Molecules*, 2nd Ed. (Oxford University Press, New York), 2005.
10. Graybeal, J. D., *Molecular Spectroscopy*, 1st Ed. (McGraw-Hill, New York), 1998.
11. Hollas, J. M., *High Resolution Spectroscopy*, 2nd Ed., John Wiley, Chichester, 1998.
12. McHale, J. L., *Molecular Spectroscopy*, 1st Ed. (Prentice Hall, New Jersey), 1999.
13. Pavia, D. L., Lampman, G. M., Kriz, G. S. and Vyvyan, J. A., *Introduction to Spectroscopy*, 4th Ed. (Brooks-Cole, California), 2009.

14. Chandra, S., *Molecular Spectroscopy* (Alpha Science International, Oxford), 2009.
15. Svanberg, S. *Atomic and Molecular Spectroscopy: Basic Aspects and Practical Applications*, 4th Ed. (Springer, Berlin), 2004.
16. Hollas, J. M., *Modern Spectroscopy*, 4th Ed. (John Wiley, Chichester), 2010.

Chapter 2

Nuclear Magnetic Resonance Spectroscopy

Preeti Gupta,[a] S. S. Das,[b] and N. B. Singh[c]
[a]*Department of Chemistry, DDU Gorakhpur University, Gorakhpur, India*
[b]*Department of Chemistry (ex-professor), DDU Gorakhpur University, Gorakhpur, India*
[c]*Department of Chemistry and Biochemistry, SBSR and RDC, Sharda University, Greater Noida, India*
ssdas3@gmail.com

2.1 Introduction

Nuclear magnetic resonance spectroscopy is a very valuable technique for determining the structures of organic compounds because atomic nuclei in different environments produce signals in different positions of the spectrum [1–6]. Since NMR spectroscopy depends on the reversal spin of a nucleus and its interaction with the external magnetic field, it will be worthwhile to discuss first the nature of spinning particles and then the interaction between spin and a magnetic field [5–7].

Spectroscopy
Edited by Preeti Gupta, S. S. Das, and N. B. Singh

ISBN 978-981-4968-32-4 (Hardcover), 978-1-003-41258-8 (eBook)
www.jennystanford.com

2.2 Nature of Spinning Nucleus

Many nuclei possess a spin and angular momentum which varies from nucleus to nucleus. The total angular momentum depends on the nuclear spin number I, which may have values equal to 0, ½, 1, 3/2..., depending on a particular nucleus [8–11].

The simplest nucleus is that of the hydrogen atom which behaves as a tiny spinning bar magnet. It has only one particle, the proton. The proton possesses electric change as well as mechanical spin and any charged body which spins produces a magnetic field. The proton has a spin of ½. Apart from the hydrogen nucleus, another particle present in all nuclei is the neutron whose mass has been equal to that of a proton and possesses no charge. It has also a spin of ½. Therefore, if a particular nucleus contains p number of protons and n number of neutrons, then its total mass will be equal to (p + n). In view of the observed spin values, some empirical rules have been formulated which are mentioned below:

(i) Nuclei with both p and n odd (i.e., charge and mass even) have zero spins. For example, ^{4}He, ^{12}C, ^{16}O, ^{32}S, etc.

(ii) Nuclei with both p and n odd {i.e. charge is odd but (mass = p + n) is even} have integral spin. For example, ^{2}H (spin = 1), ^{14}N (spin = 1), etc.

(iii) Nuclei with odd mass possess half-integral spins. For example, ^{1}H and ^{15}N have spin =1/2, ^{17}O has spin = 5/2, ^{11}B = 3/2, etc.

2.3 Effect on the Nucleus of an External Magnetic Field

The external magnetic field has a great influence on the nucleus proton. It will try to adjust itself according to the imposed external field and can adopt two types of orientations either (i) parallel to the external magnetic field or (ii) opposite to the field. The parallel orientation corresponds to a lower energy state while the antiparallel represents to higher energy level. However, in the absence of a magnetic field, the molecules are randomly oriented. This is shown in Figure 2.1.

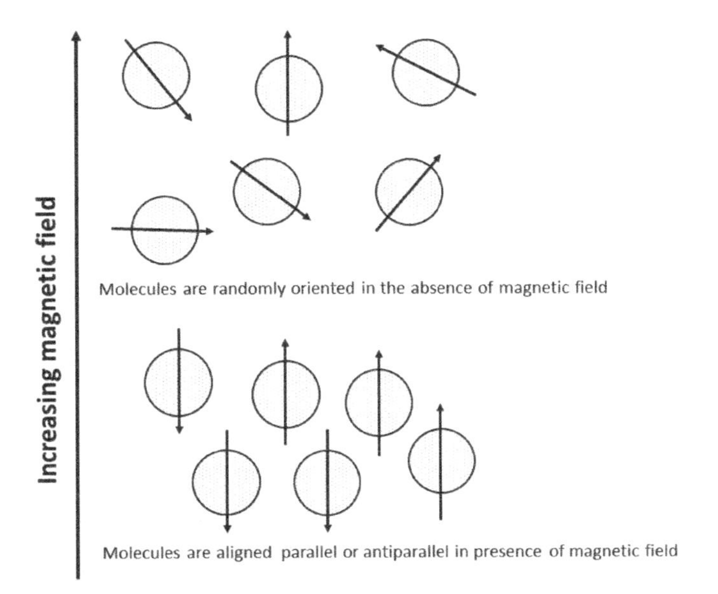

Figure 2.1 Alignment of molecule in presence and absence of magnetic field.

2.4 Precessional Motion of Nucleus

The nucleus besides arranging itself parallel or antiparallel to the external magnetic field can also perform a characteristic motion known as *precessional motion,* under the effect of the external magnetic field. We can understand this precessional motion in a better way by taking the example of a spinning top. The spinning top may perform a complete vertical motion or slower waltz-like motion under the influence of the earth's gravity. In the spinning waltz-like motion, the spinning axis of the top moves slowly around the vertical axis at a particular angle (Figure 2.2) [1].

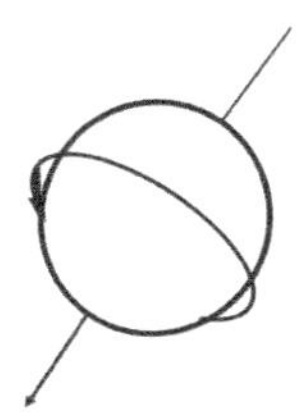

Figure 2.2 Molecule performing precessional motion in presence of an external magnetic field.

The proton can perform precessional motion in two principal orientations, i.e., (i) parallel (aligned) to the field or (ii) antiparallel (opposed) to the field. The precessional motion is shown in Figure 2.3 which is also called the *Larmor precession* [1].

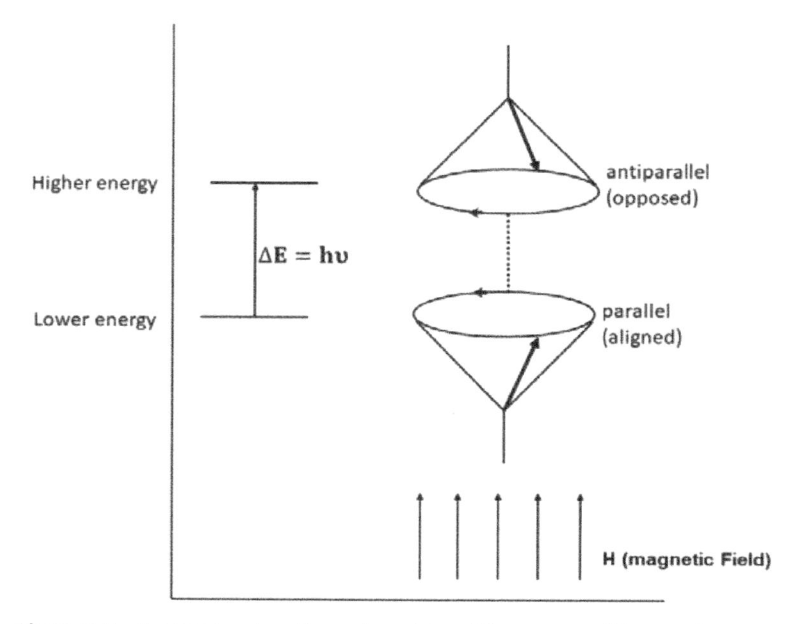

Figure 2.3 Precessional motions of nuclei and the energy difference between two orientations (aligned and opposed) conditions in presence of the magnetic field H.

2.5 Precessional Frequency: Larmor Frequency

The precessional frequency may be defined as the number of revolutions per second made by the magnetic moment vector of the nucleus in the magnetic field H. The angular precession frequency, υ, known as Larmor precession frequency, is directly proportional to the strength of the external magnetic field, H. Thus

$$\upsilon \propto H \tag{2.1}$$

This relationship is very important in NMR spectroscopy.

2.6 Energy Transitions and the Origin of NMR

A nucleus precessing in the aligned (parallel) orientation (being in the lower energy state) can absorb energy and may go into the antiparallel (opposed) orientation situated at a higher energy level. It can also go back into the original aligned state from the antiparallel state by losing the extra energy. If the precessing nuclei interact with electromagnetic radiation in the radio frequency region the low-energy precessing proton may absorb this energy and jump into the higher energy level. But this absorption of energy from the electromagnetic radiation (radio frequency region) will take place only when the precessing frequency (υ) of the spinning nucleus is similar to the frequency of the radiofrequency beam. When such absorption occurs, then the nucleus and radio frequency radiation will show a resonance. This absorption of energy is recorded in form of an NMR spectrum. This is the reason that this phenomenon is termed as the *nuclear magnetic resonance.*

2.7 Theory of NMR

2.7.1 Nuclear Spin and Magnetic Moment and Interaction Between Nuclear Spin and Magnetic Field

In atoms, nuclei are composed of protons and neutrons and like electrons both these particles spin on their own axes. Because of spin, they possess angular momentum equal to ½ ($h/2\pi$) for each particle. The resultant angular momentum for all the protons and neutrons present in a nucleus is called nuclear spin. This nuclear spin angular momentum, I is given by [1, 11]

$$I = \sqrt{I(I+1)}\left(h/2\pi\right) \tag{2.2}$$

where I is the spin number for a nucleus and can have any one value: 0, 1/2, 1, 3/2, ... The numerical value of I is related to the mass number and atomic number as mentioned in Table 2.1.

Table 2.1 Spin number with odd and even mass and atomic numbers [1]

Mass Number	Atomic Number	Spin Number I
Odd	Even or odd	½, 3/2, 5/2.....
Even	Even	0
Even	Odd	1, 2, 3

Thus, the nucleus of ^{1}H has I = 1/2, whereas ^{2}He, ^{12}C, and ^{16}O have I = 0 and are, therefore, nonmagnetic. Deuterium, ^{2}H, and nitrogen ^{14}N have I = 1. For sodium I = 3/2 and for chlorine I = 5/2. ^{11}B, ^{13}C, ^{15}N, ^{17}O, ^{19}F, and ^{31}P are other magnetic nuclei that have been studied in detail [1, 2, 4].

Protons and neutrons in a nucleus possess magnetic properties. This gives a magnetic moment μ for the nucleus. In fact, the charged nucleus spinning about an axis constitutes a circular electric current which in turn produces a magnetic dipole. Thus, the spinning nucleus behaves as a tiny bar magnet whose axis is coincident with the spin axis. The strength of this small magnet, i.e., the magnetic moment μ is related to the nuclear spin quantum number I by the equation

$$\mu = g\beta_{\mathrm{N}}\sqrt{I(I+1)} \text{ (in JT}^{-1}\text{)} \tag{2.3}$$

where g is called the *Lande's splitting factor* and is characteristic of each nucleus and β_N is known as the *nuclear magneton* which is given by

$$\beta_{\mathrm{N}} = \frac{eh}{4\pi m_{\mathrm{p}}} \text{ (in JT}^{-1}\text{)} \tag{2.4}$$

where m_p is protonic mass and e is the charge. Since the magnetic moment of a nucleus, μ, is a vector quantity with components in various allowed directions. Thus, the component μ_z in the direction of the field is given by

$$\mu_{\mathrm{Z}} = g\beta_{\mathrm{N}} I_{\mathrm{Z}} \tag{2.5}$$

where I_z is the component of angular momentum vector I in the particular reference components field direction. Along a particular direction Z, the components, I_z, may be written as

(i) $I_z = I, I{-}1, ..., 0, ..., -(I{-}1), -I$ (where I is an integer), or
(ii) $I_z = I, I{-}1, ..., 1/2, z - 1/2, ..., -I$ (where the value of I is half-integer)

So that I_z can have a total of $(2I + 1)$ values. When an external magnetic field is absent, all these orientations have the same energy. But in presence of a uniform magnetic field of strength H, these different allowed orientations of μ get split into different energy levels. The nucleus under the influence of a magnetic field then may assume any one of $(2I + 1)$ orientations with respect to the direction of the external magnetic field.

The energy, E due to this magnetic interaction is given by

$$E = -\mu H \cos\theta \tag{2.6}$$

where θ is the angle between the axis of rotation of the spinning nucleus (i.e., the axis of the dipole) and the direction of the magnetic field (Figure 2.4).

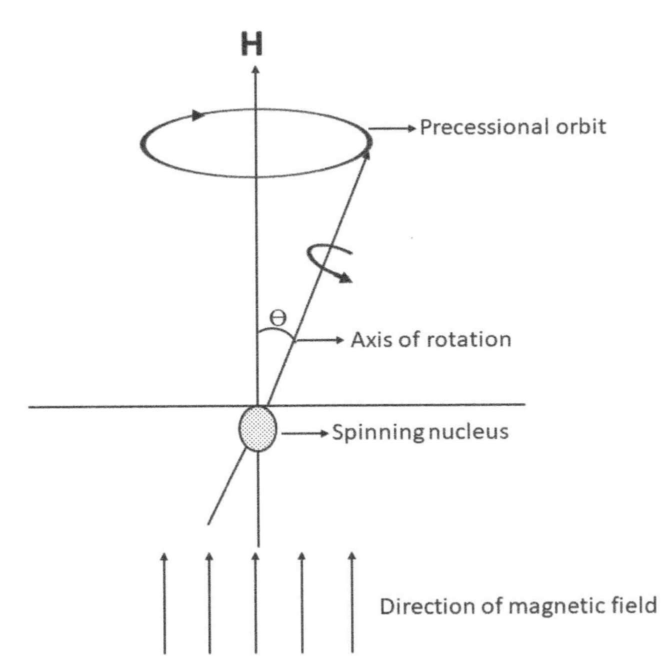

Figure 2.4 Representation of Precessional orbit and axis of rotation in the direction of the magnetic field.

If μ_z is the component of μ in the direction of the field, then $\cos\theta = \mu_z/\mu$

So that, one can write

$$E = -\mu_z H \qquad (2.7)$$

Since μ_z is given by Eq. (2.4) so on substituting its value in Eq. (2.7), the energy equation becomes

$$E = -g\beta_N H I_z \qquad (2.8)$$

The value of the energy, E, therefore, depends on the values of both I_z and H. However, for a given value of H, the energy levels arising from equation $E = -g\beta_N H_0 I_z$ will be the following.

(i) When $I = I_z = 0$, then $E = 0$. This means the nuclei with zero nuclear spins will not show any splitting in the magnetic field.

(ii) When $I = 1/2$ then $I_z = 1/2$ and $-1/2$ and then the two energy levels will be $-1/2\ g\beta_N H$ and $½g\beta_N H$. Thus, a proton ($I = 1/2$) will be able to assume only one of the two possible orientations that correspond to energy levels $\pm\mu_z H$ in an applied magnetic field. In the absence of the field these two levels are said to be degenerate (Figure 2.5).

(iii) Similarly, when $I > 1/2$, a large number of orientations (energy levels) are possible. For the case when $I = 3/2$, then $I_z = 3/2$, $1/2$, $-1/2$, and $-3/2$. Thus, there will be four energy levels.

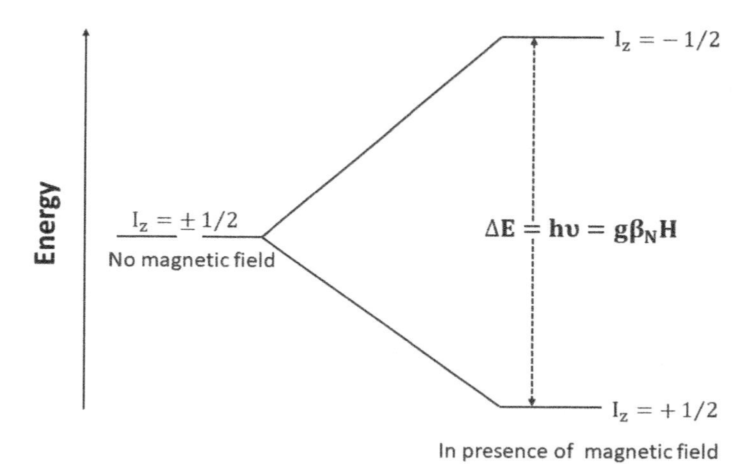

Figure 2.5 Effect of the magnetic field in the splitting of $I_z = \pm 1/2$ states.

If E_2 and E_1 are the energies in the two levels I_z and I_{z-1} then the energy difference, ΔE is given by

$$\Delta E = E_2 - E_1 = -g\beta_N H(I_z - I_{z-1}) \quad (2.9)$$

Since the transition may take place only between two adjacent levels, i.e, $I_z \rightarrow I_{z-1} = \pm 1$, therefore

$$E = +g\beta_N H \quad (2.10)$$

where the plus sign (+) refers to absorption and the minus sign (-) indicates the emission of energy.

If υ is the frequency of the electromagnetic radiation which corresponds to a transition, then the transition between any two energy levels will take place either by the discrete amount of energy in the radio frequency region. Thus, $\Delta E = h\upsilon$ of absorption or emission, and in such case

$$\upsilon = \frac{\Delta E}{\mathrm{h}} = \pm \frac{g\beta_N H}{h} Hz \quad (2.11)$$

Further, since $\mu = g\beta_{\mathrm{N}\sqrt{(I(I+1))}}$ and $I = \frac{\sqrt{I(I+1)h}}{2\pi}$,

Therefore, $$\frac{\mu}{I} = g\beta_N . \frac{2\pi}{h} \quad (2.12)$$

or $$g\beta_N = \frac{h\mu}{2\pi I} \quad (2.13)$$

Therefore, substituting its value in Eq. (2.11), we get the exact frequency given by

$$\upsilon = \frac{\mu H}{2\pi I} \text{ Hz} \quad (2.14)$$

The gyromagnetic ratio (γ) is defined as

$$\gamma = \frac{\text{magnetic moment}}{\text{angular moment}} = \frac{\mu}{I}$$

Therefore, the frequency $$\upsilon = \frac{\gamma H}{2\pi} \text{ (in Hz)} \quad (2.15)$$

Thus, when a nuclear spin interacts with a beam of electromagnetic radiation and the beam has the same frequency as that of the spinning nucleus then the interaction is coherent and energy can be exchanged. However, if the frequency is different, then there will be no interaction. The phenomenon of resonance will occur when the frequency of the magnetic field and the frequency of the spinning nucleus are equal. The electromagnetic energy, which becomes equal to the energy difference between parallel and antiparallel spin states of the nucleus at some value of field strength, can be absorbed by the molecule under investigation, causing it to spin or 'flip' from the lower energy parallel state to the higher energy antiparallel state. When this occurs, the spin is said to be in resonance with the applied electromagnetic radiation giving rise to the signal on the detector. The frequency of the energy required to induce spin-state flipping of the nuclei varies directly with the applied magnetic field. The larger the field, the larger the difference between parallel and anti-parallel spin states, and the higher will be the energy of the signal required to induce the change. A plot of signal intensity versus frequency is called a spectrum, in which the unique signal produced by each nucleus is recorded as a peak at a specific resonance frequency. A typical NMR spectrum is shown in Figure 2.6.

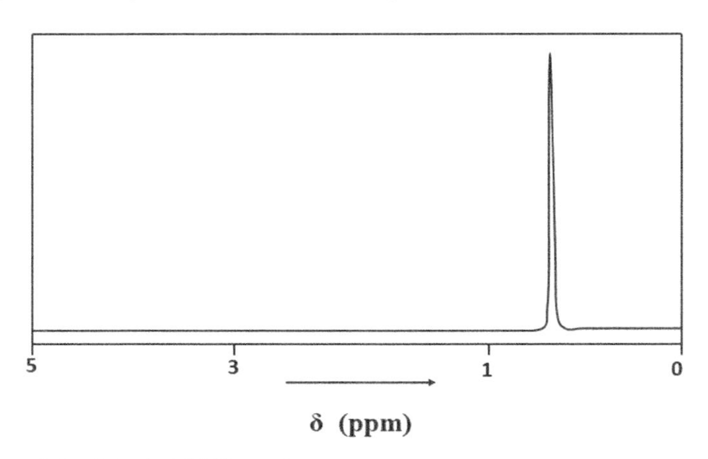

Figure 2.6 A typical NMR spectra.

2.7.2 Population of Nuclear Energy Levels

For the system containing non-interacting protons with spin ½, the spin energy levels are split in an applied external magnetic field. If the energy separation between +½ and −½ states is denoted by ΔE. The classical theory is used in calculating the populations of such levels. At a temperature T (in Kelvin), the ratio of the population of the upper state and lower state is given by the expression

$$\frac{N_{\text{upper}}}{N_{\text{lower}}} = \exp\left(-\frac{\Delta E}{kT}\right) \quad (2.16)$$

where k is Boltzmann constant. This suggests that for all temperatures above absolute zero, the upper level will always be populated to some extent. Although, for a large value of ΔE, the population may be negligible and insignificant.

When the nuclear spin is ½, the above Eq. (2.16) may also be written as

$$\frac{N_{\text{upper}}}{N_{\text{lower}}} = \frac{N_{(-1/2)}}{N_{(+1/2)}} = e^{((-g\beta_{\text{N}}H_0/kT))} \quad (2.17)$$

Since g, β_{N}, and H_0 have already been defined earlier, it can be shown that the ratio of $N_{\text{upper}}/N_{\text{lower}}$ is nearly equal to unity. This means that the population of the nucleus in the ground and excited state are equal.

2.7.3 Saturation and Relaxation

When the population of nuclei in the two spins states is equal then the probability of upward (absorption) transition becomes equal to that of downward (emission) transition. Electromagnetic radiation theory suggests that the probability of an upward transition (absorption) resulting from the absorption of energy from the magnetic field is exactly equal to the probability of a downward transition (emission) due to a process stimulated by the field. In other words, both absorption and emission are equal and the nuclear resonance effect would not be observed. In the equally populated level, no further absorption can occur. This phenomenon in which the populations of nuclei in the two spin states are equal is known as *saturation* of the resonance signal and the system is said to be saturated.

Under the ordinary condition, there is finite excess of nuclei in the lower energy state. As such due to these excess nuclei, absorption of energy in the radiofrequency region of electromagnetic radiation may take place. When these nuclei absorb radiofrequency radiation, the number of excess nuclei present in the lower energy state starts decreasing. Accordingly, the intensity of the absorption signal also diminishes. When the population of the nuclei in the two spin states becomes exactly equal, then it vanishes completely. When this happens no more energy will be absorbed. This situation is known as the saturation of the resonance signal.

When an NMR instrument is properly operated, certain mechanisms bring back the nuclei from the upper excited state to the lower energy state. The different types of radiation-less transition in which the excess spin energy is shared either with the surroundings or with some other nuclei of the upper energy level which is returning to the lower state are called a relaxation process. Further, the time taken for the dissipation of fraction $1/e = 0.37$ of excess energy is called the relaxation time. These relaxation processes may be of two types:

1. Spin–spin relaxation
2. Spin–lattice relaxation

In the spin–spin relaxation, the nucleus of one atom situated in the higher energy state transfers its energy to another atom in the lower energy state provided the nuclei are in very close proximity. Thus, spin–spin relaxation involves a mutual exchange of spin energy in which when one nucleus loses energy, the other nucleus gains that energy. This results in no net change in the population of the two spin states. This relaxation process is also known as *transverse relaxation*.

However, in the spin–lattice relaxation, the energy is transferred to the lattice, i.e., the high-energy nucleus loses its energy by transferring it to some electromagnetic vector present in the surrounding environment. The term lattice means a whole array of neighboring molecules, ions, or atoms (either in the same sample molecule or in solvent molecules, etc.) that contains or surrounds the precessing nuclei. The molecules present in the lattice are undergoing translational, rotational, and vibrational motions and thus, possess electrical and magnetic properties. Therefore, several small magnetic fields are present in the lattice. And it is quite likely that a particular small magnetic field properly oriented in the lattice might be able to absorb the energy lost by the precessing nuclei in the higher energy state so that the total energy of the system remains unchanged. The nucleus returns to the lower state. Because of this mechanism, in this relaxation process, there is an excess of nuclei in the lower energy state. This is a necessary condition for observing the nuclear resonance phenomenon. Spin-lattice relaxation is also known as *longitudinal relaxation*.

2.8 Concept of Chemical Shift

The concept of an isolated nucleus is practically not possible because all the nuclei are associated with the electron in any atom or molecule. Thus, when the nucleus is placed in an external magnetic field, it precesses (revolve) at a rate depending upon the gyromagnetic ratio and the strength of the external field. Along with this nucleus, the surrounding electrons will also revolve in the magnetic field around the nucleus. These circulating electrons generate their magnetic field (H_{induced}) in a direction opposite to the applied external field (H). This will influence the resultant field and the nucleus becomes shielded to a certain extent from the applied field. Because of this shielding, the rate at which the nucleus precesses will also change and in turn, the frequency of the resonance will also be changed. Therefore, different degrees of shielding will bring resonance at different frequencies and cause the NMR signal frequency to shift. This leads to the definition of another phenomenon known as *chemical shift*. The magnitude of the chemical shift depends upon the type of nucleus and the circulating electron motion in the atoms or the molecules. Chemical shift provides great information about chemical bonds and the structure of molecules.

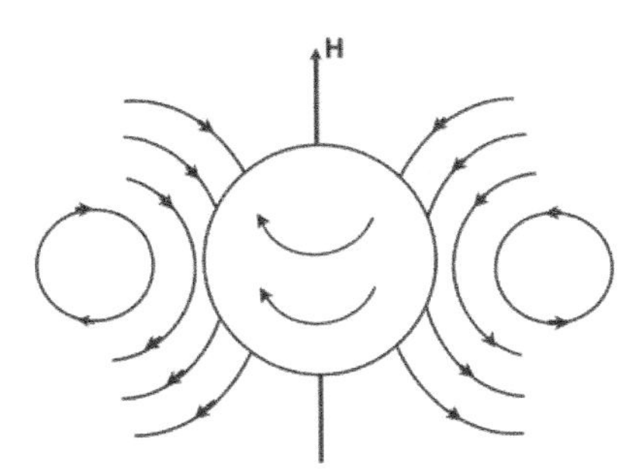

Figure 2.7 The diamagnetic circulation about a nucleus produces an opposing field of electron clouds in presence of a magnetic field, *H*.

The total field that the nucleus experiences because of this shielding phenomenon is given by

$$H_{\text{applied}} = H_{\text{effective}} - H_{\text{induced}} \quad (2.18)$$

where the induced field depends directly on the strength of the applied field. Therefore, one can write

$$H_{\text{induced}} = \sigma H_{\text{applied}} \quad (2.19)$$

where σ represents the shielding of the nucleus by the electrons and it is a dimensionless constant known as the shielding or screening constant (It is of the order of 10^{-5} for protons). Thus, we may write,

$$H_{\text{induced}} = H_{\text{applied}}(1 - \sigma)$$

$$H_{\text{eff}} = H(1 - \sigma)$$

where H_{eff} is the effective field experienced by the nucleus when it is put under a magnetic field. As such, the modified resonance condition in presence of an extranuclear electronic environment can be expressed as

$$\text{h}\nu = g\beta_{\text{N}}H(1 - \sigma) \quad (2.20)$$

But for an unshielded bare proton,

$$\Delta E = \text{h}\nu = g\beta_{\text{N}}H_{\text{eff}}$$

So that, $\nu = \dfrac{g\beta_{\text{N}}H_{\text{eff}}}{\text{h}}$

Whereas for a shielded proton

$$\Delta E' = \text{h}\nu' = g\beta_{\text{N}}H_{\text{eff}}$$

So that

$$\nu' = \frac{g\beta_{\text{N}}H(1-\sigma)}{\text{h}}$$

or

$$\nu' = \nu(1-\sigma) \quad (2.21)$$

The modified splitting conditions for isolated and shielded protons in the nucleus are shown in Figure 2.8.

A positive value of σ means that the proton is shielded by the electrons, whereas a negative value of σ indicates the deshielding of the nucleus. The value of H_{eff} is less than H when the nucleus is shielded. Therefore, the magnetic field strength H must be increased to bring the nucleus to a resonance condition. Contrary to it, when the nucleus is deshielded, the value of H_{eff} is more than H. This means that the resonance will take place at lower fields. Thus, it could be predicted that the identical nuclei (e.g., H) when shielded (or deshielded) with different chemical environments will resonate at different values of the applied magnetic field, giving NMR peaks at different values of field strengths. Since these values are characteristics of that shielded nucleus, they may greatly help in identifying different types of chemical environments present around that particular nucleus. Further, since the shifts in the position of the resonance peaks occur due to differences in the chemical environments around the nucleus, it is given the name chemical shift.

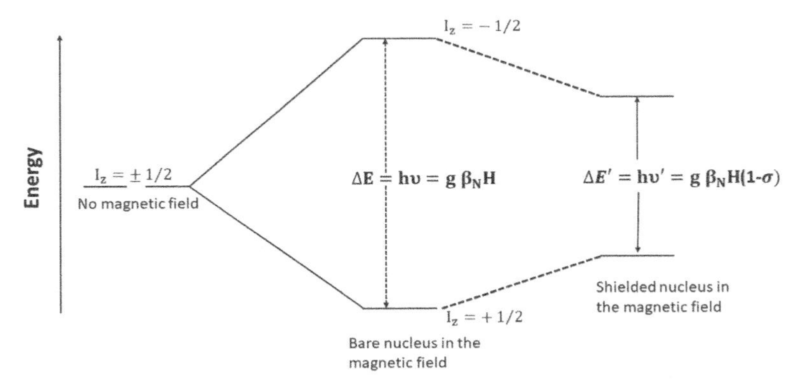

Figure 2.8 Splitting of the base and shielded nucleus in a magnetic field.

For example, let us consider the simplest case of ethanol as shown in Figure 2.9 which contains three peaks. The peak areas are directly proportional to the number of equivalent protons on the hydroxyl, methylene, and methyl groups i.e. 1 : 2 : 3.

This may be explained on the basis of the fact that oxygen is a better electron acceptor in comparison to carbon due to its higher electronegativity. Therefore, the electron density around the hydrogen atom in the C–H bond will be larger than in O–H bonds. It is then expected that the shielding constant $\sigma_{C-H} > \sigma_{O-H}$ and hence,

$$H_{CH} = H(1 - \sigma_{C-H}) < H_{OH} = H(1 - \sigma_{O-H}) \qquad (2.22)$$

The –OH hydrogen nucleus will experience a greater field due to smaller shielding constant σ_{O-H}; whereas, the more shielded $-CH_3$

nuclei will experience a smaller field. The OH molecule will resonate first and absorb energy from the beam; the CH_2 nuclei resonate next and finally, the CH_3 nuclei will resonate at a higher field. The effect of the increasing field on the energy levels of the CH_3, CH_2, and OH hydrogens in CH_3–CH_2–OH is shown in Figure 2.9.

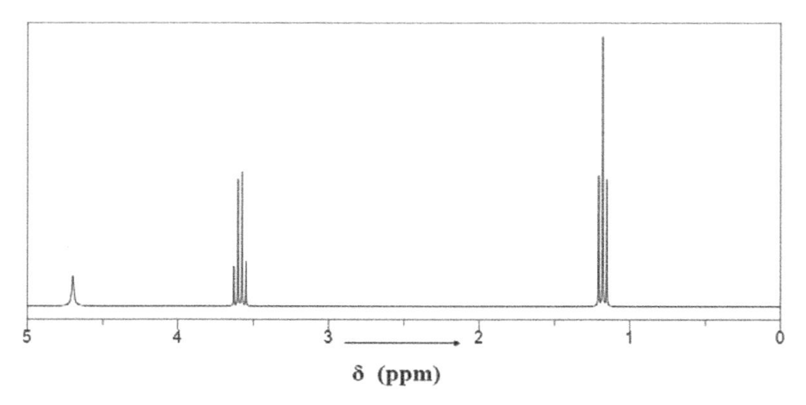

Figure 2.9 NMR spectrum of ethanol.

Another important aspect of NMR spectroscopy is the accurate measurement of the position of the field strength at which protons absorb energy in various environments from the radiofrequency probe. The position of the NMR signals helps in predicting what kinds of protons are present in the sample (aromatic, aliphatic, primary, secondary, tertiary, benzylic, vinylic adjacent to halogens or some in other atoms or groups). All these different kinds of protons have a different electronic environment that determines the position of the absorption by the proton in the spectrum.

But there is a drawback in NMR spectroscopy. The strength of the applied magnetic field, H, and the absolute position of the absorption cannot be accurately determined from the instrument. Therefore, for greater accuracy, relative proton frequencies re-determined. For obtaining relative values of absorption, a suitable reference material is used. The difference in the field strengths at which the sample nuclei and the nucleus of the reference compound absorb energy is measured.

The universally accepted reference for hydrogen resonance and ^{13}C is tetramethyl silane, $Si(CH_3)_4$ known as TMS. However, in aqueous solutions, the salt $(CH_3)Si\ CH_2CH_2CH_2SO_3^-Na^+$ known as DSS is used since TMS is not soluble in water or D_2O.

Tetramethyl silane is chosen as an internal reference because it has the following advantageous properties over other reference substances.

1. It gives an intense sharp signal even at low concentration because its all 12 hydrogen nuclei (having the same chemical environment) are magnetically equivalent. So, they absorb radiation exactly at the same position.
2. It absorbs at a higher frequency than all common types of organic protons and therefore, can easily be recognized. (The σ_{TMS} is large).
3. It is chemically inert.
4. It is magnetically isotropic.
5. It has a low boiling point (b.p. 270 °C, so that it is easily removed from the sample of a valuable organic compound after use.
6. It is soluble in most organic solvents and can be added to the sample solution (0.01–1.0 %).

For chemical shift measurement, two scales in parts per million are used. These two scales, delta (δ) and tau (τ) are related to each other by the expression,

$$\delta = 10 - \tau \tag{2.23}$$

On the τ scale, most chemical shifts have values between 0 and 10, and the resonance of standard TMS is arbitrarily set at a τ scale value of 10 ppm. The low τ value represents low field absorption while a high τ value suggests high field absorption. The τ scale suits nearly all protons as their spectra fall between 0 and 10 on this scale but due to its some disadvantages, chemical shift positions are now universally expressed in δ units. δ unit is a dimensionless number and is independent of field strength. Therefore, a signal having 4.2 δ obtained from a 60 MHz spectrum will always have the same 4.2 δ value even when 100 MHz or 600 MHz instruments are used. δ values (in ppm) are obtained from the following relation:

$$\delta = \frac{\upsilon_{\text{Sample}} - \upsilon_{\text{TMS}}}{\upsilon_0} \tag{2.24}$$

where δ is the chemical shift (in ppm), υ_S and υ_{TMS} are the frequencies (in Hz) of the signals for the sample and the reference TMS, respectively, υ_0 is the operating frequency of the instruments (in MHz).

2.9 Types of Shielding

The magnetic shielding of protons in molecules occurs due to the circulation of electrons induced by the applied field. The shielding may be classified into the following types:

1. Local diamagnetic shielding
2. Neighboring paramagnetic shielding
3. Interatomic diamagnetic shielding

Besides these, shielding effect may also arise due to the diamagnetic susceptibility of the solvents.

The *local diamagnetic shielding* arises due to the induced circulation of electrons around the nucleus itself. It generates a magnetic field opposite to the applied field. The shielding effect is controlled by the inductive and mesomeric effects that operate at the atom to which the hydrogen atom is attached. The magnitude of the local diamagnetic shielding is related to electron density at the hydrogen atom.

The neighboring paramagnetic shielding occurs due to the circulations of electrons around neighboring atoms. It is related to the anisotropy in the electron distribution around neighboring atoms. The secondary magnetic fields become effective only when the nearby protons are anisotropic, i.e., the circulation of electrons around the neighboring atoms depends on the orientation of the molecule with respect to the direction of the applied field. In this case, the electron circulation about the nucleus generates a magnetic field in the same direction as the applied field, resulting in the deshielding of the nucleus. Hence, the value of $\sigma_{\text{paramagnetic}}$ will be negative.

The *interatomic diamagnetic shielding* takes place in those molecules where electrons may circulate over two or more atoms through certain favorable paths. This effect is more important in aromatic systems in which the existence of π-molecular orbitals provides such paths for electronic circulation.

The neighboring paramagnetic and interatomic diamagnetic shielding are known together as 'long range shielding', while the diamagnetic shielding is called 'low range shielding'. The chemical shifts of a substance depend on the nature of the solvent molecules. This shielding effect is represented by $\sigma_{solvent}$ which includes all types of shieldings due to solvent. The quantity $\sigma_{solvent}$ is defined by

$$\sigma_{solvent} = \sigma_B + \sigma_A + \sigma_W + \sigma_E \tag{2.25}$$

where σ_B is the shielding contribution because of to bulk magnetic susceptibility of the solvent, σ_A is due to anisotropy of the solvent, σ_W occurs because of van der Waals interactions between the solvent and solute and σ_E arises due to an induced solvent dipole-dipole interaction.

Thus, all these three types of shielding effects will contribute to the total value of shielding constant σ, and one can write

$$\sigma = \sigma_{diamagnetic} + \sigma_{paramagnetic} + \sigma_{solvent} \tag{2.26}$$

The total magnetic field that the nucleus experience, $E_{effective}$ will change according to the total value of the shielding constant, σ.

2.10 Factors Affecting Chemical Shift

Certain factors affect the chemical shift and they provide useful information in determining the chemical structure of the substance. These factors are

1. Effect of electronegativity (inductive effect)
2. Van der Waals deshielding
3. Magnetic anisotropy
4. Effect of hydrogen bonding

2.10.1 Effect of Electronegativity (Inductive Effect)

Out of all the factors, the electronegativity of nearby groups plays an important role in predicting the chemical shift of a proton. The difference values of chemical shift and electronegativity values are mentioned in Table 2.2.

Table 2.2 Effect of electronegativity or chemical shift values for CH_3 protons attached to groups of different electronegativity

Compound	Chemical Shift (δ)
CH_3F	4.26
CH_3Cl	3.10
CH_3Br	2.65
CH_3I	2.16
CH_3-Si-	0.0

From this Table 2.2, the following questions need to be answered:

(a) How does the chemical shift of a proton vary:
   (i) with the electronegativity of the neighboring halogens?
   (ii) with the number of neighboring halogens?
   (iii) by its distance from an electronegative group? and
(b) Why does $(CH_3)_4Si$ has a lower chemical shift than the other molecules?

These could be explained in the following manner: the hydrogen nuclei are surrounded by electrons which shield the nucleus from the influence of the applied magnetic field, *H*. The greater the shielding, the lower will be the precessional frequency of the nucleus. Electronegative groups affect the chemical shift of a proton. In presence of the magnetic field, the circulation of electrons near the proton generates a local magnetic field which shields the proton from the applied magnetic field. This secondary magnetic field acts opposite to the applied field. If an electron density circulating the proton increases, the induced diamagnetic shielding effect will also increase. Consequently, the precessional frequency of the proton is lowered. But when the electron density in the vicinity of the proton is reduced, the magnitude of these types of local magnetic fields also reduces and the electron density at the nucleus decreases. The proton gets deshielded and the value of chemical shift in the NMR spectrum decreases.

As mentioned in Table 2.2, the chemical shift of the compound CH_3F is maximum $\delta = 4.26$. The fluorine in CH_3F withdraws electron density from the methyl group and deshields the proton. Methyl proton will thus experience a greater magnetic field. On the other hand, chlorine being less electronegative than fluorine will deshield the CH_3 proton to a lesser extent, and hence the protons would show a lesser chemical shift value $\delta = 3.05$. The same reasoning may conveniently be given to other compounds CH_3Br and CH_3I.

Silicon, on the other hand, is electropositive. It will give electrons to the CH_3 group of CH_3Si by the +I effect. This will cause a strong shielding of the CH_3 protons. Therefore, the protons will come to resonance at a low precessional frequency and a very low chemical shift δ value of nearly zero is observed.

2.10.2 Van der Waal's Deshielding

In densely populated molecules, such as steroids and alkaloids, some protons may be present in sterically hindered positions. In these molecules, the electron cloud of the bulky group hindering the proton will repel the electrons surrounding that proton. Thus, the proton will be deshielded and the effects of the applied external magnetic field will be large. Consequently, the deshielded proton will resonate at a lower frequency exhibiting higher chemical shift values.

2.10.3 Magnetic Anisotropy

Different functional groups present in organic molecules containing multiple bonds show anisotropic behavior. Depending upon the extent of anisotropy present in the molecule, the chemical shift values are changed. Various organic compounds containing multiple bonds may be classified into the following groups (Table 2.3).

Table 2.3 Chemical shift value of various functional groups

Functional Group	Chemical Shift (δ)
Alkene	4–8
Alkyne	1.5–3.5
Carbonyl	9.5–10
Aromatic	6–9

In the case of π-electron circulation under the influence of an external field, the effect is complex. It may cause shifts to a higher frequency (known as paramagnetic or downfield shifts), or to a lower frequency (called diamagnetic or upfield shifts). Since these effects are paramagnetic in some directions and diamagnetic in others around π electrons so they are called the anisotropic effect.

2.10.3.1 *Alkenes*

In the case of alkene, the circular motion of π-electrons under the influence of an external magnetic field generates an induced magnetic field. This induced secondary magnetic field opposes the external magnetic field in the center of the molecule. When the molecule is perpendicular to the applied field, the induced magnetic field will be diamagnetic around the carbon atom and paramagnetic near the protons. The applied magnetic field outside π-electrons is then strengthened. This is shown in Figure 2.10. Thus, the alkene protons are deshielded and experience larger fields. Therefore, the resonance will take place at a higher chemical shift value in the spectrum at $\delta = 4–8$.

However, when the orientation of the molecule is such that the induced magnetic field generated by the circulation of π-electrons is parallel to the applied external field, *H*, then the net field experienced by the proton will be greater than *H*. Thus, the protons will experience resonance at higher δ values than expected.

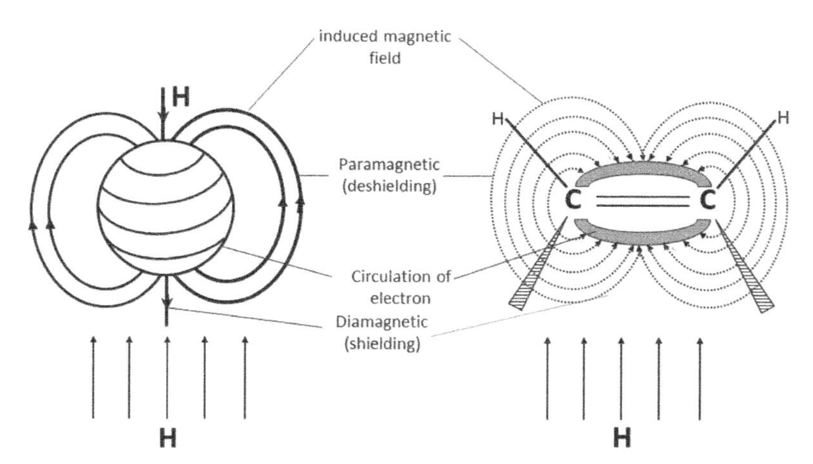

Figure 2.10 The shielding and deshielding of π-electron in alkene under the influence of magnetic field *H*.

2.10.3.2 *Alkynes*

The peaks of alkyne protons appear at low δ values (1.5–3.5) in the NMR spectrum. The electrons at triple bonds circulate in such a manner that the alkyne protons are diamagnetically shielded. This has been illustrated in Figure 2.11.

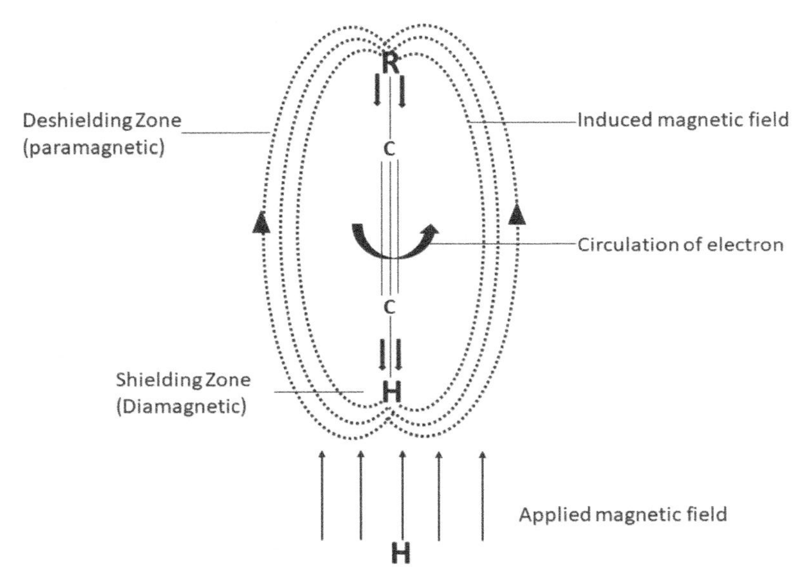

Figure 2.11 The shielding and deshielding zone in the case of alkyne under the applied magnetic field H.

If the axis of the alkyne molecule is parallel to the direction of the external field *H*, the cylindrical sheath of π-electrons of the triple bond will circulate the axis and the resultant induced magnetic field will work in the opposite direction of *H* near the proton. The strength of the field at the shielded alkyne proton becomes less causing the protons to show resonance at low δ values in the spectrum.

2.10.3.3 *Carbonyl compounds*

The shielding and deshielding zones in carbonyl groups are a little different from that of the alkene. The deshielding zone is shown in Figure 2.12 by two cone-shaped spaces which originate at the center of the oxygen atom. In between these two cone-shaped space-shielding zones exist. Thus, protons which are present in this space are deshielded, while those protons situated outside this space become shielded. The aldehyde and formyl protons of formate esters fall within the shielded cone space, so they show resonance at a high chemical shift value in NMR spectra. On the other hand, the shielded protons will appear in resonance at lower chemical shift values [1].

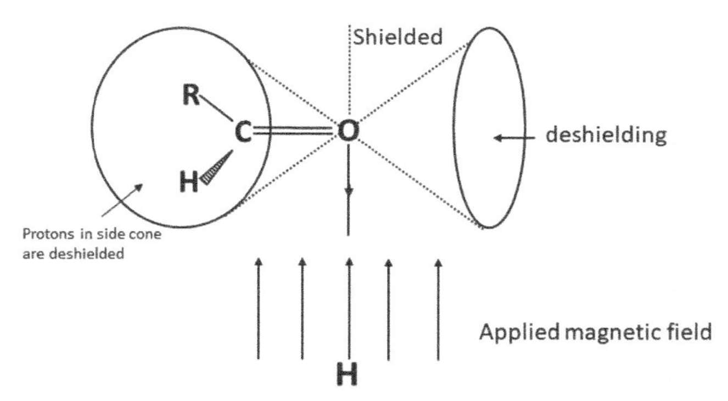

Figure 2.12 The shielding and deshielding of protons in carbonyl compounds under the applied magnetic field *H*.

2.10.3.4 *Aromatic compounds*

The aromatic protons of benzene are very similar to the protons of the double bond and the π-electrons are found to be delocalized on the benzene ring in form of loops above and below the aromatic ring. When an external magnetic field *H* is applied, these loops of electrons circulate and generate an electric current which is known as the ring current. This electric current then produces a magnetic field, whose direction and shape are shown in Figure 2.13 in the center of the ring, the induced magnetic field opposes the applied field *H*. Since it is coming toward the external field its nature will be diamagnetic. However, the returning induced magnetic field outside the ring is going in the direction of the applied field, and therefore, it becomes paramagnetic on both sides of the benzene ring. Due to this reason, the protons lying outside the aromatic ring in this periphery are deshielded and exhibit resonance at higher chemical shift values. The protons situated above and below the plane of the aromatic ring appear at low chemical shift values in the spectrum. Let us take examples of methyl and annulene.

The methyl protons (CH_3–) in toluene show resonance at $\delta = 2.4$ in the spectrum. When this methyl group is attached to an acyclic conjugated alkene, then methyl protons resonate at $\delta = 1.95$. This suggests that the ring current in aromatic compounds has a greater deshielding effect as compared to the deshielding of the conjugated alkene group.

Similarly, in the annulene molecule (Figure 2.13) a ring current is produced so its 12 protons situated outside the periphery get deshielded while the six internally situated protons are shielded. Those peripheral protons, therefore, show resonance at higher values of $\delta = 8.9$ whereas the internal protons appear, at a frequency even lower than those of the reference compound TMS, i.e., at $\delta = 1.8$.

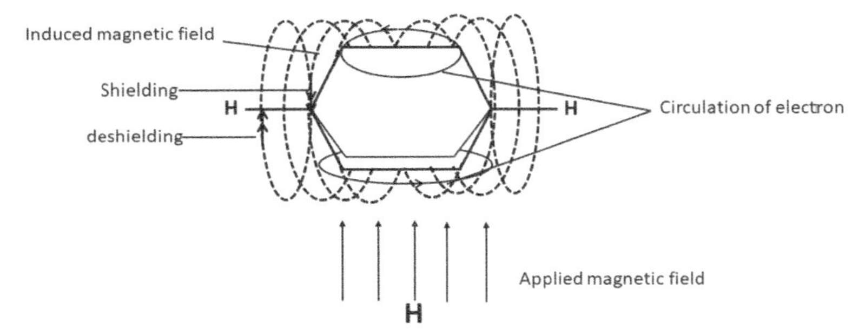

Figure 2.13 The shielding and deshielding of protons in aromatic compound annulene when a magnetic field *H* is applied.

2.10.3.5 *Effect of hydrogen bonding*

When in a molecule H-atoms are involved in hydrogen bonding then it will show resonance at a lower field in the spectrum in comparison to that molecule where H-atoms are not involved in hydrogen bonding. When a hydrogen bond is present, the H-atom attached to an electronegative atom pulls away the electron of the H-atom toward itself. So, this H-atom becomes deshielded and therefore, it will experience a greater external magnetic field. The resonance will appear in a downfield (high δ value). The greater the strength of H-bonding larger will be the magnitude of chemical shift. The amount of hydrogen bonding depends upon the concentration and temperature of the medium.

2.10.3.6 *Effect of solvents (types, concentration, temperature, and hydrogen bonding)*

Since the polarity of the solvent is different, the peak positions in the spectrum of a compound using a particular solvent may be slightly different when another solvent of different polarity is used. Therefore, in the spectrum, the name of the solvent used must be mentioned. Generally, the NMR signals for protons attached to carbon may shift by changing the solvent to a very little extent. But if significant bonding or dipole-dipole interaction exists, then greater chemical shifts are noticed. For example, in the NMR spectrum, chloroform dissolved in cyclohexane gives a signal at $\delta \sim 7.3$, but when benzene is used as a solvent, then in the spectra, the NMR signals shift to $\delta \sim 5.74$. In fact, benzene acts as a Lewis base to chloroform, and the electron density around the proton of the chloroform is changed due to considerable charge–transfer which causes a shift in the NMR signal of chloroform in benzene solution to a lower δ value (~ 5.74). The ring current in benzene is also responsible for this shift.

In simple compounds with –OH, –NH, and –SH functional groups, the hydrogen bonding is mainly responsible for shifts in NMR signals when solvents of different polarities are used. In hydrogen bonding, there is a transfer of electron cloud from the hydrogen atom to a neighboring electronegative atom, i.e., O, N, or S. Thus, in a concentrated solution when hydrogen bonding is strong, the deshielding of the hydrogen nucleus is greater and the NMR signals for –OH, –NH, and –SH protons are obtained at higher δ values than that at a lower concentration. In dilute solutions, the hydrogen bonding is less as such the hydrogen nucleus is more shielded, and the NMR signals appear at lower δ values. When the temperature is raised, the intermolecular H bonding reduces and consequently, chemical shift values are lowered giving NMR signals at low δ values. Thus, intermolecular H-bonding increases the δ values while increasing the dilution or temperature of the system lowers the chemical shift values.

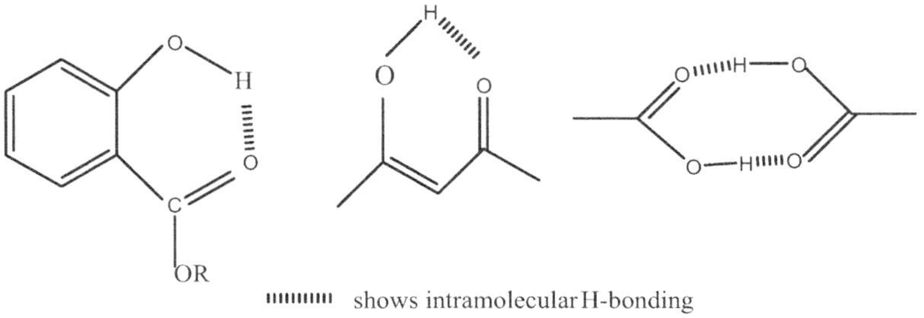

shows intramolecular H-bonding

Figure 2.14 The dashed line shows intramolecular H-bonding.

In those molecules such as salicylates, enols of β-dicarbonyl compounds, and carboxylic acids where intramolecular hydrogen bonding is present, the signals in the NMR spectrum remain unchanged at different solution concentrations. This is because the dilution has no effect on intramolecular hydrogen bonding. In salicylates, the signal for the OH occurs at δ (10–12) while for the enol OH, the NMR signals appear at a higher δ value (11–16). But in carboxylic acids due to their stable dimeric association, a special case of hydrogen bonding is present which exists even in a very dilute solution. As such NMR signals of carboxylic OH are obtained between $\delta = 10$ and $\delta = 13$ mostly near $\delta = 10$–12.

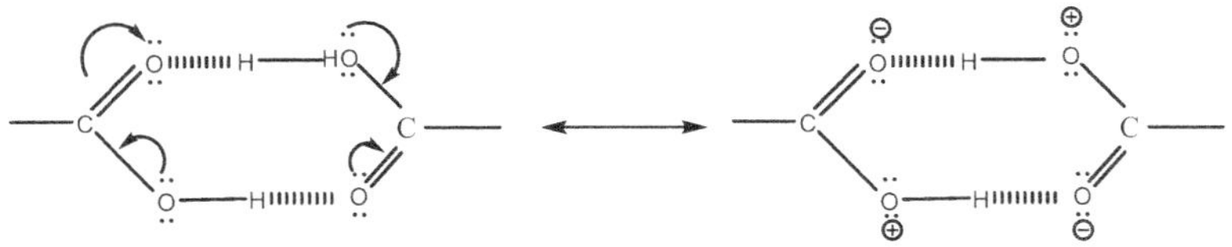

Figure 2.15 Stable dimeric association in dicarboxylic acid showing intramolecular H-bonding.

2.10.3.7 *Solvents used in NMR*

The spectra of organic compounds are recorded in the solution state because the non-viscous samples give sharp NMR spectra. However, if the compound itself is non-viscous then a neat liquid can be used without dissolving it in any solvent. Solvents are chosen on the basis that at least 10% of the compound under investigation should be soluble in that solvent. Further, the solvent should be free of protons. The common solvents mostly used in NMR spectroscopy are CCl_4,

CS_2, $CDCl_3$, C_6D_6, D_2O, $(CD_3)_2SO$, $(CD_3)_2CO$, or $(CCl_3)_2CO$, etc. Trifluoro acetic acid is a good solvent for the NMR spectrum of amines as its spectra do not interfere with those of amines.

2.11 Spin–Spin Coupling

2.11.1 The Splitting of Signals in Proton NMR Spectra

It is a well-known fact that in the NMR spectrum every resonance signal gives information about only one kind or one set of protons in a molecule. But under high-resolution NMR spectra appearance of fine structures (multiplets) was observed [1]. In certain molecules, instead of a singlet, a group of peaks (multiplets) appeared in the spectra. These multiplets or fine structures are produced due to the spin–spin coupling of the nuclei connected, or even separated by several covalent bonds. In those molecules, where the protons are separated by more than three covalent bonds, the spin–spin coupling is known as long-range coupling.

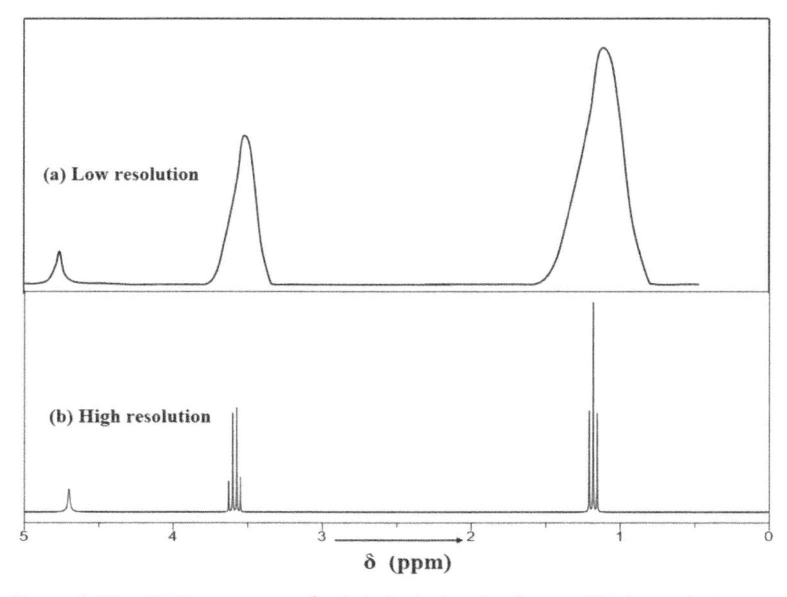

Figure 2.16 NMR spectrum of ethyl alcohol under low and high resolution.

For example, let us look at the NMR spectrum of ethyl alcohol CH_3CH_2OH under low and high resolution. Both the spectrum is shown in Figures 2.16(a) and (b).

This molecule has three kinds of protons and thus, three signals are expected to appear in its NMR spectrum. A singlet is observed for the –OH proton whereas, the $-CH_2$ and $-CH_3$ protons give a group of peaks instead of a single peak. The $-CH_3$ proton peak splits up into a triplet while $-CH_2$ protons show a quartet in the NMR spectrum under high resolution (as shown in Figure 2.16b). The appearance of these fine structures as multiplets is due to a phenomenon known as spin–spin splitting.

Consider 1,1,2-trichloroethane molecule as another example [1]. This molecule has got two types of protons labeled 'a' and 'b'. In its spectrum, for 'a' kind of proton a triplet and for 'b' kind of proton a doublet is obtained, as shown in Figure 2.17.

Cl
a
C
Cl
Cl
C
b

1,1,2 trichloroethane

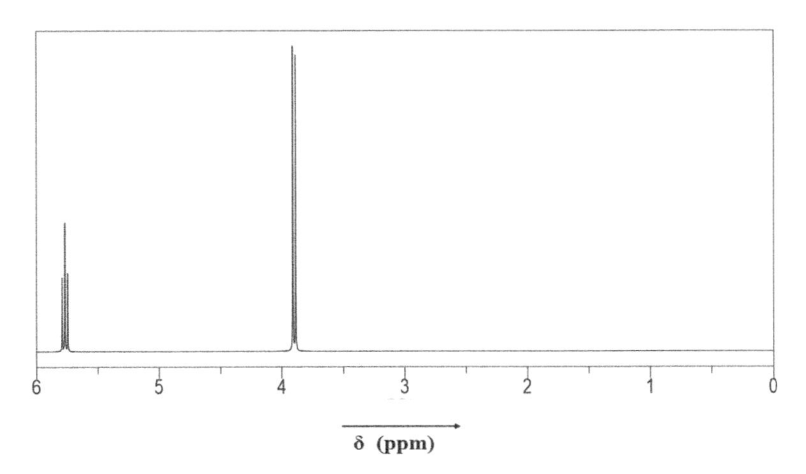

Figure 2.17 NMR spectrum of 1,1,2-trichloroethane molecule.

In all the examples discussed above, the splitting of spectral lines takes place due to a coupling interaction between neighboring protons. The protons which are on adjacent atoms can indirectly interact will each other depending on the nature and number of the bond between them.

2.11.2 Theory of Spin–Spin Coupling

In order to explain and understand the phenomenon of spin–spin coupling, let us consider a molecule having two neighboring protons H^a and H^b in different magnetic environments. These two protons will show resonance at different positions in the NMR spectrum at different values of chemical shifts. Instead of showing a single peak (singlet), they split into two lines and yield doublets. The spacing between the lines of the doublets is called a *coupling constant* and denoted by the symbol 'J'. The separation between the lines of each doublet in the spectrum will be equal.

The splitting of singlet peaks into doublets for the two different protons (a and b) could be explained in a simple manner. The peak position for the proton, H^a depends on the overall magnetic environment of protons, which gets affected due to the neighboring protons H^b. For the proton H^a, there are two possibilities of spin orientations of the protons H^b, either parallel or antiparallel. So two different magnetic fields will be experienced by the proton H^a due to the two-spin orientation of proton H^b. As a result, the proton H^a can align (↑↑ parallel) with H^b and increase its net magnetic field or may be opposed (anti-parallel (↑↓)) by the proton H^b and decreases its net magnetic field. Thus, the proton H^a will show resonance twice to give a doublet instead of a singlet (Figure 2.18). A similar explanation can also be given for the other proton H^b which also exhibits a doublet in the spectrum.

In molecules, such as ethyl bromide and 1,1,2-trichloroethane, $-CH_3$ protons of ethyl bromide and –CH proton of 1,1,2-trichloroethane splits into three peaks (triplets). In both the molecules, the $-CH_3$ protons and –CH protons contain adjacent $-CH_2$ protons. The splitting of $-CH_3$ protons of ethyl bromide and the –CH proton of 1,1,2-trichloroethane due to the presence of neighboring $-CH_2$ protons in the respective molecules could be explained in the following manner:

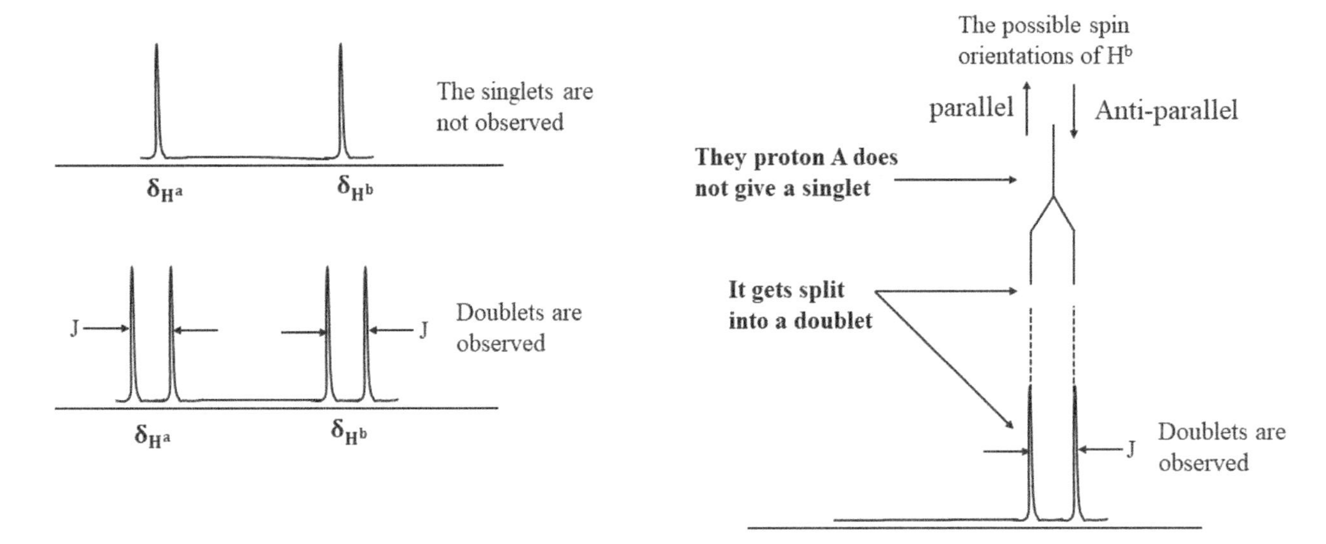

Figure 2.18 Schematic representation of splitting of singlets into doublets due to spin–spin coupling.

(i) If we label the protons of $-CH_3$ group of ethyl bromide and $-CH$ proton of 1,1,2-trichloroethane as H^a and the neighboring protons of $-CH_2$ group as H^b and $H^{b'}$, then, there are three possible orientations of spin due to protons H^b and $H^{b'}$. The proton A of $-CH_3$ group or $-CH$ group may experience the two nuclear spins of proton H^b and $H^{b'}$parallel (↑↑) to it.

(ii) These two nuclear spins may be antiparallel (↓↓) to H^a.

(iii) One spin may be parallel while the other spin may be antiparallel. But this situation will lead to two possibilities:

(a) The spin of proton H^b is parallel and that of $H^{b'}$ is antiparallel to the proton H^a (↑↓).

(b) The spin of proton H^b is antiparallel and that of the proton $H^{b'}$ is parallel to H^a (↓↑).

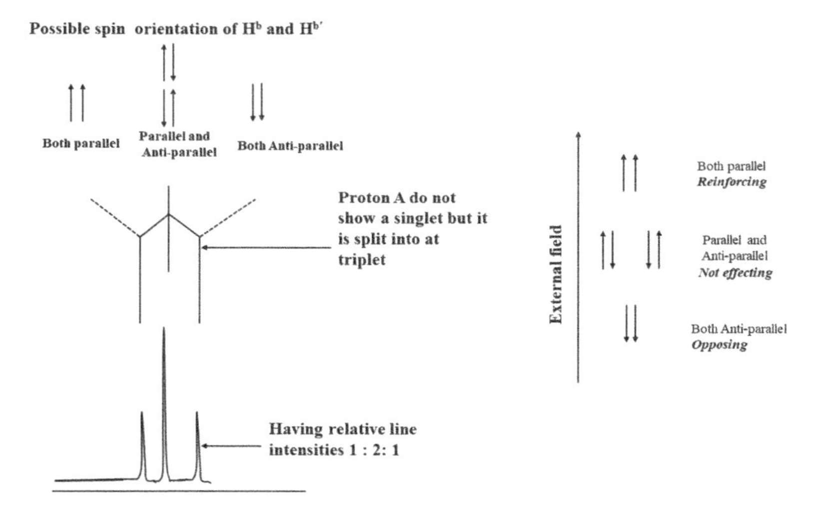

Figure 2.19 Schematic representation of three different spin arrangements due to two neighboring protons (Hb and Hb') of the $-CH_2$ group and the splitting of $-CH_3$ proton Ha into a triplet in the spectrum.

Thus, three different spin arrangements will exist. Due to this, the proton H^a of $-CH_3$ group of ethyl bromide and $-CH$ group of 1,1,2-trichloroethane exhibit triplets in their spectrum (Figure 2.17). In the spectra, it is clearly seen that the central peak of the triplet is nearly double the other two peaks. The reason for this is the equal occurrence of parallel and antiparallel spin arrangements. Whereas the possibility of the third nuclear spin arrangement is just double due to its two possible spin arrangements (↑↓ and ↓↑). This leads to an increase in the intensity of the signal of the middle peak nearly equal to double the other two peaks of the triplets in both cases. Thus, the relative intensities in the triplet are in the ratio of 1:2:1. This situation has been clearly explained with the help of Figure 2.19.

In the NMR spectra of 1,1,2-trichloroethane, in addition to a triplet, a doublet is also seen in Figure 2.17. This is due to the two possible spin orientations of proton H^a with respect to the proton of the $-CH_2$ group (i.e., H^b and $H^{b'}$). Since proton H^a exhibit two magnetic spins around $-CH_2$ protons (one is parallel and the other is antiparallel). Therefore, the $-CH_2$ protons show resonance two times in the spectrum.

On the other hand, the spectra of ethyl bromide show a quartet along with a triplet (Figure 2.20). This quartet is obtained because of the splitting of methylene proton into four peaks due to their three neighboring protons of the $-CH_3$ group. These four peaks will have relative peak intensities in the ratio of 1:3:3:1 in the spectrum. Four different possible spin arrangements for the CH_2 protons can experience are

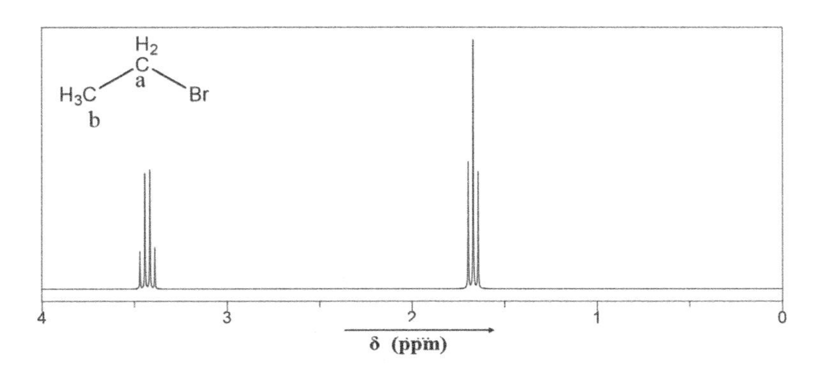

Figure 2.20 NMR spectrum of ethyl bromide.

(i) All three spins of CH_3 protons are parallel (↑↑↑).
(ii) All the three spins of CH_3 may be antiparallel (↓↓↓).

(iii) Two spins of the $-CH_3$ protons are parallel and the third is one antiparallel. This may occur in three ways, i.e.,

(↑↑↓), (↑↓↑), (↓↑↑)

(iv) Out of three spins of $-CH_3$ protons, one spin is parallel, and the other two antiparallels. This also occurs in three ways, i.e.,

(↑↓↓) (↓↑↓) (↓↓↑).

Thus, four different spin orientation takes place and the $-CH_2$ protons show four resonances as a quartet in the spectrum. Also, spin arrangements (iii) and (iv) can occur in three possible ways, therefore, the intensities of these peaks of two adjacent central lines will be triple of those two peaks appearing due to spin arrangements (i) and (ii) as shown in Figure 2.20.

2.12 The General Rule of Spin–Spin Splitting: The (*n* + 1) Splitting Rule

The simple splitting of NMR peaks is known as a first-order splitting pattern. Spin–spin splitting is governed by the following rules:

1. The splitting of lines of proton absorption depends upon the number of protons present in the neighboring group in a molecule. The simple rule for obtaining the multiplicity of proton absorption is to count the number of neighboring protons (n) and then add 'one.' This is called as $(n + 1)$ rule. According to this rule, one neighboring proton splits the NMR peaks into a doublet, two equally coupled neighboring protons produce a triplet, and so on. Thus, the general formula of this rule which covers all the nuclei, is $(2nI + 1)$, where I is the spin number of the nuclei.
2. The relative intensities of the resonance peaks of a multiplet depend on the number of neighboring protons. In simple cases

of interacting nuclei, the relative intensities of a multiplet are symmetric about the midpoint and are numerically proportional to the coefficients of the terms in the binomial expansion of $(x + 1)^n$. Thus, when $n = 1$, the binomial expression becomes $(x + 1)' = x + 1$ and for this doublet, the peaks are in the ratio 1:1; when $n = 2$, $(x + 1)^2 = x^2 + 2x + 1$. Hence the triplet peaks will be in the ratio 1:2:1; similarly, when $n = 3$: $(x + 1)^3 = x^3 + 3x^2 + 3x + 1$ and the quartet peaks will appear in the ratio 1:3:3:1, and so on. The multiplicity and the relative intensities can be obtained from the Pascals triangle as shown in Table 2.4 and multiplicity in some compounds are given in Table 2.5.

Table 2.4 Pascal diagram showing a number of peaks and their relative intensities due to neighboring protons

Number of neighbouring proton	Number of peaks	Relative intensities (Pascal diagram)
0	Singlet	1
1	Doublet	1 1
2	Triplet	1 2 1
3	Quartet	1 3 3 1
4	Quintet	1 4 6 4 1
5	Sextet	1 5 10 10 5 1

Table 2.5 The multiplicity, the ratio of intensities of peaks, and their pattern in the spectrum because of the number of nearest protons in some compounds

Compound	Multiplicity	Number of nearest neighbouring protons	Ratio of intensities	Spectral pattern
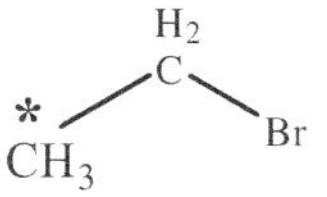 * CH_3–CH_2–Br	Doublet	1	1:1	
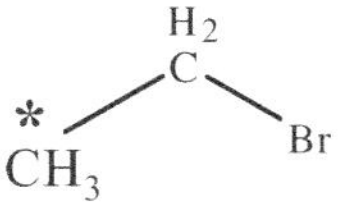 * CH_3–CH_2–Br	Triplet	2	1:2:1	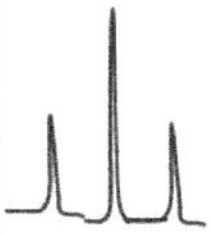
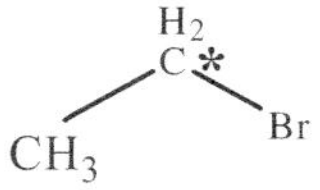 CH_3–CH_2*–Br	Quartet	3	1:3:3:1	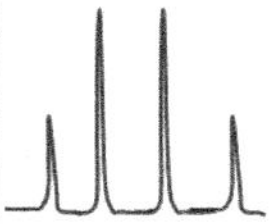
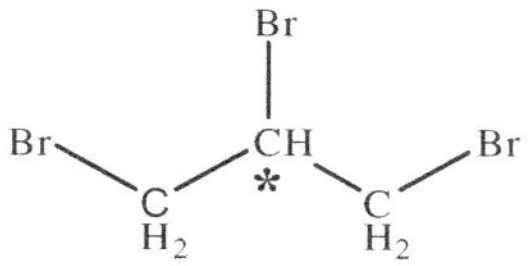 Br–CH_2–CH*(Br)–CH_2–Br	Quintet	4	1: 4:6:4:1	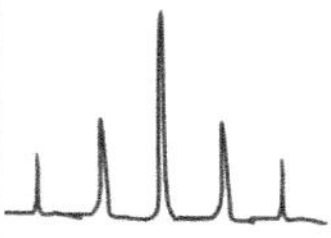

* indicates the proton getting split due to spin–spin coupling.

2.13 Complex NMR Spectra: Breakdown of (*n* + 1) Splitting Rule

So far now we have discussed spin–spin splitting which obeys the $(n + 1)$ rule where n is the number of protons in the neighboring group of the molecule. Let us now extend this rule to more complex systems where the spectra of certain compounds contain such splitting patterns which do not appear to be simple ones and cannot be predicted by the general $(n + 1)$ splitting rule. In such complex spectra (i) multiplicative splitting and (ii) in certain cases the breakdown of the $(n + 1)$ general rule takes place.

2.13.1 Multiplicative Splitting

Multiplicative splitting means the splitting of one set of protons by more than one other set of protons. When the coupling constants are equal and multiplicative splitting occurs, the numbers of splitting are then predicted by the general rule $(n + 1)$ but when the coupling constants involved are different the complex multiplicative splitting takes place. This could be explained by taking the example of the spectrum of vinyl pivalate whose spectra are shown in Figure 2.21.

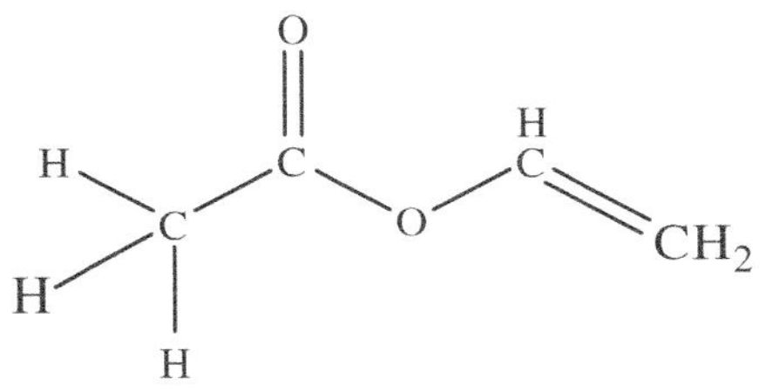

Structure of Vinyl Pivalate

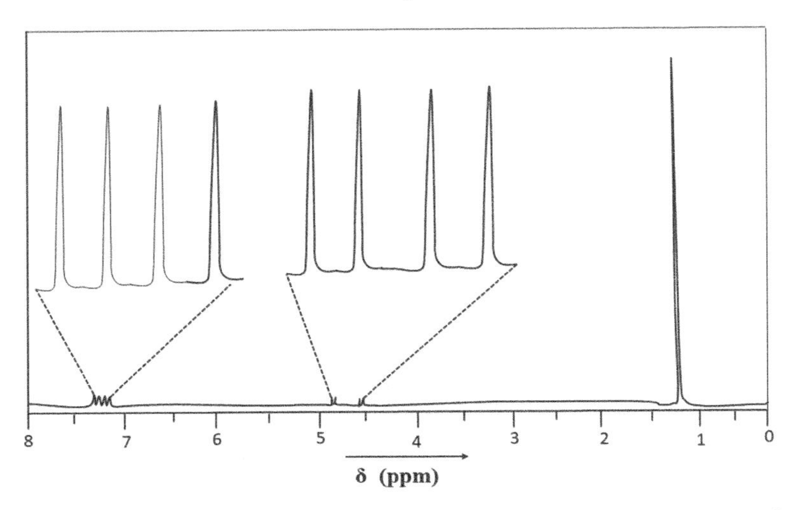

Figure 2.21 NMR spectrum and multiplicative splitting pattern in vinyl pivalate.

In this compound, nine equivalent tert-butyl protons labeled as H^A exhibit a structure at chemical shift $\delta = 1.26$ giving a sharp singlet. The protons H^B and H^C are quite far from the electronegative atom oxygen and therefore, they show the smallest chemical shifts nearly at δ = 4.5–5.0. There are three alkenes' protons labeled as H^B, H^C, and H^D that are non-equivalent protons. Therefore, the absorption of all these three protons will take place at different values of chemical shifts. Further, each proton gets split by the other two with a different coupling constant. Therefore, the splitting will be multiplicative in nature.

Multiplicative splitting can be illustrated with the help of a splitting diagram (Figure 2.22) for the alkene protons of vinyl pivalate. The single line of protons H^d gets split by the proton H^c into a doublet and the coupling constant J_{cd} = 14 Hz. Each line of the doublet has half the intensity of the original peak, the distance from the original is the same. The next proton H^b then further splits these two lines into two doublets each with coupling constant J_{bd} = 7 Hz. The intensity of each line further reduces to half. As such, the signal for proton H^d exhibits four lines of equal intensity, having one-fourth of the intensity of the original one. The resonance of proton H^d is then called a *doublet of doublets*.

Similarly, the splitting of H^b and H^c protons can also be explained. Each of these two protons resonates to exhibit a doublet of doublets with coupling constant value J_{bc} = 1.5 Hz. In the spectrum, a total of 12 lines are obtained for alkene protons of vinyl pivalate. This yields a complex splitting pattern due to multiplicative splitting. This suggests that a complex splitting pattern is obtained due to multiplicative splitting.

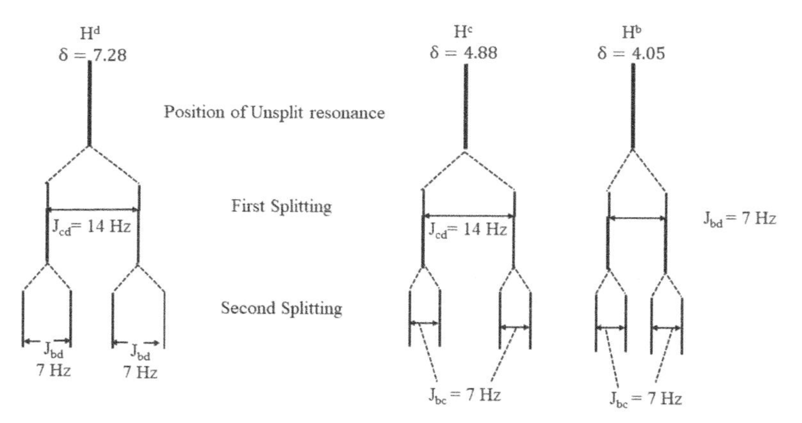

Figure 2.22 Schematic representation of multiplicative splitting in vinyl pivalate.

The case of multiplicative splitting with an identical coupling constant can be explained with the help of the NMR spectrum of 1-nitropropane [1].

2.13.2 Non-first-order Spectra: Breakdown of the (*n* + 1) Rule

It is difficult to explain the spectra of those compounds which contain more complex splitting with the help of the (*n* + 1) rule. Thus, (*n* + 1) rule breaks down. The condition of a first-order spectral behavior is expressed as $\Delta\upsilon_{ab}$ (in Hz) >>J_{ab}. Fairly unperturbed spectra are obtained when $\Delta\upsilon \geq 6J$. Under this condition, the chemical shift difference between coupled protons is greater than 60 Hz. But when the chemical shift difference of the coupled protons is noticeably less than 60 Hz, the spectra exhibit non-first under splitting patterns. Therefore, a complex spectrum is observed which can be explained with the help of the (*n* + 1) rule.

The spectra of some compounds with more complex splitting patterns cannot be predicted by the general (*n* + 1) rule for the first-order spectra. In general, the first order spectra occur only when the separation between multiplets, i.e., the chemical shift difference $\Delta\upsilon$ (in Hz), between two coupled protons is much greater than the value of coupling constant J. If the difference in the chemical shift of two resonances of protons a and b is $\Delta\upsilon_{ab}$ and if J_{ab} is the value of their coupling constants, then the spectra follow the condition.

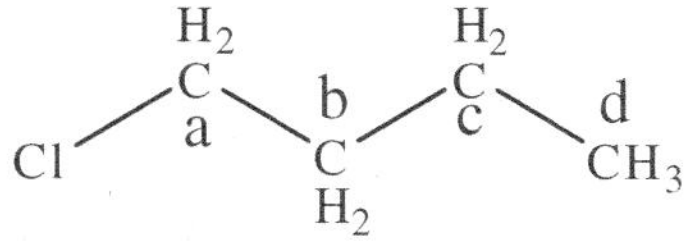

Cl–CH_2(a)–CH_2(b)–CH_2(c)–CH_3(d)

1-chlorobutane

If the value of $\Delta\upsilon \geq 6J$ then fairly unperturbed spectra will be obtained. But practically, the value of $\Delta\upsilon$ should be greater than the coupling constant J by a factor of nearly 10 to show a clear first-order spectrum.

When the difference in the chemical shift $\Delta\upsilon$ of the coupled protons is substantially less than the ~10 times, non-first order splitting patterns are obtained. The general (*n* + 1) rule breaks down and a complex NMR spectrum is observed. For example, in the NMR spectrum of 1-chlorobutane, the differences in the chemical shift values of H^b, H^c, and H^d protons are less. So, the $(n + 1)$ rule will not be followed and one would observe non-first order splitting patterns for these protons. However, H_a proton will show a triplet (first-order splitting) according to the $(n + 1)$ rule.

In a non-first-order spectrum, the signals from the coupling protons are observed close to each other and the difference in the chemical shift value is also small. Therefore, distorted signals will be observed.

Figure 2.23 illustrates this in the compound of AX type (1-chlorobutane). As the NMR signals come closer, the inner peaks enlarge at the expense of outer peaks and the positions of the lines also change. In the end, when the two protons have the same chemical shift so that $\Delta\upsilon = 0$, the interaction between the spins will not be seen in the spectrum and a single peak is obtained.

Many non-first order spectra can be simplified by utilizing the fact that coupling constants do not vary with the operating frequency. Thus, the NMR spectra could therefore be obtained at different operating frequencies and magnetic field strengths. The chemical shifts (in Hz) change is proportional to the operating frequency of the instrument. Therefore, very large operating frequencies (large magnetic field) are used for obtaining the NMR spectrum. The chemical shift value increases but the coupling constants remain unchanged. As a result, the difference in the chemical shift value $\Delta\upsilon$ increases in comparison to coupling constant value J, and the condition for the first order spectra is obeyed again. For example, if the NMR spectrum of 1-chlorobutane is recorded by a 360 MHz instrument then the spectra will be completely first-order spectra and obey the $(n + 1)$ splitting rule. However, when its spectrum is recorded on a 60 MHz instrument then non-first order will be obtained in which the $(n + 1)$ rule is broken down. The value $\Delta\upsilon = 108$ Hz and J = 7 Hz is obtained with the help of 360 MHz instruments. Since, $\Delta\upsilon >> J$, therefore, it obeys the condition of the first order spectra as shown in Figure 2.24a. But, the values of $\Delta\upsilon$ and J, obtained with the help of a 60 MHz instrument, were found to be 18 Hz and 7 Hz, respectively. This suggests that the difference between the value of $\Delta\upsilon$ and J is not large and a non-first order splitting pattern will be obtained in the NMR spectrum of 1-chlorobutane [7].

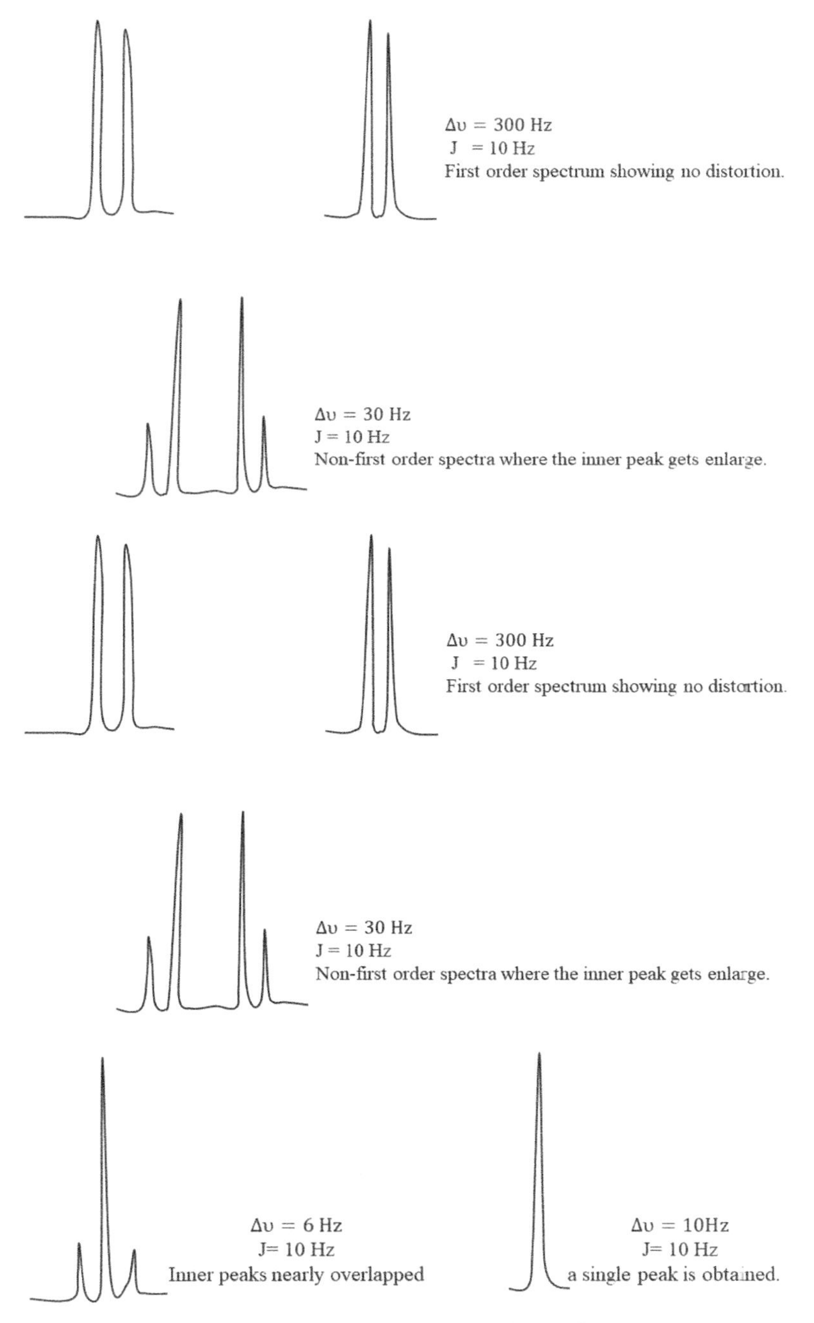

Figure 2.23 Schematic representation of the effect of $\Delta\upsilon$:J in the spectral pattern.

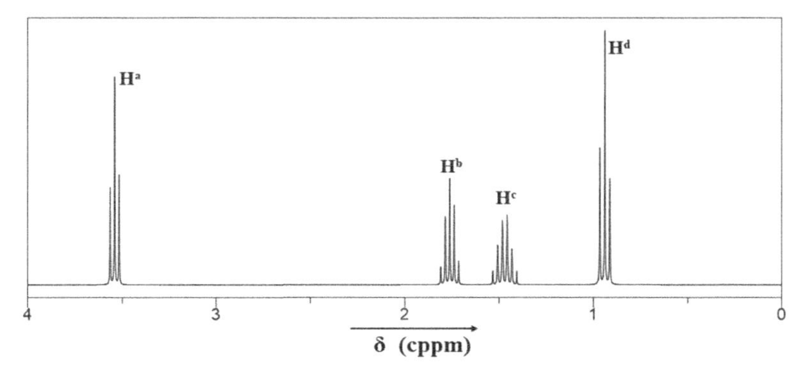

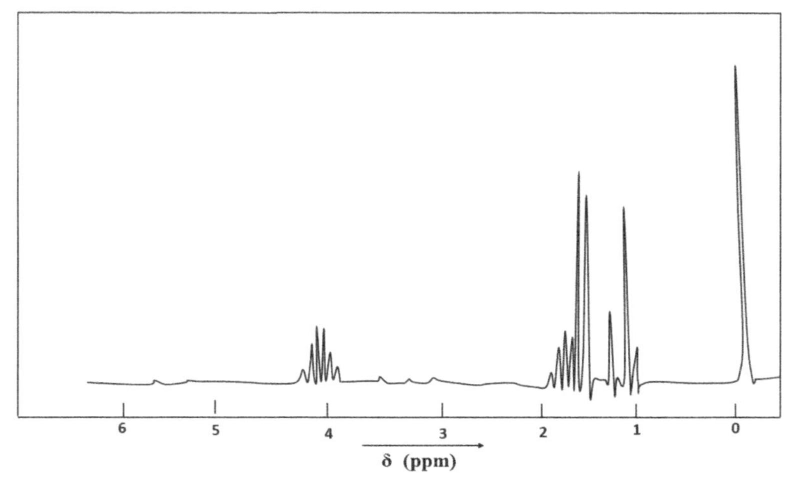

Figure 2.24 (a) NMR spectrum of 1-chlorobutane at 360 MHz. (b) NMR spectrum of 1-chlorobutane at 60 MHz.

2.14 The Coupling Constant (J)

In a multiplet, the distance between the centers of the two adjacent peaks in the spectrum is usually constant. This is called the *coupling constant* denoted by the symbol 'J.'

The magnitude of the separation of the peaks in the multiplets depends on the spin–spin interactions of the neighboring protons, but it is entirely independent of the strength of the applied magnetic field strength. The coupling constant value J which is always constant does not depend on the strength of the applied field. The coupling constant of the two nonequivalent hydrogen atoms, designated as H^a and H^b is represented as J_{ab} (either in Hz and c.p.s).

It is possible to distinguish between the two singlets and one doublet, or between a quartet and two doublets with the help of the values of the coupling constant, J. The spectrum is recorded at two field strengths and if the separation between the lines of the peak (J value) does not change, then the NMR signals show a doublet. However, if the separation between the lines of the peaks increases with the increasing frequencies of the applied field, then the signals indicate the presence of two singlets in the NMR spectrum. The value of J usually lies between 0 and 20 Hz. Similarly, a quartet can also be distinguished from the two doublets.

Let us take a compound as mentioned below which gives two signals in the NMR spectrum.

$$>\overset{a}{CH}—\overset{b}{CH_2}—$$

The peak signals for CH^a proton type show a triplet due to the influence of two equivalent protons of the $-CH^b{}_2-$ group. This triplet formed due to spin–spin coupling is shown in Figure 2.25a.

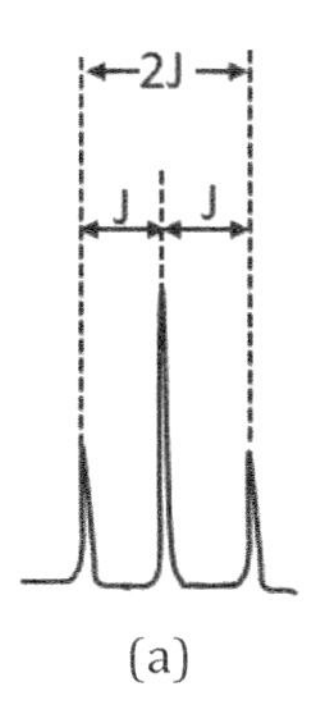

(a)

Figure 2.25 *(Continued)*

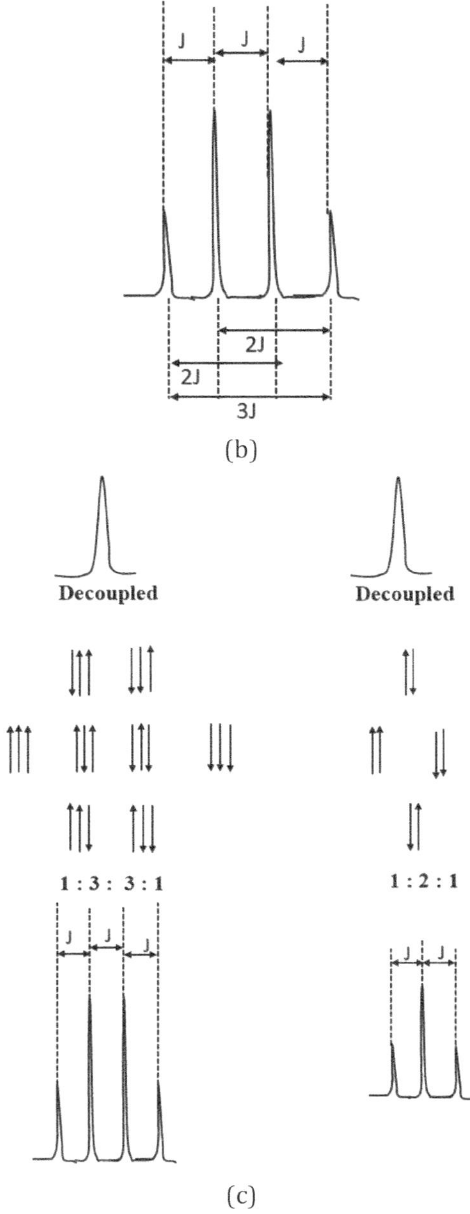

Figure 2.25 (a) Schematic representation of a triplet showing similar J values between its two adjacent peaks. (b) Schematic representation of a quartet showing equal J values between its two adjacent peaks. (c) Schematic representation of decoupled and coupled with the compounds of proton H^a and H^b in CH^a_3-CH^b_2-X.

Figure 2.25 clearly shows that the distance between the two adjacent peaks (J) in the triplet is exactly equal. Similarly in the NMR spectrum of this compound, the signals from the proton a ($>CH^a{-}CH^b{}_2{-}$) will exhibit a quartet due to the interactions by three equivalent $-CH_3$ group protons of type H^b. This quartet is shown in Figure 2.25b. The intensity ratio of the peaks, in this case, will be 1:3:3:1. Here also, the J values are the same and constant.

In the compound $CH^a{}_3\text{-}CH^b{}_2\text{-}X$, the splitting pattern is shown in Figure 2.25c. The coupling of protons in the adjacent carbon atoms is responsible for the splitting patterns. The figure suggests that the spin–spin splitting takes place due to the presence of the proton on the adjacent and (d) carbon atom. The proton H^b of $-CH_2$ group splits into four in the ratio 1:3:3:1 whereas the proton H^a $-CH_3$ group splits into three peaks in the ratio 1:2:1. In this figure, the sets of arrows show the possible alignments of adjacent nuclear spin under the influence of applied magnetic field [7].

2.14.1 Factors Influencing Coupling Constant, J, and Various Types of Couplings

The J values depend upon the number of bonds present between the coupling protons. Because the coupling takes place via the electrons of the bonds that connect the two coupled protons. As per convention, the number of bonds involved in the coupling is mentioned as a superscript to the coupling constant J. Therefore, it may be written as (i) ^1J represents a one bond coupling called the direct coupling ($^{13}C{-}^1H$), (ii) the coupling between protons of a $-CH_2$ group will be denoted as ^2J, and (iii) when the coupling takes place between protons present on adjacent carbon atoms then it is represented as ^3J and so on. Different types of couplings important for the change in the values of the coupling constant are mentioned below.

2.14.1.1 *Geminal coupling*

The study of the geminal coupling constant gives important information about the structure, conformation, and fine electronic interaction in a compound. The magnitude of geminal coupling constants, J_{Gem} generally varies in the range of 10–18 Hz. Geminal protons are attached to the same carbon atoms of a particular >CH2 group having different chemical environments. In the case of saturated compounds, due to the presence of geminal protons, the J value depends upon the bond angles between the two protons attached to a C atom (—∢). J values can be positive or negative.

The geminal coupling depends upon the electronegativity of atoms or group to which the protons are attached. J values increase with the increase in the electronegativity of the attached substituents. However, electronegative substituents like F, O being a π-acceptor withdraw electron density from the π-bond of vinyl compounds, and the value of the J decreases. For example, in the case of ethylene, the value of germinal coupling is nearly equal to +2.5 Hz but decreases to -3.2 Hz of vinyl fluoride (CH_3F). In vinyl compounds, the value of geminal coupling is quite low nearly in the range of 1–3 Hz. On the other hand, when electropositive substituents (Si, Li, etc.) are present on the neighboring carbon atom, they behave as electron donors, and the value of the coupling constant increases. For example, the J value of $C_2H_3Si(Me)_3$ is 3.8 Hz and for $C_2H_3Li(CH_2{=}CHLi)$, J = 7.5 Hz. Also, if there are two substituents out of which, one is a π-donor and the other one a π-acceptor, the J value will be too small to measure ($J \approx 0$). The geminal coupling also varies with ($C-\hat{C}-C$) bond angles and the value is highest ($J_{Gem} \approx$ 10–14 Hz) in strain-free cyclohexane and cyclopentane. However, when the angular strains in the molecules are increased, then the value of J_{Gem} decreases. For example, in cyclobutanes, the value of J_{Gem} is ≈ 8–14 Hz and in cyclopropane, the value of J_{Gem} is ≈ 4–9 Hz.

2.14.1.2 *Vicinal coupling*

Vicinal coupling occurs when that three bonds separate the protons which are present on adjacent carbon atoms in a molecule. The magnitude of vicinal coupling constant $^3J_{vic}$ varies widely in the range from 0 to 25 Hz depending upon the molecular structure and its conformations [12] but mostly it has been reported that the $^3J_{vic}$ values range from 0 Hz to 12 Hz. Values of vicinal coupling in *trans* form (≈ 11 to 19 Hz) are more than the *cis* form (≈ 6 to 14 Hz) for olefinic compounds.

Trans form $^3J_{vic} \approx$ 11 to 18 Hz cis form $^3J_{vic} \approx$ 6 to 14 Hz

The $^3J_{vic}$ values depend on the dihedral angle between the protons. $^3J_{vic}$ values can be calculated from the values of dihedral angle ϕ between the two protons.

$^3J_{vic,H-H} = 8.5 \cos^2\phi - 0.28$ (where ϕ varies from 0 to 90°)

$^3J_{vic,H-H} = 9.5 \cos^2\phi - 0.28$ (where ϕ varies from 90° to 180°)

The magnitude of vicinal coupling changes with the increasing or decreasing electronegativity of the attached substituent on a group. If the substituent group is more electronegative, the $^3J_{vic}$ value will be smaller. For example, in unhindered ethanes, the value of $^3J_{vic}$ is nearly equal to 8Hz, whereas, in halogenated ethane, the value of $^3J_{vic}$ value falls in the range of 6–7 Hz. The dihedral angle ϕ between two vicinal protons C–C bonds also affects the value of J_{vic}. The observed values are in good agreement with the "KARPLUS equation."

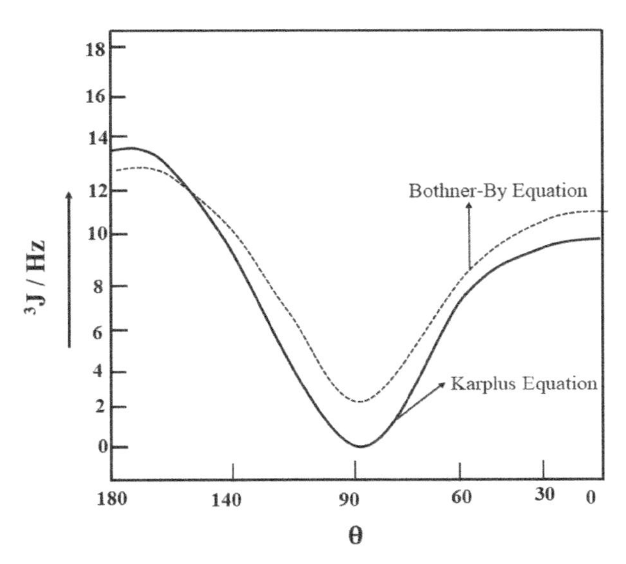

Figure 2.26 Two curves are plotted between the predicted coupling constant for the coupled proton as the function of the dihedral angle.

The Karplus equation based on observation and supported by theoretical considerations is written as

$$^3J_{vic,H-H} = J_0\cos^2\phi - K \quad (2.27)$$

where the constants J_0 and K have been used to correct the effect of substituent groups.

Similar to the Karplus curve, another Bothner-by equation also yields a curve in which the different J_0 values (as mentioned in the Karplus equation for 0–90^0 and 90–100^0 dihedral angles) are not required. The Bothner-by equation is mentioned below as

$$^3J_{H-H} = 7 - \cos\phi + 5\cos2\phi \quad (2.28)$$

The typical Karplus and Bothner-by equation curves are shown in Figure 2.26.

The maximum vicinal coupling occurs when the protons are in the trans-co-planner position (ϕ = 180^0). For cis coplanar protons (ϕ = 0^0), $^3J_{vic}$ values are also large but when protons are at 90^0 to each other very low $^3J_{vic}$ values are observed. This variation is ^3J values have been illustrated as follows:

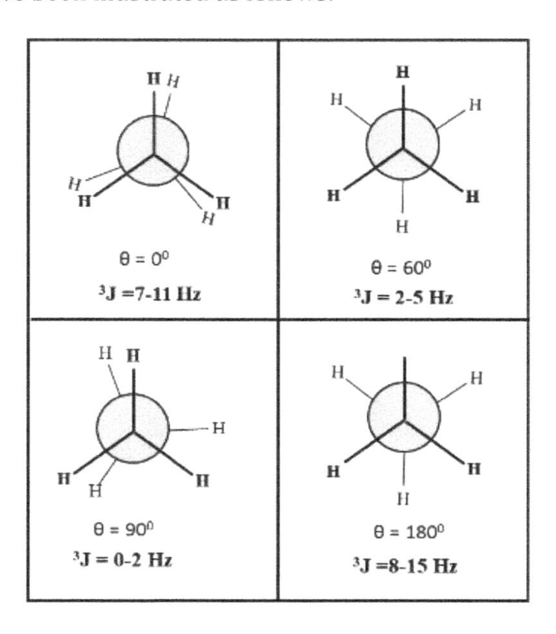

2.14.1.3 *Long-range proton–proton coupling*

When proton–proton coupling extends over more than three bonds long-range coupling occurs. It is represented as ^4J coupling. These values are too small to detect ($^4J_{LR}$ is ≤ 1 Hz). ^4J coupling is mostly observed in the aromatic compounds (meta couplings), such as allylic, propargylic, and allenic compounds. In saturated carbons or heteroatom ^4J coupling is very rare.

Table 2.6 Long-range proton–proton coupling constant (J_{LR}) for different compounds

Structure	J_{LR}, Hz
H—C=C—C—H, Allyl	1–4
H—C—C=C—C—H, Homoallyl	0.5–5
H—C=C=C—H, Alkene	6–7
H—C=C=C—C—H, Alkene	3–4
H—(C≡C)$_n$—H, Acetylene	up to 1
H—(C≡C)$_n$—C—H, Acetylene	~ 2 (n=2)
H—C—(C≡C)$_n$—C—H, Acetylene	2–3 (n=1)
H—C=C—C=C—H, 1,3-diene	up to 2
H—C—X—C—H, Carbon chain including heteroatom	0–3

2.14.1.4 *Heteronuclear coupling*

When organic molecules, if other magnetic nuclei apart from hydrogen are also present, then it may cause an additional complication in the NMR spectrum. This spin–spin splitting caused by other magnetic nuclei under an applied magnetic field is known as heteronuclear. This phenomenon could be explained with the help of NMR spectra of 2,2,2-trifluoroethanol CF_3–CH_2–OH. The NMR spectrum of this compound shows that the –CH_2 group protons gives a singlet. This observation is very much similar to the splitting of the –CH_2 group of –CH_3 proton in ethanol. The coupling constant $J_{H\text{-}H}$ and $J_{H\text{-}F}$ are approximately equal [1]. The spectra of 2,2,2-trifluoroethanol are presented in Figure 2.27.

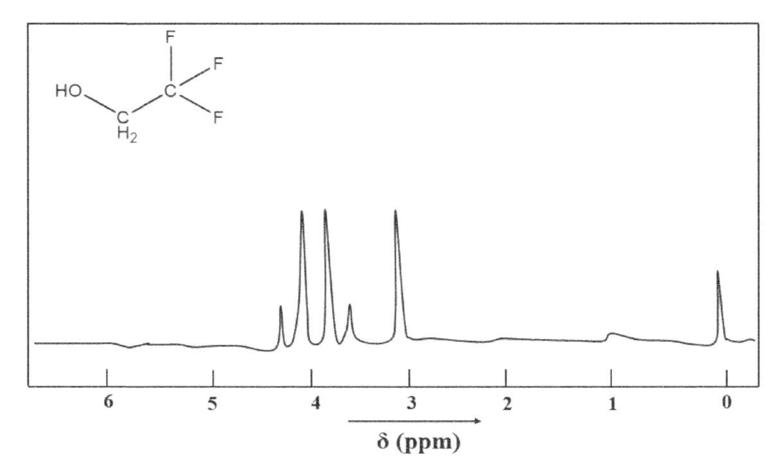

Figure 2.27 NMR spectrum of 2,2,2-trifluoroethanol.

2.14.1.5 *^{1}H–^{13}C spin–spin coupling*

Isotope ^{13}C with a magnetic moment and spin of ½ is present in natural carbon only to an extent of 1.08–1.1 % [1, 4]. These ^{13}C nuclei split the ^{13}C-H group protons to yield an additional weak signal (doublet) along with the main signal in the NMR spectra. These weak signals are produced by splitting the peaks of protons connected directly to this carbon isotope and are known as ^{13}C satellites. These two signals are separated from each other with the coupling constant, $J_{H-^{13}C}$ by ≈ 120 Hz. The value of $J_{H\text{-}13C}$ gets affected by the character of the hybridization of the carbon orbital along the C-H bonds [13, 14].

2.14.1.6 Deuterium coupling

In the NMR spectroscopy, three isotopes of hydrogen viz, ^{1}H, deuterium (2H or D), and tritium (^{3}H or T) can be used but tritium being radioactive is less commonly measured. Each isotope of hydrogen yields different resonance frequencies. For example, ^{1}H shows resonance at 400 MHz whereas in deuterium (^{2}H) the resonance is observed at 61.4 Hz. ^{2}H-^{2}H coupling is nearly 40 times smaller than the ^{1}H–^{1}H coupling and hence, it is very difficult to obtain. But, ^{1}H–^{2}H coupling can be observed. Therefore, any group containing one deuterium nucleus will exhibit resonance in presence of an applied magnetic field. Deuterium having spin quantum number, I = 1 can exhibit three orientations corresponding to –I, 0, and +I. Thus, if the NMR spectrum of ^1H proton is recorded for >[a]CHD group, three peaks resonance at different frequencies are observed leading to and in the NMR spectrum, a triplet is observed. The coupling constant is denoted as J_{HD}.

The three spin states of deuterium are nearly equal to each other as such the line intensities of this triplet are found in ratio 1:1:1. However, if 'H' is coupled with a particular group that contains more than one deuterium then, in that case, the multiplicity of ^1H–^2H coupling will occur according to the general formula (2n + I) = 6. Thus, a coupling due to two deuterium will yield a quintet while a coupling due to three deuterium will show a septet in the NMR spectrum.

2.15 Proton Exchange Reaction in NMR Spectrum of Alcohol

The chemical shift of proton of a particular –OH group in pure and dry alcohol depends upon the involvement of that proton in H-bonding when the NMR spectrum is obtained. In the case of pure ethanol, CH_3CH_2OH, three resonance peaks must be observed. The first peak is a triplet for –CH_3 protons that appears at nearly 1.3 ppm due to spin–spin coupling with protons of the –CH_2 group. (ii) the second peak is a multiplet for the protons of the –CH_2 group appears at nearly 3.6 ppm due to the coupling with –CH_3 protons and the protons of the –OH group. This multiplet consists of eight lines because the –CH_2 protons are under the influence of two different types of proton environments. The (n+1) splitting rule predicts a quartet due to –CH_3 protons but because –OH group each line of the quartet gets further split into two yielding a total of eight lines in the multiplet. (iii) The third peak is triplet for –OH proton due to its spin–spin coupling with –CH_2 group protons appearing at nearly 5.3 ppm in the NMR spectrum.

The NMR spectrum of pure and very dry ethanol exhibits these three types of resonance peaks as shown in Figure 2.28.

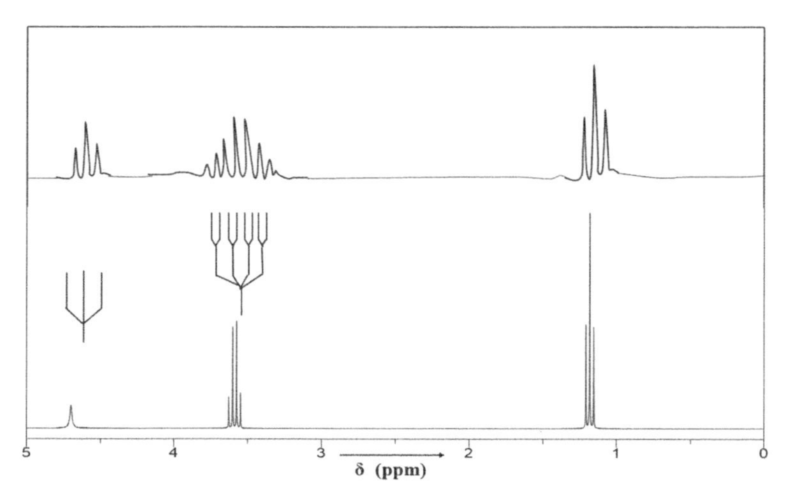

Figure 2.28 Schematic representation of ^{1}H NMR spectrum of (a) pure dry ethanol and (b) ethanol with acidic impurities.

However, if the NMR spectrum of commercial-grade ethanol or water-containing ethanol is observed then the resonance peak due to the proton of the –OH group appears as a singlet. This clearly shows that the –OH proton is not involved in the coupling with the protons of the $-CH_2$ group. In fact, the presence of water, acid, and base in the ethanol restricts the -OH resonance to yield a single line and erase all other splittings associated with this proton. Further, the resonance of CH_2 protons shows a quartet because of splitting only by CH_3 protons. This type of behavior is quite common for alcohols, amines, and other compounds with a proton bonded to an electronegative atom. The NMR spectrum of commercial group ethanol is also shown in Figure 2.28. This phenomenon is known as chemical exchange or proton exchange reaction. This chemical proton exchange is basically due to equilibrium involving chemical reactions which takes place very rapidly when the NMR spectrum is being observed. The chemical reaction involves a proton exchange between the protons of alcohol and those of water or other alcohol molecules as shown below:

$$R{-}\ddot{O}(H) + H{-}O{-}H \rightleftharpoons R{-}\overset{\oplus}{O}(H){-}H + {:}\overset{\ominus}{O}{-}H$$

$$R{-}\overset{\oplus}{O}(H){-}H + {:}\overset{\ominus}{O}{-}H \rightleftharpoons R{-}\ddot{O}{-}H + H{-}O{-}H$$

The exchange of –OH proton among ethanol molecules takes place very rapidly. In fact, it is so rapid that one particular proton (H) does not stay for a sufficiently long time on a particular oxygen atom. Hence, nuclear coupling is not observed. Rapidly exchanging protons do not show spin–spin coupling with neighboring protons. The presence of acid and base also catalyzes this exchange reaction. They accelerate it to such an extent that splitting is no longer visible.

The proton exchange reactions are absent when (i) the sample under investigation is pure and dry (ii) when the spectrum is recorded at a low temperature and (iii) when the sample has been dissolved in a highly polar solvent such as DMSO, dimethyl sulfoxide Me_2SO. Low-temperature and highly polar solvents prevent the coupling of –OH protons with the neighboring protons.

The proton exchange reaction also occurs very rapidly in some other compounds like carboxylic acids, phenols, amines, thiols, etc. in which the proton is attached to nitrogen, sulfur, or oxygen. So, no coupling takes place between the protons of these functional groups with other protons on adjacent neighboring groups.

2.16 Restricted Rotation

Rotation about a single bond at room temperature is so rapid that, it is not possible to isolate the rotational isomer. Also, the presence of such isomers in the solution cannot be detected by NMR spectroscopy. But at a temperature much below room temperature, the rate of rotation is greatly reduced. Hence, NMR absorptions resulting from each isomer can be observed. The rotation in a compound is restricted by the presence of a double bond in it and because of the restricted rotation cis and trans isomers are formed. The cis and trans forms have different properties and thus give different NMR spectra. The most thoroughly investigated example of this phenomenon is the hindered rotation which exists about the –CN bonds has considerable double bond character.

The NMR spectra of simple amides generally suggest that an alkyl one group attached to N may either be cis to O or can be trans to O.

For example N,N′-dimethyl formamide exists in two resonating forms as mentioned below.

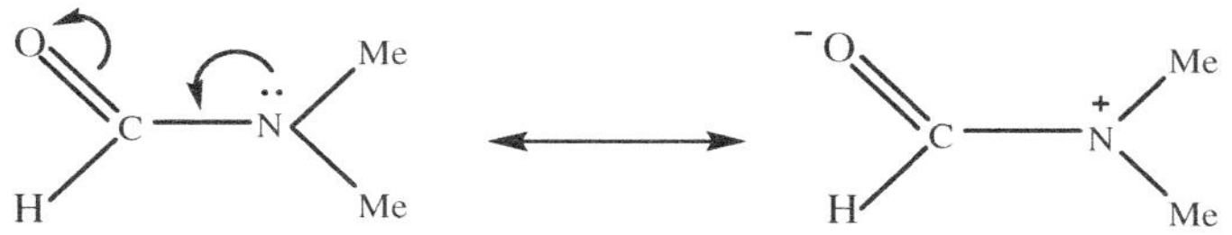

The conjugation between the carbonyl and the non-bonding electron pairs of N increases the double bond character of the –CN bond in this compound. At room temperature, this restricts the rotation which leads to the formation of *cis* and *trans* isomers, in which one methyl group ($-CH_3$) is cis to the C=O oxygen while in the other methyl group ($-CH_3$) is trans to O. Its NMR spectrum, when recorded at room temperature yields two NMR peaks for methyl groups (for cis appears at $\delta = 2.881$ and for trans at $\delta = 2.97$). Actually, this is against the expectation that there should have been only one resonance peak in the NMR spectrum of the compound since both the methyl groups are magnetically equivalent. But at higher temperature ~130°C, the NMR spectrum of this compound exhibit only one peak as expected. At a higher temperature, since the rotation around the –CN bond is very rapid and one particular methyl group is unable to stay for a sufficiently long time at one position, therefore, the NMR spectra are not able to differentiate the cis and trans methyl groups and only one peak is obtained instead of two peaks.

In another example of 1,2-dibromoethane, there are four chemically equivalent protons and being also magnetic equivalent, should yield the same chemical shift positions. But the three rotational isomers of 1,2-dibromoethane clearly suggest that the four protons are present in different conformations and, in a real sense, are magnetically non-equivalent. In one conformation, the encircled proton lies in between H and Br, in the second conformation, it lies between H and H. Therefore, if the NMR spectrum is recorded, one should get two different resonance peaks in the spectrum. But, this can only be observed at a very low temperature. When the rapid motion around C–C freezes and these two different conformations can be detected. But when the NMR spectrum is recorded at room temperature, the rotation around the C–C bond is so fast that the proton does not stay at one position for a long time and experiences a similar magnetic environment around it. Therefore, an intense resonance peak around $\delta = 3.65$ ppm appeared [1].

2.17 Deuterium Exchange in NMR Spectroscopy

Deuterium (^{2}H, D) exchange reaction has special importance in NMR spectroscopy. ^{2}D NMR and ^{1}H NMR both operate at different frequencies in a magnetic field. The ^{2}H NMR resonance peaks cannot be detected under the conditions of ^{1}H proton NMR spectroscopy. Deuterium substitution is used to simplify the NMR spectra and identify the resonance peaks of a particular group in a compound.

When deuterium is used as a solvent, then the labile proton (H atom) of –OH, –NH, –SH, etc. groups are exchanged by D_2O. The proton-exchange reaction mechanism is applicable in this case also and ROH is converted in ROD, RCOOH into RCOOD and $RCONH_2$ in $RCOND_2$, etc. The exchange reaction is written as

$$R{-}OH + D{-}O{-}D \rightarrow R{-}O{-}D + H{-}O{-}D$$

As a result of the above reaction, the resonance peaks for proton (H-atom) are diminished and a new peak corresponding to H–O–D is observed in the spectra at around $\delta = 5$.

This process, known as *deuteriation*, is particularly important for the detection of -OH, -SH, and -NH groups in the compounds. Simply shaking the solution of such compounds with little deuterated water (D_2O) causes a very quick exchange of the -OH, –SH, and –NH protons to form –OD, –SD, or –ND group, etc. in the sample compounds. Thus, in the NMR spectrum of such deuterated compounds, the resonance peaks expected for –OH, –NH, or –SH groups are not found. This deuterium exchange eliminates the –OH, –NH, or –SH group resonances completely. This process is known as D_2O exchange or D_2O shake.

Thus, the presence of such groups in the sample is confirmed by running two NMR spectra; one dissolving the sample in a solvent other than D_2O, and in another case, the solution of the sample is prepared in the same solvent and then a few drops of D_2O is mixed smoothly into it.

2.18 Simplification of Complexities of Proton NMR Spectra

The complex NMR spectra can be simplified for more accurate determinations of chemical shift and coupling constant values. This could be done with the help of following methods:

2.18.1 By Increase in the Field Strength

The position of the chemical shift depends upon the strength of the applied magnetic field, whereas the coupling constant values are independent of it. This fact could be utilized in the simplification of complex NMR spectra of certain compounds. For example, when the NMR spectrum of a compound is recorded by a low field instrument (say 60 MHz) two multiplets due to proton coupling are observed which overlap with each other. These overlapping multiplets can be separated by increasing the field strength of the NMR instrument nearly to ≈100 MHz or 600 MHz. In this way, the originally observed complex spectrum becomes easy to analyze. By this technique, the difference in the chemical shift values, $\Delta\upsilon$ increases with respect to the coupling constant, J.

2.18.2 By Spin Decoupling Method

When the neighboring protons contain more than one spin orientation, the resonance peaks will appear as multiplets making the spectrum more complex for interpretation. In the complex spectrum, it is not easy to identify the mutually coupled protons from one another as they give the coupling constant values. It also becomes difficult to determine directly the multiplicity and the coupling constants in the spectrum.

This problem could be solved with the help of the spin decoupling technique. In this method, the observed coupling of the neighboring protons is fully eliminated by irradiating the proton with the radiofrequency source. In doing so, a rapid transition between irradiated protons takes place and the coupled proton partner experiences only the average of its two spin states. Therefore, the coupled proton partner show resonance at a single frequency, and the coupling between the protons vanishes. In other words, the two protons are decoupled from each other and make the NMR spectrum simple.

In this technique, in addition to the basic NMR instrument, another tunable radiofrequency source is required to irradiate the neighboring proton at the necessary frequency. Since two radiofrequency sources are used simultaneously in the process, this spin decoupling technique is also known as *double resonance* or *double irradiation*. Due to this process, the observed multiplets are converted into a singlet.

Let us understand this process more clearly by taking a specific compound of the type in which the protons H and H′ have different environments. In its spectrum, each proton will show two doublets in different field strengths as shown in Figure 2.29.

Now one of the protons H′ is irradiated with an appropriate radiofrequency source. As a result, a rapid transition between the two spin states of H′ occurs, and due to upward and downward transitions, the lifetime of the nucleus (in any one spin state) becomes extremely less for coupling with H. So, the proton H will resonate only once and not twice (it sees only one time-averaged view of H′). The time (Δt) required to resolve the two lines of a doublet for a pair of coupled protons is related to the coupling constant. The proton H will show a doublet when each spin state of H′ has a lifetime greater than Δt. Double irradiation reduces this lifetime and therefore, the proton H does not resonate to show a doublet and appears simply as a singlet in the spectrum.

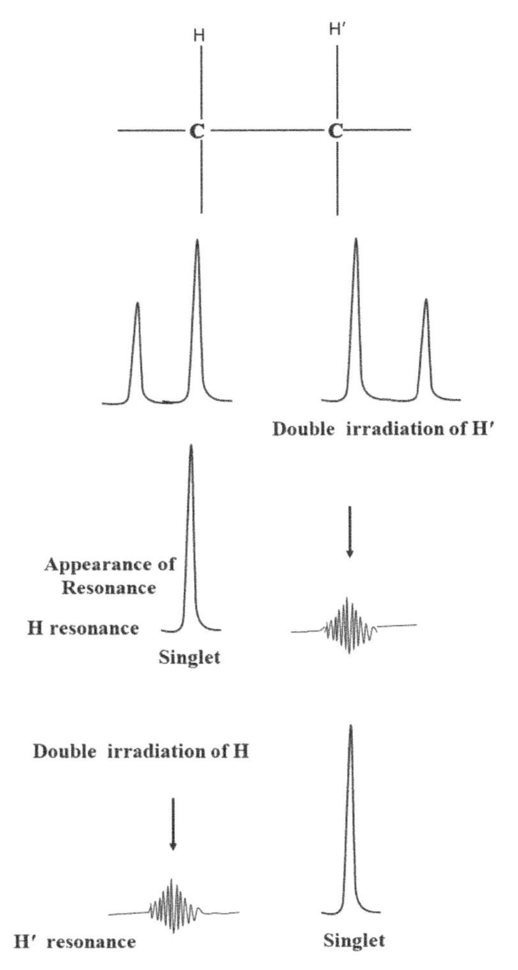

Figure 2.29 Representation of spin-decoupling.

2.18.3 By Using Chemical Shift Reagents

Chemical shift reagents simplify a complex NMR spectrum into a simple one for easy identification and analysis. The overlapping resonance peaks in complex NMR spectra can be separated from each other to make the spectra into a simple first-order one. β-diketone has been used as a potential extractor and complexing agent in the spectrophotometric determination of metal ions in dilute solutions and chromatographic separations. Lanthanide β-diketone could therefore as a useful shift reagent [2]. Also, several complexes of Eu and lanthanides when added to the sample show marked improvement and simplicity in the spectrum. The spectrum almost becomes of the first order, when it is stretched over a much wider range of frequencies on the addition of such reagents in the solution of compounds under investigation. The lanthanides ions are found to cause a shift in the NMR signals with a line broadening effect. The LSR should show good solubility in a common NMR solvent. Few commonly used LSRs are lanthanide complex with two enolic, β-diketones, FOD, and FDM. These complexes possess protons and are soluble in CCl_4. In doing so, the spectrum is found to be separated into a wide range of frequencies and gets simplified nearly into a first-order spectrum.

The effect of the addition of lanthanide shift reagent [Eu (DPM)$_3$] on the ^{1}H NMR spectrum of 6-methyl quinoline at 60 MHz in $CDCl_3$ is shown in Figure 2.30. The upper spectrum is taken without adding LSR in the normal manner while the lower portion of the spectrum was recorded after the addition of soluble europium (III) complex to the solution. This spectrum is stretched to a wide range of frequencies and has been simplified nearly to first order spectrum.

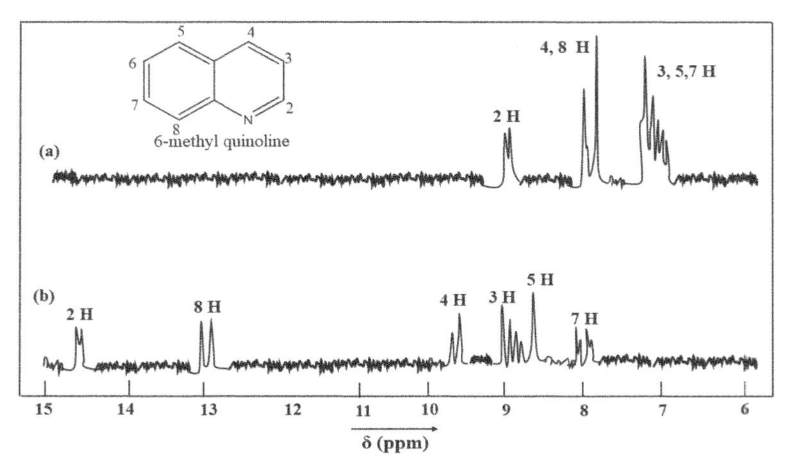

Figure 2.30 NMR spectrum of 6-methylquinoline (a) in $CDCl_3$ solution, (b) on the addition of LSR [Eu (DPM)$_3$].

2.18.4 NMR of Nuclei Other Than Protons

Besides ^{1}H NMR spectroscopy, any other nucleus possessing a nuclear spin can also be studied by NMR spectra. Among various nuclei, ^{13}P, ^{19}F, and ^{13}C find great attention in NMR. Both ^{31}P and ^{19}F have 100% abundance and in them, the value of $I = 1/2$. ^{31}P and ^{19}F compounds can be studied in terms of their characteristic chemical shifts and spin–spin coupling to neighboring nuclei.

2.18.4.1 *^{13}C NMR*

Although the principles of ^{13}C NMR and ^{1}H NMR are basically the same, one problem which arises is that the ^{13}C nucleus gives intrinsically weaker resonance than that of a proton due to its magnetic properties. The ratio of ^{13}C and ^{1}H NMR resonances is nearly equal to 1/62.9 means the resonance intensity of ^{13}C is 0.159 times less than the ^{1}H proton NMR spectroscopy. Some characteristic features of ^{13}C NMR are unique. Firstly, like a ^{1}H NMR, the coupling (splitting) between carbons is not observed. The reason for this is the low natural abundance of ^{13}C. The probability of occurrence of ^{13}C at a given carbon atom is only 0.0110, so the probability of occurrence of ^{13}C atoms within the same molecule will be so low that no coupling is observed (for coupling to occur two ^{13}C atoms must be present in the same molecule). However, isotopically enriched ^{13}C compounds may be prepared and then in those compounds, the regular splitting rules may apply.

A second important feature of ^{13}C NMR is that the range of chemical shifts is very large (approximately 40 times greater) than in ^{1}H NMR. The chemical shifts of ^{13}C nuclei in organic compounds mostly range from $\delta = 0$ to about $\delta = 260$. Thus, chemical shifts in ^{13}C NMR spectroscopy can very conveniently be measured (in ppm) using tetramethyl silane (TMS) as a reference where TMS is arbitrarily given the value $\delta = 0$. The chemical shift of a particular carbon atom depends on its structural environment and the chemical shifts are quite sensitive to small changes that occur in its chemical environment. Due to this unique characteristic of ^{13}C NMR, it is often possible to obtain clear distinct resonances even for two carbons in very similar chemical environments.

The third important aspect of ^{13}C NMR is that coupling of ^{13}C by the neighboring protons (^{13}C–^{1}H splitting) is quite large. Strong spin–

spin coupling occurs not only between a ^{13}C nucleus and the nuclei of hydrogen atoms bonded to it but also between a ^{13}C atom or the nuclei of hydrogen atoms separated by two, three, and more bonds. Due to such extensive coupling a complicated spectrum is obtained which is very difficult to interpret. Such splitting may sometimes be very useful in the interpretation of ^{13}C NMR spectra. For example, an H_3C-H group in a molecule will give rise to a doublet (due to ^{13}C=^{1}H coupling) in the ^{13}C NMR spectrum with the coupling constant $J_{C\text{-}H}$=125 Hz, the group >C=C–H will yield a coupling in which J_{CH} = 170 Hz while a group –C≡C–H shows coupling in which J_{CH} = 250 Hz. Since the difference in J values are large, these groups can be easily distinguished in the spectrum.

But now with the help of FT instruments which permit simultaneous irradiation of all ^{13}C nuclei, ^{13}C NMR spectrometry is extensively used. In this technique, a short powerful radiofrequency of the order of a few microseconds simultaneously excites all the nuclei. Each ^{13}C nucleus then shows free induction decay (FID) which is represented as an exponentially decaying sine wave. The frequency of this wave is equal to the difference between the applied frequency and the resonance frequency of that nucleus. If in a compound more than one ^{13}C nuclei are present then FID may be represented by superimposed sine waves of different characteristic frequencies in form of an interference pattern. These data are then fed into a computer to build up the signal thus, Fourier transformation by the computer converts this information into a conventional ^{13}C NMR spectrum.

2.18.4.2 *^{19}F NMR spectroscopy*

Fluorine-containing compounds can easily be identified and analyzed with the help of ^{19}F NMR technique. The isotope of fluorine ^{19}F has got 100% natural occurrence and possesses a nuclear spin I = 1/2 with a very high gyrometric ratio. As a result, the ^{19}F isotope of fluorine gives a very good response in NMR analytical techniques [15–17]. Besides ^{1}H, the nucleus ^{19}F possesses certain properties which are very favorable to nuclear magnetic resonance. In the ^{19}F NMR technique, the range of chemical shifts of ^{19}F is nearly 20 times greater than those of ^{1}H proton. ^{1}H, ^{19}F, and ^{31}P nuclei are highly sensitive NMR active nuclei with spin ½. The chemical shift of ^{19}F NMR ranges up to 800 ppm. This range is a little less in organofluorine compounds ≈50–70 ppm CF_3 groups and 200–220 ppm CH_2F groups containing compounds.

The NMR spectrometer for ^{19}F needs a little modification in the conventional ^{1}H spectrometer. In this case, the appropriate frequency source of the magnetic field strength of ≈56.5 MHz at 1.4 Tesla is required [1]. ^{19}F being a sensitive nucleus gives sharp resonance signals and possesses a wide chemical shift range. The analysis of the ^{19}F NMR spectrum is done in a similar manner to that of the ^{1}H proton. The analysis of a number of fluorine atoms in the pure sample can be obtained from its spectrum. The number of multiplets and spin–spin coupling present gives information about the fluorine environment and molecular structure. The coupling constant in ^{19}F NMR is usually large as compared to those with ^{1}H. Besides homonuclear F–F coupling in most fluorinated compounds containing hydrogen, the H-F coupling also takes place. The positions of chemical shift in ^{19}F NMR are generally measured by taking $CFCl_3$ or (CF_3COOH) as reference compounds [1]. Chemical shift ranges for ^{19}F in some common organic environments are discussed [1]. It is found that the range of the chemical shift for ^{19}F is from 0 to ≈200 ppm which is ~20 times larger than the range of chemical shift of ^{1}H proton NMR in which the range is only from 0 to ~10 ppm. As such, the fluorine resonance peak position in the spectrum shows proper separation from each other. The value of the coupling constant between fluorine nuclei ^{19}F-^{19}F is larger than the value ^{1}H-^{1}H coupling constant. Long range ^{19}F-^{19}F couplings (^{2}J, ^{3}J, ^{4}J, and ^{5}J) over four bonds (through F–C–C–C–C–F bonds) are possible and its J values lie in the range 0–18 Hz. The longer the range of coupling, the smaller will be the value of J [1]. For geminal F–F coupling, the values of coupling constant lie between 43 and 370 Hz whereas in vicinal coupling J values range between 0 and 39 Hz. The coupling constant value, J in cis and trans F ranges between 0 and 58 Hz and between 106 and 148 Hz, respectively. ^{1}H coupling with fluorine, ^{19}F is also strongly observed between geminal hydrogen and fluorine. In this case, large values of coupling constants are obtained usually in the range of 42–80 Hz. However, in the case of vicinal coupling, the J values for ^{1}H-^{19}F are less (~1.2–29 Hz). In the case of cis and trans ^{1}H-^{19}F couplings, the J value ranges from between 0–22 Hz and 11–52 Hz, respectively [1].

When the benzene ring contains an F atom then fluorine can also couple with ^{1}H proton on the ring, in that case, the values of coupling constant $J_{H\text{-}F}$ depend upon the ortho, meta, and para positions. The coupling constant values for ortho, meta, and para positions are 7.4 to 11 Hz, 4.3 to 8 Hz, and 0.22 to 2.7 Hz, respectively [1].

Figure 2.31 explains the ^{1}H NMR and ^{9}F NMR spectrum of 1-bromo-1-fluoroethane. The ^{1}H proton spectrum was recorded at 60 MHz and the ^{19}F nuclei spectrum was at 56.46 MHz using a magnet of 1.4 T strength.

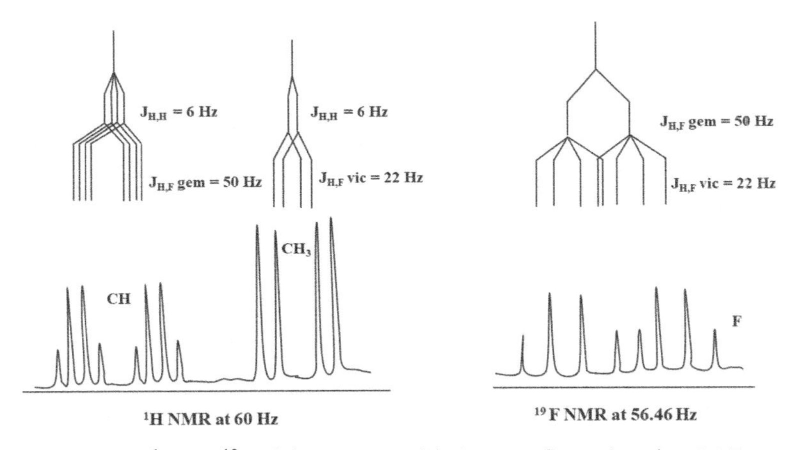

Figure 2.31 ^{1}H and ^{19}F of the compound (1-bromo-1fluroethane) at 1.4 T.

In the ^{1}H spectrum, the H-H coupling and the H-F coupling both take place (Figure 2.31a). In this case, the quartet and the doublet obtained due to CH_3 and CH protons further get split up due to the presence of the 19F atom. Therefore, the doublet of the CH_3 proton becomes a quartet due to vicinal H-F coupling (J_{vic}). In the spectrum, a double doublet is observed for the CH_3 group. However, the quartet obtained by the splitting of CH proton by CH_3 protons further gets split into two due to geminal H-F coupling. Thus, one observes a doublet of the quartet in the 1H spectrum 1-bromo-1-fluoroethane.

On the other hand, in the ^{19}F spectrum of this compound, the signal first splits into a doublet due to H-F coupling and then each line of this doublet gets further split into four because of the presence of vicinal CH_3 protons. Finally, the spectrum shows a doublet of the two-quartet overlapping on each other (Figure 2.31b).

2.18.4.3 *^{31}P NMR*

^{31}P NMR spectroscopy is a useful means for characterizing inorganic compounds and complexes containing phosphorus. ^{31}P, ^{19}F, and ^{1}H NMR spectroscopy are very much similar to each other in terms of natural abundance and spin. The importance of isotropic ^{31}P NMR increases because the isotopic abundance of ^{31}P is also 100% with a high gyromagnetic ratio (17.23). Further, the spin of the ^{31}P nucleus is also ½, therefore its spectra are easy to analyze. The multiplicity of resonance peaks in ^{31}P NMR and ^{1}H NMR can be predicted with the help of the formula (2n + I) which was applied to ^{1}H proton and ^{19}F fluorine. The sensitivity of ^{31}P NMR, which is less than ^{1}H proton and greater than ^{13}C NMR, yields sharp lines with a wide range of chemical shifts in δ its spectrum. The overlapping of peaks in the ^{31}P spectrum is rarely found. In the ^{31}P NMR spectrometer, an 11.7 T magnet frequency source is used and the resonance signal is observed at nearly 202 MHz. The standard reference for ^{31}P NMR is 85% H_3PO_4 but it is always used as an external standard due to its very high reactivity. Chemical shift values are quite high and range between +250 and −250 ppm. ^{31}P is more important because of its presence in nucleic acids. Chemical shift values in ^{31}P NMR depend upon the number of hydrogen substituents. Tertiary phosphine such $P(Me)_3$, $P(et)_3$, PR_3, Pbu_3, show continuous downfield shift as the length of the alkyl chain increases.

2.19 Instrumentation of NMR Spectroscopy

A modern spectrometer whose schematic diagram is shown in Figure 2.32 generally contains the following parts [11].

A Magnet

Generally, three types of magnets are used in NMR spectrometers:

(a) Permanent magnet with a field strength of 14000 G: This type of magnet is used in 60 MHz instruments for the detection of ^{1}H proton spectra.

(b) Electromagnet having a field strength of 23000 G: Such a magnet is used in 1000 MHz instruments.

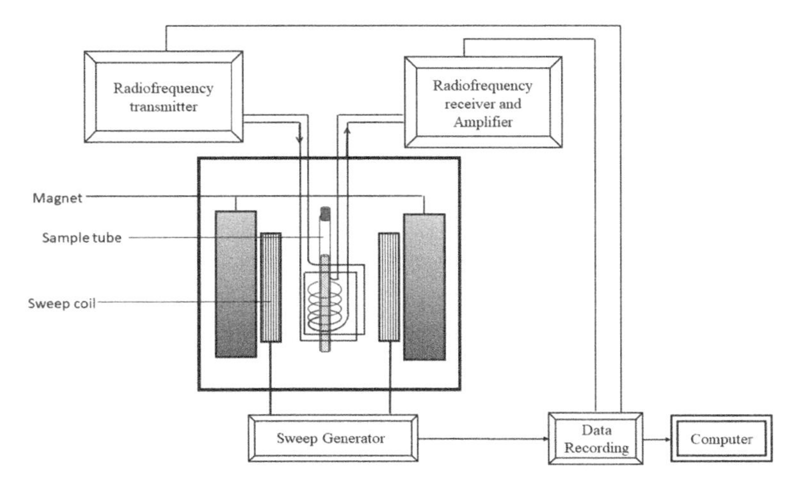

Figure 2.32 Schematic representation of NMR spectrometer.

(c) Superconducting magnet: This is used with a field strength of 51000 G and 71000G. The 51000 G magnet is used in 220 MHz NMR instruments and the 71000 G magnet is used in 300 MHz NMR instruments.

(d) The superconducting magnets are made by immersing electromagnets in liquid He at a temperature of ≈4 K. By doing so, the electrical resistance disappears making the metal behave like a superconductor. The most important requirement of an NMR spectrometer is to have a homogenous magnetic field around the sample. Therefore, additional coils, known as Golay coils, are used to remove this inhomogeneity in the field of the main magnet.

Radiofrequency power source

This is called a radiofrequency transmitter which is a crystal oscillator controlled at a single frequency such as 60 MHz for a 14000 G magnet,100 MHz for a 23000 G magnet, and 300 MHz for a 71000 G magnet. This radiofrequency power is passed through a coil inside which the sample tube is kept.

Radiofrequency receiver

The NMR signal is filtered amplified and detected by an RF receiver. The receiver uses electronic filters to separate the desired radiofrequency signal from all other signals passing through it.

Sweep generator

The strength of the magnetic field is changed with the help of a sweep generator. Its output is then synchronized with the axis of an oscilloscope or XY-recorder. Generally, the variable electromagnetic coils are used as sweep coils. The double coiled type probe which holds the sample tube is fitted into the magnetic gap. The arrangement of the sample tube and the coils in the gap of the magnets is also shown in the figure. The coils of the radiofrequency receiver are oriented in such a way that its axis is perpendicular to the direction of the magnetic field and also to the axis of oscillator coils. The sample tube is mounted in such a way that it may provide a steady pinning rate (around 30Hz).

Sample tube

The cylindrical symmetry and uniform thickness of the sample tube used in NMR spectroscopy are very important. Any imperfections in the thin glass-walled sample tubes may give inhomogeneity in the magnetic field strength and may cause significant variations in the measurement of the chemical shift and coupling constant values. Several commercial, microcells and sample tubes are used. The standard sample tube consists of a small glass bulb having a capacity of 32–100 mL for holding the sample, and a teflon holder for holding the sample bulb in the right position inside the standard NMR sample tube with the help of a KLF rod. This rod is removed after when the microcell is placed in the probe of the instrument. The space below the teflon holder is filled with CCl_4 in order to minimize the bobbling of the sample bulb when the tube is rotated at a regular rate of around 30 Hz in the spectrometer.

Data recorder

The data recording system viz, oscilloscope and X-ray recorder are completely synchronized with the output of the sweep generator.

Operational details

The appropriate magnetic field needed for the energy absorption by the sample is supplied from the magnet. RF transmitter (radiofrequency power) which provides energy at the operating frequency is then absorbed by the coil connected to the radiofrequency source.

The strength of the magnetic field over the range of chemical shift is then changed with help of a sweep generator and sent to the probe through the sweep coils. The energy absorption is detected with the help of a radiofrequency receiver coil is amplified and then obtained as a spectrum through the x-y chart recorder and stored in a computer attached to the NMR spectrometer.

2.20 Nuclear Overhauser Effect

In 1953, a new original phenomenon was discovered by Albert Overhauser which describes that the intensities of NMR signal can significantly be enhanced by irradiation of one of the two nuclei which are located very close to each other within the molecule and moving in the magnetic field. This is known as NOE. By irradiating one of these two nuclei leads to a change in the intensity of the resonance signal of a proton by altering the Boltzmann equilibrium population distribution of the nearby proton. Therefore, this leads to an opening of a new relaxation pathway and the intensity of the resonance signal of the nearby nuclei is also get enhanced. No coupling constant is needed to be present between the two nearby nuclei for observing this phenomenon. Overhauser discovered this phenomenon between nuclei and unpaired electrons. However, when both are nuclei then this Overhauser effect is more useful in determining the structural details and molecular geometry of the compounds. NOE is noticeable more when the distance between the two nuclei is very short (~ 2–4 Å). In the simple experiment, to observe the phenomena of NOE, first a normal NMR spectrum of the compound is recorded and the spectrum is obtained. Now, one of the nuclei is then irradiated and the NMR spectrum is recorded again then the intensity of the resonance signals of the nuclei situated closest in space to the irradiated nucleus gets enhanced more in comparison to those nuclei situated at a larger distance from the irradiated one [9]. For example, if the proton of isovanillin is recorded in a normal way and then irradiated at the $-CH_3O$ group, the integral for the ortho proton (doublet) is significantly increased i.e., this doublet then appears as a more intense signal. This method is used fully in determining stereochemical relationships in molecules.

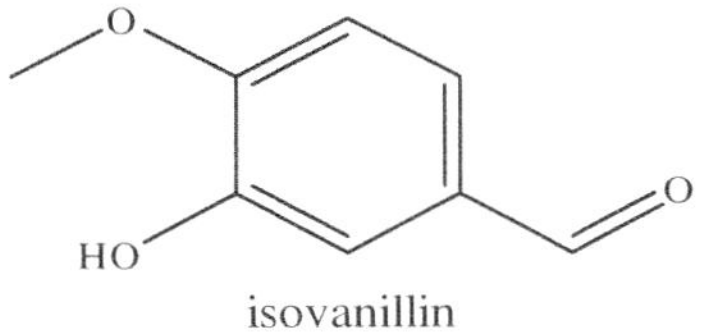

O

HO

O

isovanillin

The phenomenon of NOE can be understood simply by considering a molecule containing two protons H^a and H^b situated close to each other. In the normal ^1H NMR spectra, two peaks will be obtained. Now, on irradiation of H^a proton, the peak for H^a proton, the peak for H^a will subside but on the other hand, the peaks for the proton H^b will become more intense and enhanced. Theoretically, proton ^1H-^1H NOE is only 50 % while ^1H-^{13}C NOE is 200%.

2.21 FT-NMR

In Fourier transform NMR spectroscopy, a series of highly intense pulses of radiofrequency radiation (~30 microseconds) is used for the irradiation of the sample instead of a continuous radiofrequency pulse that is used in normal NMR technique. Fourier transform is a mathematical operation in which a time domain spectrum is converted to a frequency domain spectrum and vice versa. Thus, the complex NMR spectrum breaks into a simpler one. In ^1H proton FT NMR, the sample is irradiated periodically in a fixed magnetic field with highly intense pulses of radiofrequency radiation containing all frequencies for ^1H proton. In every environment, the protons absorb their specific frequencies from the strong pulse, and then these frequencies couple with one another to form coupling energies sub-levels. After switching off the pulse radiation, the relaxation process of proton nuclei occurs. During this relaxation process, the protons simultaneously re-emit the absorbed energies due to coupling with one another. Due to reemitted energies, a characteristic radiofrequency signal is recorded as a function of time which decays rapidly. This rapidly decaying signal is a time domain (a complex interacting pattern) (Figure 2.33). The output of this decay signal is transformed into the frequency domain spectrum where each individual frequency is separated from the complex interacting pattern by Fourier transformation with the help of a digital computer. These signals are then converted into a normal NMR spectrum. This technique is also called as pulse FT-NMR technique. But it is very different from continuous wave NMR (CWNMR).

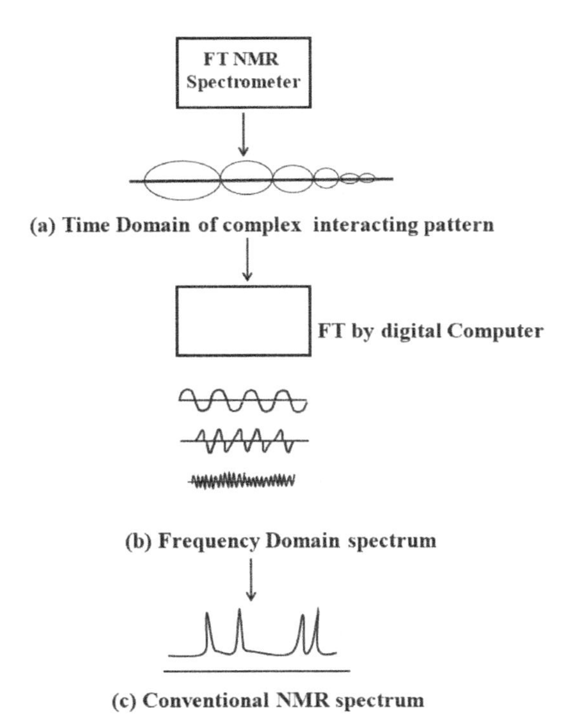

Figure 2.33 Schematic diagram for representing Fourier transformation in NMR spectroscopy, (a) time domain, (b) frequency domain, and (c) normal conventional NMR.

Advantage of FT-NMR

The entire FT-NMR spectrum can be carried out in a few seconds (≈30 μs) only during which recording, computerization, and transformation, all are done. On the other hand, the normal NMR spectroscopies for several minutes (~13 min) are needed to record the spectrum. FT-NMR spectroscopy is also advantageous for (i) the detection of weak signals, (ii) analyzing and recording spectra of samples even at very low concentrations (10–50 mg), and (iii) studying the samples of those nuclei which have very low natural abundance with small magnetic moments. It is used in engineering, and quality control in industries and medical institutes. MRI is the most important application of this tool.

References

1. Kemp, W., *Organic Spectroscopy*, 3rd Ed. (Palgrave, London), 1991.
2. Kellar, J., *Understanding of NMR Spectroscopy* (John Willey), 2002.
3. Pavia, D. L., Lampman, G. M., Kriz, G. S. and Vyvyan, J. R., *Introduction to Spectroscopy*, 4th Ed. (Cengage Learning, USA), 2009.
4. lonin, B. I. and Ershov, B. A., *NMR Spectroscopy in Organic Chemistry*, 1st Ed. (Plenum Press, New York), 1970.
5. Harwood, L. M. and Claridge, T. D. W., *Introduction to Organic Spectroscopy* (Oxford University Press, New York), 1996.
6. Schmid, G. H., *Organic Chemistry* (McGraw-Hill Education, Europe), 1995.
7. Loudon, G. M., *Organic Chemistry*, 3rd Ed. (Cummings Publishing, San Francisco), 1995.
8. Silverstein, R. M. and Webster, F. X., *Spectrometric Identification of Organic Compounds*, 6th Ed. (Wiley, New York), 1997.
9. Singh, N. B., Gajbhiye, N. S. and Das, S. S., *Comprehensive Physical Chemistry* (New Age International, New Delhi), 2014.
10. Lambert, J. B., Gronert, S., Shurvell, H. F., Lightner, D., and Cooks, R. G., 2nd Ed. *Organic Structure Spectroscopy* (Pearson, UK), 2013.
11. Rabek, J. F., *Experimental Methods in Polymer Chemistry* (John Wiley, Bristol), 1980.
12. Reich, H. J., Geminal proton-proton couplings ($^{2}J_{H\text{-}H}$), https://www.chem.wisc.edu/areas/reich/nmr/05-hmr-04-2j.htm
13. Muller N. and Pritchard, D. E., C-13 splittings in proton magnetic resonance spectra. 2. Bonding in substituted methanes, *J. Chem. Phys.*, 31(6), 1471–1476, 1959.
14. Shoolery, J. N., Dependence of C13-proton spin coupling constants on s character of the bond, *J. Chem. Phys.*, 31, 1427, 1959.
15. Claridge, T., *High Resolution NMR Techniques in Organic Chemistry* (Elsevier, UK), 2016.
16. Martino, R., Gilard, V. and Malet-Martino, M., *NMR Spectroscopy in Pharmaceutical Analysis* (Elsevier, Boston), 2008.
17. Friebolin, H. *Basic One- and Two-Dimensional NMR Spectroscopy* (Wiley-VCH, Weinheim), 2011.
18. Pretsch, E., Bühlmann, P. and Badertscher, M., *Structure Determination of Organic Compounds*, 4th Ed. (Springer, Germany), 2009.

Chapter 3

Carbon-13 NMR

Tamanna Yakub,[a] Bhawana Jain,[a] Anupama Asthana,[a] Ajaya K. Singh,[a,b] and Md. Abu Bin Hasan Susan[c]
[a]*Department of Chemistry, Govt. V. Y. T. PG Autonomous, College, Durg, Chhattisgarh 491001, India*
[b]*School of Chemistry & Physics, Westville Campus, University of KwaZulu-Natal, Durban 4000, South Africa*
[c]*Department of Chemistry, University of Dhaka, Dhaka 1000, Bangladesh*
ajayaksingh_au@yahoo.co.in

3.1 Introduction

Nuclear magnetic resonance (NMR) is a type of spectroscopy in which radio frequency pulses of electromagnetic radiation are involved. Holding the nuclei in a magnetic field establishes the magnetic energy levels. The spin states of nuclei degenerate in the absence of a magnetic field, i.e., possess the same energy and so transitions between energy are not possible. As a magnetic field is applied, radio frequency radiation can induce transitions between these energy levels. NMR spectroscopy is most frequently used on nuclei having a half-integral value of spin angular momentum, i.e.,

Spectroscopy
Edited by Preeti Gupta, S. S. Das, and N. B. Singh

ISBN 978-981-4968-32-4 (Hardcover), 978-1-003-41258-8 (eBook)
www.jennystanford.com

I = ½. Examples of such nuclei are proton (^{1}H), carbon-13 (^{13}C), oxygen-17 (^{17}O), fluorine-19 (^{19}F), phosphorous-31 (^{31}P). Nuclei with I = 0 cannot produce NMR spectra, whereas nuclei with I > 1 can only produce spectra in a few instances. Table 3.1 shows some nuclei with non-zero spin [1].

Table 3.1 Properties of some nuclei with non-zero spin

Nucleus	Spin	Natural abundance (%)	Magnetic moment μ	Magnetogyric ratio ϒ (10^7 rad T s^{-1})	NMR frequency ν (MHz)
^{1}H	1/2	99.99	2.793	26.752	400.00
^{2}H	1	0.015	0.857	4.107	61.40
^{7}Li	3/2	92.58	3.256	10.398	155.45
^{13}C	1/2	1.081	0.702	6.728	100.58
^{14}N	1	99.63	0.404	1.934	28.89
^{15}N	1/2	0.37	- 0.283	- 2.712	40.53
^{17}O	5/2	0.037	- 1.893	- 3.628	54.23
^{19}F	1/2	100	2.627	25.181	376.31
^{23}Na	3/2	100	2.216	7.080	105.81
^{27}Al	5/2	100	3.639	6.976	104.23
^{29}Si	1/2	4.70	- 0.555	- 5.319	79.46
^{31}P	1/2	100	1.131	10.841	161.92
^{59}Co	7/2	100	4.639	6.317	94.46
^{77}Se	1/2	7.58	0.533	5.12	76.27
^{195}Pt	1/2	33.8	0.600	5.768	85.99
^{199}Hg	1/2	16.84	0.499	4.815	71.31

In particular, different chemical environments of the various forms of hydrogen present in a molecule are revealed by the NMR, from which we can determine the structures [2]. The NMR spectra of numerous compounds of ^{1}H, ^{19}F, ^{31}P, ^{11}B, ^{14}N, ^{17}O, and ^{29}Si have been investigated. The ^{1}H, ^{19}F, and ^{31}P NMR spectra, in particular, have been quite effective in solving structures, identifying molecules, and analyzing.

The first NMR observed regarding ^{13}C nuclei was reported in 1957. The first experimental ^{13}C NMR was achieved in 1965. The advancement of ^{13}C NMR for analytical and research tools for organic chemistry has been available since the 1970s. This isotopic abundance of ^{13}C, the sole stable isotope of carbon with a non-zero spin, is only 1.081% in nature. This has restricted the widespread application of ^{13}C NMR despite its immense practical and theoretical significance. Despite its low abundance, ^{13}C is a good candidate for NMR spectroscopy [3]. The magnetic moment is high enough to prevent quadrupole effects, and the spin is half that of the magnetic moment, so the resonances fall in a useful frequency range. The advances in technology and the use of suitable techniques and instruments have now rendered it feasible to produce ^{13}C NMR spectra in natural abundance in a wide range of substances. For the investigation of very large molecules and mixtures of compounds, the signal-to-noise (S/N) ratios are sufficient to ensure significant analysis. Chemical shifts are significantly large and occur within well-defined limits for many different types of molecules. They can often identify even nearly related molecules due to their sensitivity to minor changes in structure. Within a series of related compounds, the shifts change smoothly as well, and may thus predict structures with a high degree of precision.

Carbon shows indirect spin–spin coupling with hydrogen directly bonded to it to a large extent to give rise to distinctive multiplets that are highly useful in the interpretation of the spectra. The S/N ratio is, however, not high enough to allow the spectra for use in qualitative analysis as liberally as ^{1}H, ^{19}F, and ^{31}P spectra. This leaves scopes for substantial development in this area to widen practical uses. The availability of a considerable amount of ^{13}C data, on the other hand, may provide a knowledge base and theoretical background of chemical shifts for further development. The database of a complete series of related simple compounds can be created through studies of many types of compounds. The aromatic, polysubstituted methane, methyl derivatives, and heterocyclic compounds are some examples.

Since the spin number of the ^{12}C nucleus is zero, it does not exhibit magnetic activity. But ^{13}C like the ^{1}H nucleus has a spin number of ½ (i.e., $I = ½$) (Figure 3.1). Thus the number of orientation of spinning of nucleus is $(2I + 1)$ i.e. $2 \times ½ + 1 = 2$. ^{13}C is less sensitive than ^{1}H, about 1/4th of the proton.

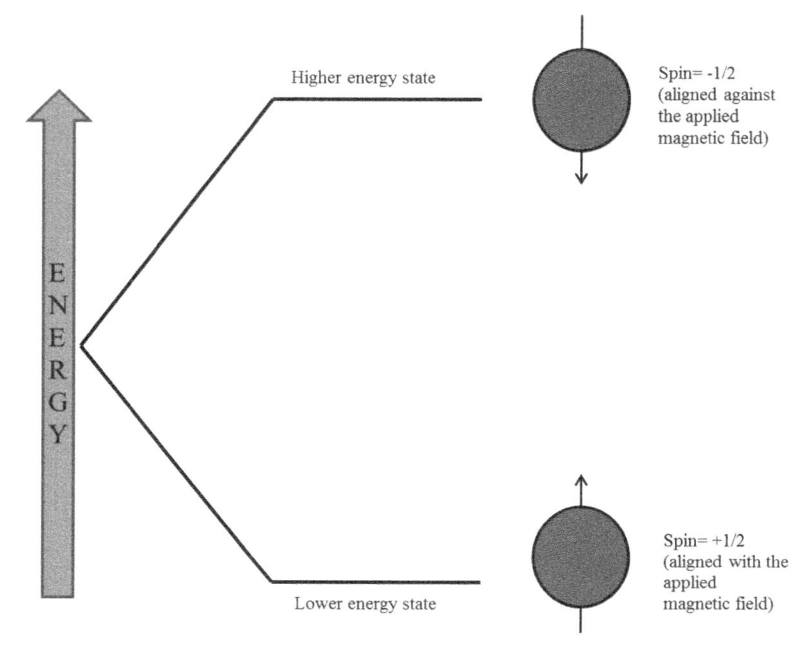

Figure 3.1 Relative energy of both spin states of I = ½ nucleus as a function of the intensity of the external magnetic field.

3.2 Comparison with ^{1}H NMR Spectroscopy

^{1}H NMR spectroscopy gives indirect information about the carbon skeleton of an organic compound since most carbon atoms have at least one hydrogen. On the other hand, ^{13}C NMR spectra display signal arising from all the carbon atoms and thus gives direct information on the carbon skeleton. In ^{13}C spectra, signals are spread over the chemical shift range of 200 ppm, compared with a range less than 20 ppm for proton.^{13}C spectra are generally much simpler than the corresponding ^{1}H spectra. In ^{1}H NMR, the peak area is directly proportional to the number of equivalent protons, but in the case of ^{13}C spectra number of peaks does not depend on the number of equivalent protons, it depends on carbon. Instrumentations for ^{1}H NMR and ^{13}C NMR are quite similar. A few components are common to both, such as (1) a magnet to apply the magnetic field, B_0, (2) devices for generation of B_1, pulse and receiving the resultant NMR signal, (3) a probe to place the sample in the magnetic field, (4) hardware to stabilize the B_0 field and optimize the signal, and (5) a computer to control most of the operations as well as to process the NMR signals.

3.2.1 Characteristic Features of ^{13}C NMR Spectra

As compared to ^{1}H NMR (which has a chemical shift of 0–12 ppm relative to tetramethylsilane, TMS), ^{13}C NMR has a larger chemical shift (0–220 ppm). Since the natural abundance of ^{13}C is low, ^{13}C–^{13}C coupling is insignificant in a molecule. As a result, each magnetically non-equivalent carbon in the ^{13}C NMR spectrum (decoupled proton) results in a single strong signal which experiences further subsequent splitting. The integrated area of the peak in a ^{13}C NMR spectrum doesn't have to be in proportion to the number of carbon atoms responsible for the signal. As a result, the area beneath the ratio does not need to be considered.

For proton coupling for each carbon or a group of magnetically equivalent carbon, splitting of the signal occurs by the proton directly attached to the carbon, and the spectra follow the (n + 1) rule. The frequency necessary to observe resonance in the ^{13}C nucleus is approximately one-fourth of that required to view proton resonance in other nuclei. Since the electronegative atom is directly attached to ^{13}C, the chemical shift is higher for ^{13}C than it is for proton. Table 3.2 shows the basic characteristics of ^{13}C.

Table 3.2 Basic physical characteristics of ^{13}C NMR

Isotope	Natural abunda-nce (%)	Nuclear spin (I)	Magnetogyric ratio, ϒ (10^7 rad/T^{-1} s^{-1})	Quadrupole moment, Q (10^{-28} m^2)	Resonance frequency (MHz at a field of 1 T)	Relative sensitivity at constant field
^{13}C	1.081	½	6.728	0	10.705	0.016

3.2.2 Referencing ^{13}C NMR Spectra

TMS is primarily used as the reference for recording ^{13}C spectra. Since ^{13}C NMR is associated with relatively low sensitivity, substantial amounts of TMS need to be added. It is a common practice to use a secondary reference where solvent peaks serve the purpose. The choice of TMS as the standard is based on several advantageous features:

1. It comprises four equivalent carbon atoms in the same environment. This results in a single and sharp peak (since lots of carbon atoms contribute to the same).
2. The C-Si bonds in TMS have electrons closer to the carbon atoms as compared to most other molecules. Due to the largest shielding of the carbon nuclei from the external magnetic field, the magnetic field needs to be increased to a greater extent to ensure the carbons back into resonance. The overall result is the observation of a peak on the right-most side of the spectrum. Almost all other signals are observed at the left side of this peak [4].

Table 3.3 Chemical shifts of several common solvents used in NMR spectroscopy

Solvent	Protio compound (δ)	Per–deutero compound (δ)
Cyclohexane	27.51	26.06
Acetone	30.43	29.22
Dimethyl sulfoxide	40.48	39.56
Dicholromethane	54.02	53.61
Dioxane	67.40	–
Tetrahydrofuran	68.12 26.30	–
Ether	66.23 15.58	–
Chloroform	77.17	76.91
Carbon tetrachloride	95.99	–
Benzene	128.53	127.96
Acetic acid	178.27	–
Carbon disulphide	192.8	–
Carbon disulfide external	193.7	–

NMR spectroscopy deals with a set of solvents, the most common ones are listed with their chemical shifts in Table 3.3. Since the isotopic effects in chemical shifts in ^{13}C NMR are very pronounced, the values are reported separately for solvents under deuterated and protonated conditions. The magnetic field strength is, in general, stabilized by deuterium absorption of the solvent and this is the basis of computation in a modern NMR spectrometer. The observation frequency depends on the magnetic field and thus the deuterium receiver detects a fluctuation in the field fluctuation through variation in the frequency. The observation frequency, also otherwise called lock frequency, can accordingly adjust the field strength.

3.3 Instrumentation

The primary NMR spectrometers were based on electromagnets and worked in the continuous wave mode in resemblance to infrared and ultraviolet-visible spectrophotometers of the modern days. The magnets had low sensitivity and poor stability, but still, these have been used by a generation of chemists. Permanent magnets are maintainable with ease, although their sensitivity is still poor. Today a superconducting magnet is used in most research-grade NMR spectrometers and these operate in the pulse Fourier transform NMR (FT-NMR) mode. These magnets provide high sensitivity and stability and have field strengths of 7.0–23.5 T (300–1000 MHz for proton). Chemical shifts and consequently differences in chemical shifts are increased with increasing field strength. This subsequently allows greater separation of resonance resulting from the very high fields produced by superconducting magnets. A double Dewar filled with liquid helium sits atop the superconducting magnet. Liquid helium is used for immersion of the solenoid coils and the temperature is maintained at 4.2 K. The central bore tube has a diameter of 53 or 89 mm used in the current solenoid. The bore tube is stored at ambient temperature, and the B_0 field (Z) is directed parallel to the cylinder axis. The majority of materials are investigated as liquids in solvents in cylindrical NMR solvents or, more aptly, as solids. For liquid samples, the tube needs a parallel placement to the superconducting cylinder (z-axis), whereas solid samples are spun at a magic angle. A console separates the magnet and consists of transmitter channels (two or more), among other modules. A proton channel is usually coupled with a specialized carbon channel or a broadband channel, both of which have a power amplifier and a frequency synthesizer. A broadband channel can be tuned to one of several lower frequency nuclei such as ^{15}N, ^{31}P, and ^{29}Si in addition to ^{13}C.

An adjustable probe is used to insert the sample and place it in the most homogeneous area of the magnetic field. The probe consists of several components: (1) a sample holder, (2) a mechanical device for adjustment of its position in the magnetic field, (3) transmitter coils (two or more) for the supply of the double resonance fields, B_1 and B_2, as well as for receiving signals, (4) frequency/field lock circuitry coils, and (5) devices to reduce any inhomogeneity of the magnetic field.

A probe that is most commonly used for observation of proton is associated with an inner coil for ^1H detection, located closer to the sample to maximize sensitivity and an outer X-nucleus coil. For nuclei other than protons, such as ^{13}C, the X-nucleus coil is placed on the interior while the ^1H–coil is on the exterior. The 5 mm is the most frequently used size of the sample tube, which needs a solvent volume of 500–60 μL. Only biological samples require larger sample tubes, usually >10 mm.

When the sample amount is adequate and solubility in NMR solvents is sufficiently high, a small volume has to be used for measurement. Microtubes with a required volume of 120–150 μL and submicrotubes with that of 25–30 μL are appropriate for this purpose. The NMR signals of samples in small quantities or large dilutions (especially for high molecular weight) may be enhanced in a number of ways. The older approach is to increase the strength of the magnetic field since sensitivity increases with B_0. More recently, cryoprobes have been developed in which liquid helium or liquid nitrogen is used to keep the receiver coils as well as preamplifiers at a desired low temperature to reduce noise and thereby, increase the S/N ratio.

3.4 Interpretation of ^{13}C Spectra

In ^{13}C nuclei, whose relaxation time is constant over a wide range and is not equally relaxed between the pulses, the resulting peak does not integrate to give the correct number of C atoms. It is possible by inspection of the ^{13}C spectrum to recognize the nuclei that do not bear proton by their low intensity. The multiplicity of the signal is one greater than the number of protons bonded to the carbon atom giving the signal. There is substantial coupling between carbon and the attached hydrogen atom. Consequently, the proton-coupled ^{13}C NMR spectra of organic compounds are quite complex. The simplification of ^{13}C spectra is performed by the process of decoupling.

3.4.1 Broadband Decoupling

In broadband decoupling, the radio frequency generator provides a band of frequencies extending over the entire ^{1}H NMR spectrum. When $J_{C\text{-}H}$ couplings are eliminated, however, it becomes very difficult to assign ^{13}C peaks with proper carbon atoms in the molecule. Broadband decoupling increases the signal intensity and simplifies the spectrum. However, it becomes much more difficult to identify the specific carbon atoms and there may be some distortions in signal intensities [5].

3.4.2 Off-Resonance Decoupling

This is accomplished by offsetting. The central frequency of a broadband proton decoupler may be offset and off-resonance decoupling may be accomplished. The TMS proton frequency may thus be offset by approximately 1000–2000 Hz upfield or 2000–3000 Hz downfield. Protons directly bonded to the ^{13}C atoms thus, experience residual coupling although the loss of long-range coupling is observed [6]. Since the number of protons and neutrons is equal, the overall nuclear spin of ^{12}C is zero. Although the ^{13}C isotope has spin ½, it has an abundance of only 1.081%. The following features are characteristics of ^{13}C NMR spectra:

1. Using TMS as an internal standard, the chemical shift ranges up to approximately 220 pm.
2. The values of the longitudinal relaxation time, T_1, and the transverse relaxation time, T_2 give rise to a natural linewidth of about 1 Hz.

3. Common spectrometers involve a Larmor frequency of 20–100 MHz.
4. For recording a good quality spectrum, the sample amount required is 5–20 mg in 0.4–2.0 mL of solvent (for instance, $CDCl_3$) and the number of scans should be 64–6400.

During the measurement of ^{13}C FID, the whole proton resonance area of a molecule (i.e., –2 to 10 ppm) is subjected to irradiation with "white noise" radio frequency. The rate of transition between the spin states, from low to high, increases, and an individual nucleus of carbon experiences a magnetic field which is the simple average of the total magnetic fields. This subsequently causes the manifestation of coupling to vanish. As long as the concern is with carbon, all protons cease to exist. The Nuclear Overhauser Effect (NOE) becomes problematic since the intensity of the carbon line is usually double or even raised further. The integration of ^{13}C spectral signals thus becomes unreliable to render into a useful variable. Recording of decoupled as well as coupled spectra is a normal practice for ^{13}C, but for coupled spectra slight modification termed "off-resonance proton decoupling" becomes essential.

^{13}C NMR spectra provide the following useful information:

1. If the spectrum is proton decoupled, it can be used exclusively for the determination of the number of carbon atoms that are unique in a molecule. The example includes a mono-substituted phenyl group, which has four unique carbon atoms.
2. When a spectrum is off-resonance, it gives the number of hydrogen atoms bonded to each unique carbon atom. If a singlet is observed, no of hydrogen atoms of this kind is zero, while it is 1, 2, and 3 for doublet, triplet, and quartet, respectively.
3. Chemical shift of each carbon provides a lot of information about the environment of the carbon. This helps in determining the different functional groups present in a molecule. The typical chemical shift ranges of carbon nuclei associated with different functional groups are listed in Table 3.4.

Table 3.4 Ranges of chemical shift of carbon nuclei

Molecule	Chemical shift
C (alkane)	~ 0–30 ppm
C (alkene)	~ 110–150 ppm
C–N	~ 50 ppm
C–O	~ 60 ppm
C–F	~ 70 ppm
Aromatic	~ 110–160 ppm
Ester, Amide, Acid	~ 160– 170 ppm
Ketone, Aldehyde	~ 200–220 ppm

3.5 Chemical Shift

^{13}C chemical shift is somewhat parallel to ^{1}H chemical shift. It is expressed in (δ) ppm. It ranges up to 200 ppm from TMS, while in ^{1}H its range is ~20 ppm. ^{13}C shifts are related mainly to hybridization and substituent electronegativity. Proton-decoupled ^{13}C-NMR spectra represent different carbons present in molecules [7].

3.5.1 Origin

NMR chemical shifts are controlled by three key effects: diamagnetic, adjacent group anisotropy, and paramagnetic:

1. Diamagnetic effect (σ_d): The local nucleus may experience an upfield shift (low frequency, shielding) by circulating electrons in the *s* orbitals. The σ_d changes with electron density and hybridization in a predictable fashion to allow chemists to rationalize many of the observed effects and thus it dominates proton chemical shifts.
2. Neighboring group anisotropy (σ_n): Local magnetic fields, created by diamagnetic circulation, influence the neighboring nuclei. Irrespective of the nucleus to be affected, these influence at the same extent and on the order of a few ppm a constant effect is observed. Neighboring group anisotropy is thus not significant in ^{13}C NMR (δ up to 200 ppm), although in proton NMR (δ up to 20 ppm) it is significant due to a pronounced effect.

3. Paramagnetic effect (σ_p): The external magnetic field produces significant downfield (high-frequency, deshielding) shifts of nuclei in the vicinity by inducing the circulation of electrons between the ground state and the excited state of *p* orbitals. The forces exerted on electrons by the magnetic field are small in comparison with the energy differences between ground and excited states, but the changes are very large, and even a small extent of paramagnetic circulation may result in huge chemical shifts. Although ^{6}Li and ^{7}Li are likely exceptions, the chemical shift of all nuclei heavier than ^{1}H is mainly affected by the σ_p.

3.5.2 Scale

The chemical shifts for ^{13}C NMR in most cases cover 0–220 ppm (TMS = 0.0). Hybridization dictates the range of chemical shifts. For instance, sp^3 carbons have the lowest shifts in the range of 0–70 ppm, while sp carbons of the acetylene type have the value in the range of 70–100 ppm. Similarly, δ has a value in the range of 100–150 ppm for sp^2 carbons, which are bonded to C and H at 160–220 ppm for sp^2 carbons of carbonyl groups, and at 210–220 ppm for sp carbons of the allene type. Figure 3.2 summarizes the chemical shift ranges for different hybridizations. It should be noted that substituents may cause shifts outside the provided ranges.

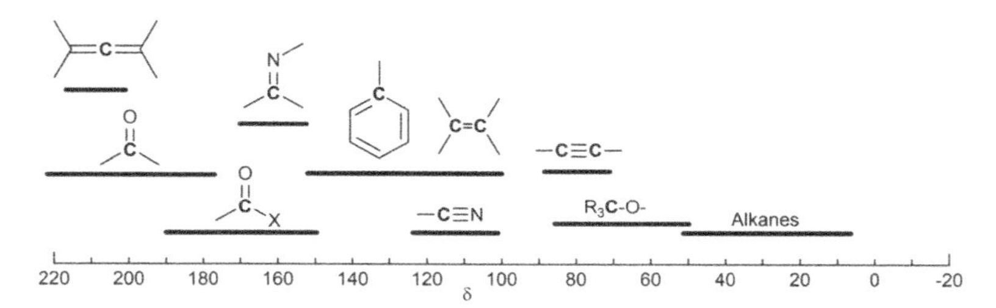

Figure 3.2 Ranges of chemical shift.

3.5.3 Effects on sp^3 Carbons

The introduction of a substituent group causes a change in chemical shift. Such a change, usually quantified by substituent perturbation

($\Delta\delta$), is used to systematically discuss the influence of substituents on model compounds with a simpler structure. The effect is the most prominent for substituent modifications at the carbon itself (α effect). A significant substituent effect is also noticeable at substitution at the β, γ positions and occasionally even at the δ position. Table 3.5 summarizes the substituent effects of a number of typical functional groups.

Table 3.5 Substituent effects of typical functional groups

X	A	β	γ
	(H–C→X–C)	(H–CC→X–CC)	(H–CCC→X–CCC)
CH_3–	+9.4 (α_c)	+9.5 (β_c)	−2.4
CH_3CH_2–	+18.8 ($\alpha_c+\beta_c$)	+7.0 ($\beta_c+\gamma$)	−2.1
CH_2=CH–	+20.4	+6.3	−2.9
HC≡C–	+4.5	+5.2	−3.5
HO_2C–	+20.8	+2.7	−2.3
N≡C–	+3.6	+2.0	−3.1
H_2N–	+29.1	+11.8	−4.5
O_2N–	+64.5	+3.1	−4.7
HO–	+49.4 (α_o)	+11.2 (β_o)	−4.7
CH_3O–	+58.7 ($\alpha_o+\beta_c$)	+6.5 ($\beta_o+\gamma_c$)	−6.0
HS–	+11.4 (α_s)	+12.1 (β_s)	−2.7
CH_3S–	+20.9 ($\alpha_s+\beta_c$)	+6.4 ($\beta_s+\gamma_c$)	−3.0
F–	+70.1	+7.4	−6.7
Cl–	31.3	+10.2	−4.6
Br–	+20.0	+10.4	−3.3
I–	−6.1	+11.1	−0.8
$(CH_3)_3Sn$–	−1.7 ($\alpha_{sn}+3\beta_c$)	+4.1	+0.9
Li–	−1.4	+6.9	+5.4

3.5.4 α-Substituent Effects

The α-effect is observed when a directly bonded H is replaced by an X group, i.e., C-H → C-X. α-substituent effects, in most cases, are influenced by the electronegativity of the attached atom. Electronegative atoms exhibit significant high-frequency shifts (e.g., CH_3OH), whereas electropositive substituents exhibit low-frequency shifts. Table 3.6 shows α-substituent effects on the ^{13}C chemical shift of some compounds.

Table 3.6 α-substituent effects on ^{13}C chemical shift

Substi-tuents	Chemical Shift	Substituents	Chemical Shift	Substi-tuents	Chemical Shift
Me–H	-2.6	Me–O–Me	61.2	Me|4c	31.5
Me–F	75.2	Me–S–Me	18.2	Me|4Si	0.0
Me–Cl	24.6	Me–Se–Me	10.4	Me|4Ge	-1.9
Me–Br	10.2	Me–Te–Me	-21.5	Me|4Sn	-9.1
Me–I	-20.2	–	–	Me|4Pb	-3.1

3.5.5 Heavy-Atom α-Effect

The electronegativity effect on chemical shift works well for first-row atoms. It also explains the chemical shift for the second-row atoms, albeit to a lower extent. A counter-effective influence is the "heavy-atom" effect. For example, all carbons bearing iodine are strongly shifted to a lower frequency. C-Te also exhibits a similar effect. Table 3.7 shows the heavy-atom effect on α-substituent on chemical shifts of various compounds.

Table 3.7 Heavy-atom α-effect on ^{13}C chemical shifts

$H\text{–}CH_3$	$I\text{–}CH_3$	$Te(CH_3)_2$	Cl_4	CBr_4	CCl_4	Ph I I
-2.1	-24.0	-21.5	-292.4	-29.4	98.6	-25.8

3.5.6 α-Effect of Triple Bonds

An unexpectedly large low-frequency shift is also caused by triple bonds (X = acetylene, nitrile) as substituents. For example, the δ value for CH_3-CN is 0.3, for CH_3C=CH is –1.9, as compared to δ = 18.7

for $CH_3CH{=}CH_2$. This might be due to the presence of a triple bond which enhances diamagnetic circulation to a greater extent.

3.5.7 α-Effect of Double Bonds

Significant shifts of allylic protons are triggered by a double bond in ^{1}H NMR in contrast to ^{13}C NMR, which shows a comparatively much smaller shift for carbons directly linked to double bonds. Small high-frequency shifts are caused by terminal vinyl groups or trans double bonds, while low-frequency shifts are caused by cis-substitution.

3.5.8 β-Substituent Effects

The β-substituent effect is observed when H atoms on a neighboring atom are replaced by an X group, i.e., C-C-H $\rightarrow$ C-C-X. Almost all substituents produce significant high-frequency shifts (typically approximately 9 ppm, but for crowded cases, it may be lower) are apparent for substituents of almost all kinds. The electronegativity of the perturbing substituent has no significant effect. The origin of β shifts is not well understood.

3.5.9 γ-Substituent Effects

When an H atom is replaced by X on the second atom (C-C-C-H$\rightarrow$ C-C-C-X), γ-effect is observed. It is the case for almost all X-substituents where the γ-carbon has attached hydrogen. Syn γ-effects and γ-gauche effects are strong proximity components to bring about stereochemistry dependence. The change in chemical shift, $\Delta\delta$ for syn γ-effects is negative to cause a shift to low frequency and the nature of the intervening groups do not essentially affect such effects. For X = CH_3, an upfield shift is observed by ~6 ppm if X and *C are either with gauche or eclipsed conformation. The effect is ~2 ppm for acyclic systems as dictated by the extent of gauche conformation. The effect is extensively used for assignments of stereochemistry.

When X and the γ-carbon are antiperiplanar, some substituents also cause anti-γ-effects. Very small antiperiplanar γ-shifts are shown by alkyl substituents, but significant effects are observed for X = O, N, or F. As compared to gauche γ-effects, anti-γ-effects are almost always smaller. They can be either lower frequency or

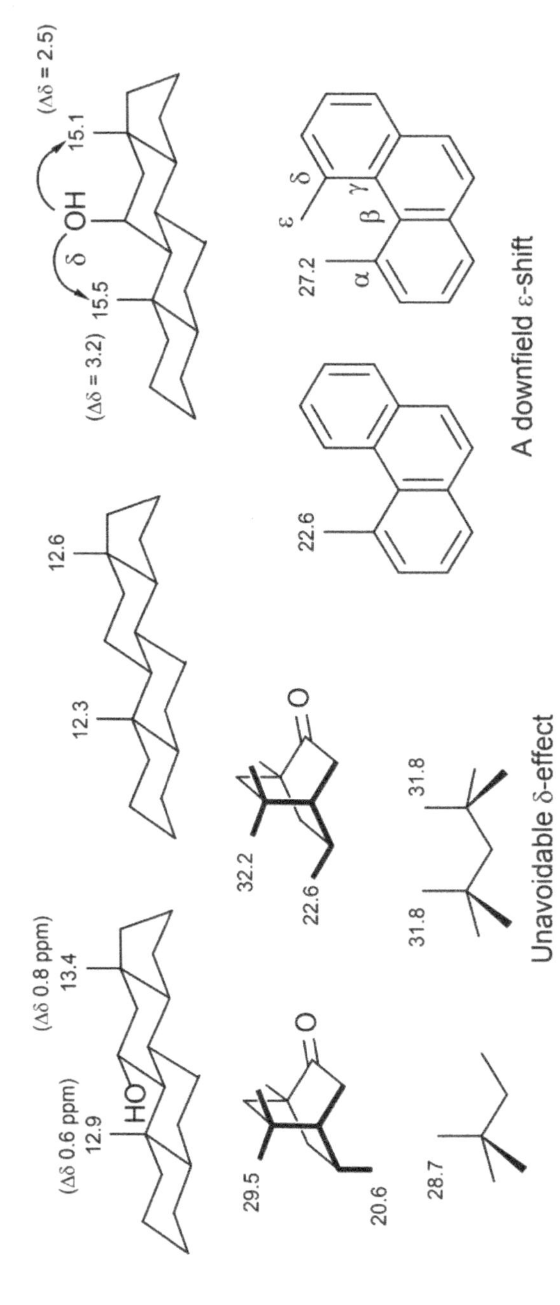

Figure 3.3 Effect of δ-substituent on chemical shift.

higher depending on a number of structural factors. For example, for 1-methyl-1-X-cyclohexanes, a downfield shift to high frequency is observed due to anti-γ-effects since the substituent is at the quaternary center of the cyclohexane. On the other hand, an upfield lower frequency shift is marked when the perturbing substituent at C-1 is H [8].

3.5.10 δ-Substituent Effects

Remote effects are usually small across single bonds with δ = 0.2 ppm, ϵ < 0.1 ppm unless substituent groups are jammed into each other. *Cis*-1,3-diaxial, for example, exhibits a downfield shift of several ppm. δ-substituent effects in some compounds are shown in Figure 3.3 [9].

3.5.11 Membered Rings

The ring size effect is noticeable for a chemical shift. For instance, an upfield shift is pronounced for 3-membered rings like epoxides, aziridines, cyclopropanes, and cyclopropenes. A similar effect is not observed for cyclobutanes and four-membered heterocycles. The ring size effect is shown in Figure 3.4.

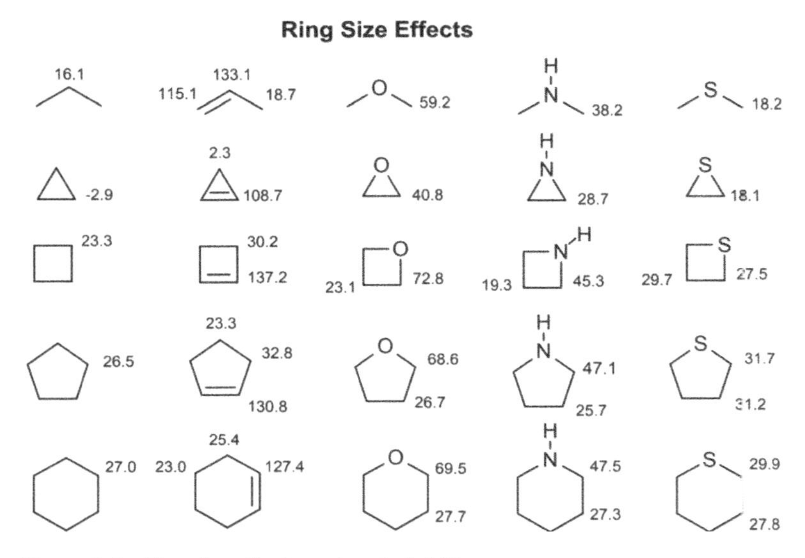

Figure 3.4 Ring size effect on chemical shift.

3.5.12 Effects on sp^2 and sp Carbons

Carbons of double and triple bonds display the α, β, and γ effects in reminiscence of saturated carbons. Furthermore, large density effects are common for such double and triple bonded carbons due to the partial charges, both positive and negative in the π-system. Carbons bonded to iodine mostly show heavy-atom effects (Figure 3.5).

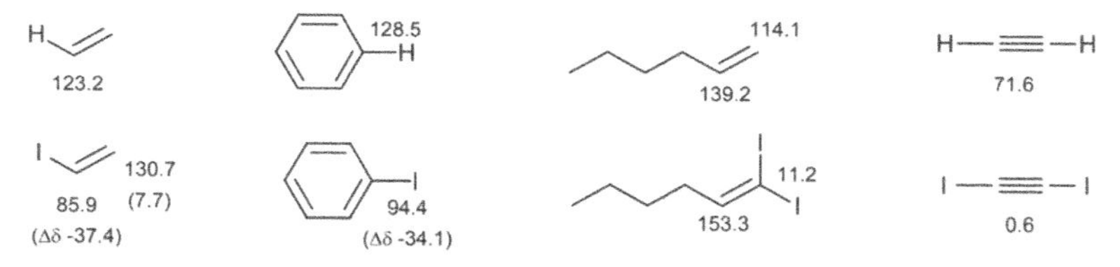

Figure 3.5 α-effect of iodine on sp^2 carbons.

3.5.13 Conjugation with π-Acceptors and π-Donors

The charge densities in π-polarized double and triple bonds dictate the chemical shifts and resonance structures may be qualitatively predicted. Figure 3.6 shows the effect of conjugation with π-acceptors and π-donors on chemical shifts. A downfield shift is a case for the β carbon of α,β-unsaturated carbonyl compounds, while a strong upfield is observed for enamines and enol ethers.

The polarization takes place in an identical fashion for alkynes with the substituents of first-row elements. But chemical shifts of alkynes with the substituents of second and third-row elements are more complicated.

3.5.14 Strongly Charged Systems

The localization of charge (sp^3 systems, σ-charge) causes interesting variation in the effect and often shows the opposite effect to those observed in π-systems. Examples include C-1 of PhLi with δ = 171.9 and CH_3Li with δ =13.2. Conversely, chemical shifts closely follow charge density (160 ppm/e), when the charge is an indispensable part of a π system. The chemical shifts of monocyclic aromatic

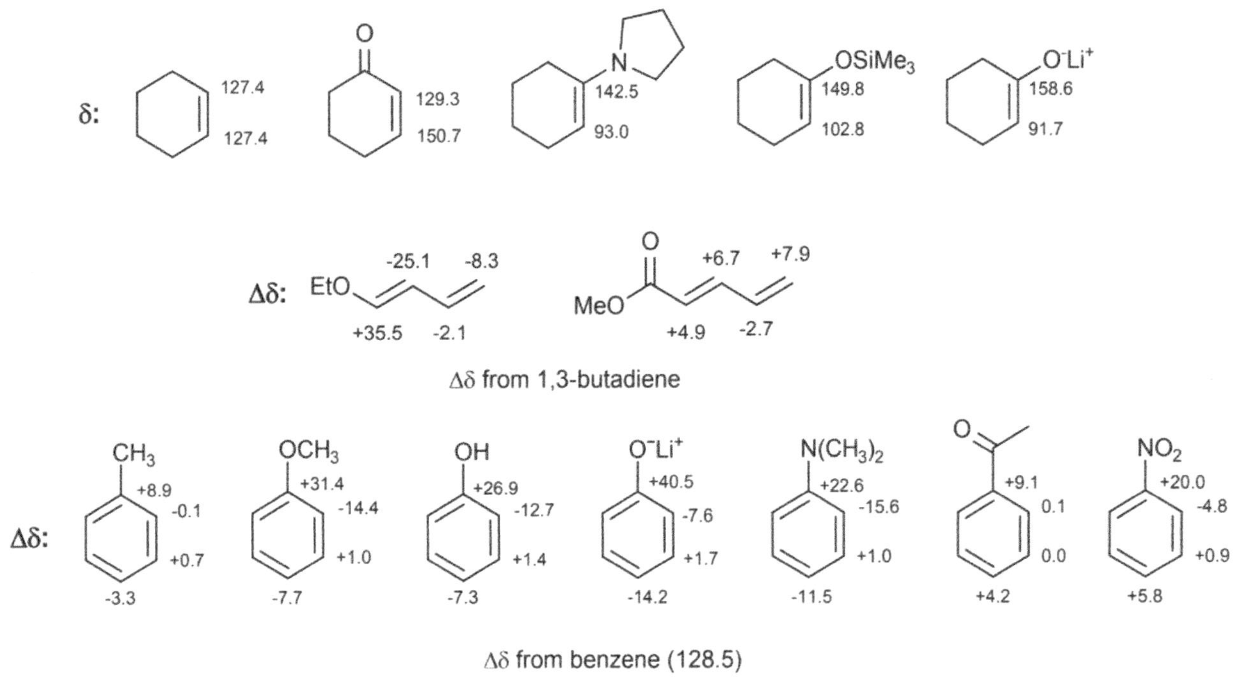

Figure 3.6 Effect on chemical shift by π-acceptors and π-donors in the conjugated system.

anions and cations show remarkable consistency, and the excess charge density at each carbon may be predicted with accuracy using the formula as shown in Figure 3.7.

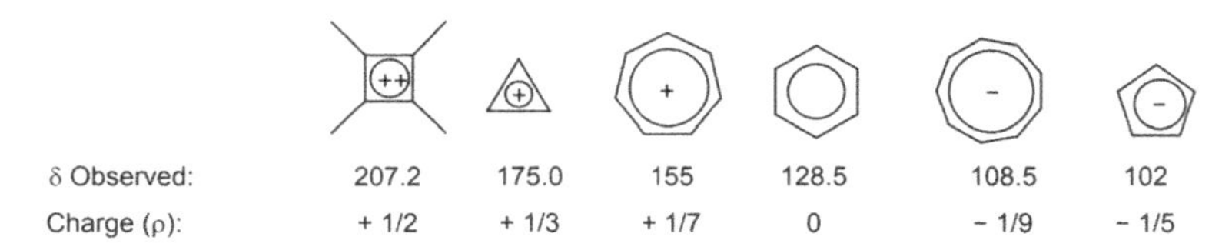

Figure 3.7 Effect of localized charge on aromatic anions and cations.

3.5.15 Carbonyl Groups

Two regions can be identified for carbonyl groups: 190–220 ppm for aldehydes and ketones, and 150–175 ppm for carboxylic acids, esters, amides, and other carbonyl functions. The ketone region is distinctively different, and the only common function that reasonably emerges for the central carbon of allenes is at about 200 ppm. Different carboxylic acid derivatives do not have unique ranges for chemical shift making the ready identification of carboxylic acids, acid chlorides, amides, esters, and anhydrides difficult. However, carbonyl stretch in the IR spectra is usually studied to solve the problem. It should be noted that carbonates, urea, and carbamates have chemical shifts very close to carboxylic acid derivatives and resolution is quite difficult. Figure 3.8 represents the chemical shift effects of numerous carbonyl groups useful for structural assignment.

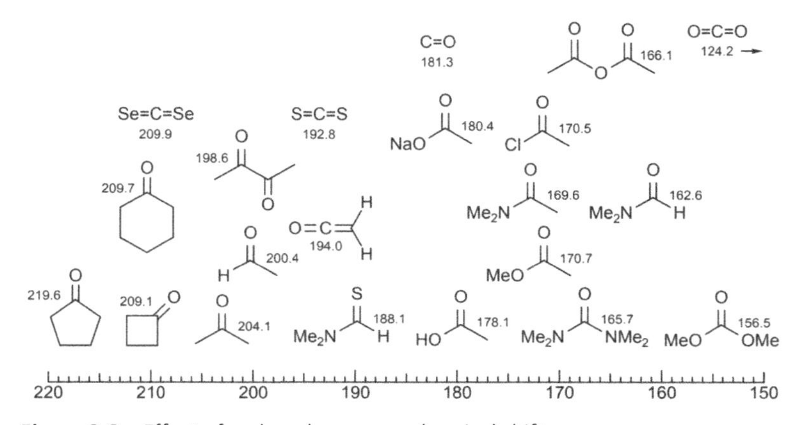

Figure 3.8 Effect of carbonyl group on chemical shift.

3.5.16 Conjugation Effects

The conjugation causes upfield shifts (6–10 ppm) for carbonyl compounds of all kinds. In such a case, conjugation takes place to a double bond or aromatic ring. Nitriles, on the other hand, have a smaller impact, but the chemical shift is still upfield. Conjugation effects in some compounds are shown in Figure 3.9.

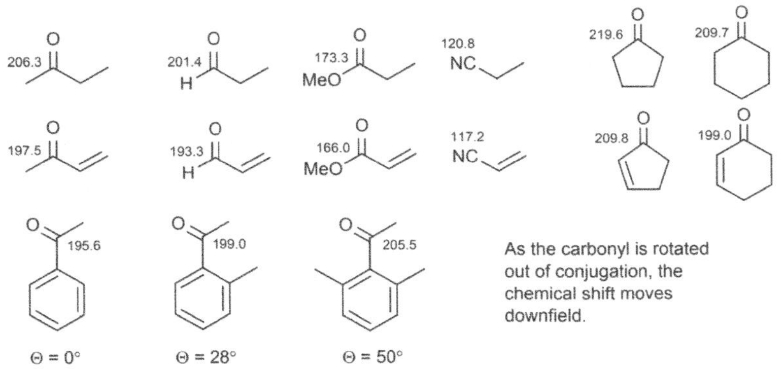

206.3 O 201.4 O H 173.3 O MeO 120.8 NC 219.6 O 209.7 O

197.5 O 193.3 O H 166.0 O MeO 117.2 NC 209.8 O 199.0 O

O 195.6 O 199.0 O 205.5

As the carbonyl is rotated out of conjugation, the chemical shift moves downfield.

Θ = 0° Θ = 28° Θ = 50°

Figure 3.9 Effect of conjugation on chemical shift.

3.5.17 Hydrogen Bond Effects

Intramolecular hydrogen bond causes significant downfield. With the exception of carbonyl groups, most carbon signals do not show solvent effects. Carbonyl groups show a downfield shift in protic solvents due to the influence of the hydrogen bond. Figure 3.10 depicts the hydrogen bond effects of various compounds on chemical shift.

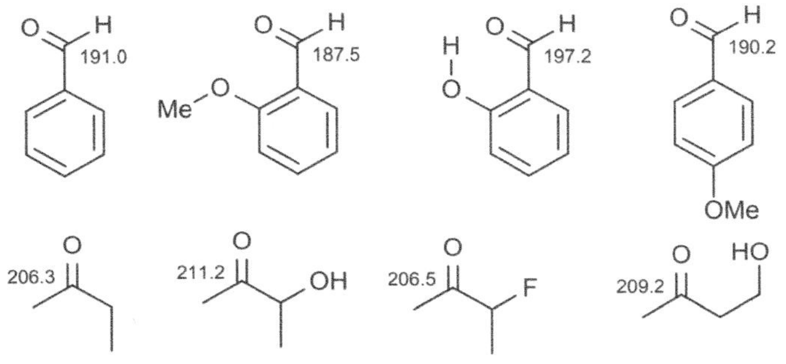

O H 191.0 Me O O H 187.5 H O O H 197.2 O H 190.2 OMe

206.3 O 211.2 O OH 206.5 O F 209.2 O HO

Figure 3.10 Effect of hydrogen bond on chemical shift.

Tables 3.8–3.11 show the ^{13}C chemical shift of alkanes, cycloalkanes, ^{13}C substituent parameter ($\Delta\delta$ ppm), and ^{13}C chemical shift of carbonyl compounds and other functional groups [10].

Table 3.8 C^{13} chemical shift for common alkanes and cycloalkanes [11]

Name	Structure	C_1[a]	C_2	C_3
Methane	CH_4	-2.3		
Ethane	CH_3CH_3	5.7		
Propane	$CH_3CH_2CH_3$	15.8	16.3	
Butane	$CH_3CH_2CH_2CH_3$	13.4	25.2	
Pentane	$CH_3CH_2CH_2CH_2CH_3$	13.9	22.8	34.7
Hexane	$CH_3CH_2CH_2CH_2CH_2CH_3$	14.1	23.1	32.2
Cyclopropane	Cyclo-$(CH_2)_3$	-3.5		
Cyclobutane	Cyclo-$(CH_2)_4$	22.4		
Cyclopentane	Cyclo-$(CH_2)_5$	25.6		
Cyclohexane	Cyclo-$(CH_2)_6$	26.9		
Cycloheptane	Cyclo-$(CH_2)_7$	28.4		
Cyclooctane	Cyclo-$(CH_2)_8$	26.9		

[a] numbering start at a terminal carbon

Table 3.9 ^{13}C substituent parameter ($\Delta\delta$ ppm) [12]

	Terminal X X–C_α–C_β–C_γ			Internal X C_γ–C_β–C_α(X) –C_β–C_γ		
X						
	α	Β	Γ	Α	Β	γ
–F	68	9	–4	63	6	–1
$-NO_2$	63	4	–	57	4	–
–OR	58	8	–4	51	5	–4
–OC(=O)R	51	6	–3	45	5	–3
–OH	48	10	–5	41	8	–5
$-NR_2$	42	6	–3	–	–	–
–NHR	37	8	–4	31	6	–4
–C(=O)Cl	33	–	–	28	2	–
–Cl	31	11	–4	32	10	–4

	Terminal X $X{-}C_\alpha{-}C_\beta{-}C_\gamma$			Internal X $C_\gamma{-}C_\beta{-}C_\alpha(X){-}C_\beta{-}C_\gamma$		
-C(=O)H	-	31	-2	0	-	-2
-C(=O)R	30	1	-2	24	1	-2
$-NH_2$	29	11	-5	24	10	-5
$-N^+H_3$	26	8	-5	24	6	-5
$-C(=O)O^-$	25	5	-2	20	3	-2
-Phenyl	23	9	-2	17	7	-2
$-C(=O)\ NH_2$	22	-	-0.5	2.5	-	-0.5
-C(=O)OH	21	3	-2	16	2	-2
$-CH{=}CH_2$	20	6	-0.5	-	-	-
-C(=O)OR	20	3	-2	17	2	-2
-Br	20	11	-3	25	10	-3
-SR	20	7	-3	-	-	-
-SH	11	12	-4	11	11	-4
$-CH_3$	9	10	-2	6	8	-2
-C≡C-H	4.5	5.5	-3.5	-	-	-
-C≡N	4	3	-3	1	3	-3
-I	-6	11	-1	4	12	-1

Table 3.10 ^{13}C chemical shift ranges of carbonyl compounds [13]

Compound class	Structure	δ (ppm)
Ketone	R-C(=O)-R	195–220
Aldehyde	R-C(=O)-H	190–200
Carboxylic acid	R-C(=O)-OH	170–185
Carboxylic ester	R-C(=O)-OR	165–175
Anhydride	R-C(=O)-O-C(=O)-R	165–175
Amide	$R{-}C(=O){-}NR_2$	160–170
Acid halide	R-C(=O)-X (X=Cl, Br, I)	160–170
Metal- coordinated CO	M-C=O	150–285
Carbon monoxide	CO	183.4[a]
Carbon dioxide	CO_2	124.8[b]

[a]CO terminally coordinated to a metal in an organometallic complex (M-C=O) is usually desheiled and appears in the range δ 180–250. [b]CO_2 – saturated $CDCl_3$.

Table 3.11 ^{13}C chemical shift of other functional groups [14]

Functional group	Structure	δ (ppm)
Oxime	$R_2C=N-OH$	145–170
Isocyanate	R-N=C=O	110–135
Isothiocyanate	R-N=C=S	120–140
Metallo-carbene	$M=CR_2$	180–400
Carbenium ions	R_3C^+	212–320
Acylium ions	$R-C^+=O$	145–155
Carbon disulfide	S=C=S	192.3
Ketenes[a]	$Z_2C_\beta=C_\alpha=O$	Z=H, Alkyl, Aryl: α, 194–206; β, 2.5–48
ketenes[a]	$Z_2C_\beta=C_\alpha=O$	Z=heteroatom: α, 161–183; β, -20–125

3.6 Application of ^{13}C NMR

^{13}C NMR spectroscopy, like ^1H NMR, is used in a range of applications, which include:

1. determination of the structure of molecules,
2. identification of molecules and molecular weight determination,
3. MRI technique same as proton although proton MRI is most common, and
4. agriculture application.

Furthermore, ^{13}C NMR provides a number of advantageous features over ^1H NMR for elucidating structures of organic and biological molecules. The fact that the backbones of molecules are identified with clarity using ^{13}C NMR rather than extracting information from the periphery offers the technique an obvious advantage over ^1H NMR. Furthermore, chemical shifts for ^{13}C for most of the organic compounds of our interest are about 200 ppm, compared to ca. 10–15 ppm for ^1H. Individual resonance peaks for each carbon atom may conveniently be assigned in compounds with molecular weights ranging from 200 to 400. In addition, in unenriched samples, the probability of two ^{13}C atoms occurring in the same molecule is low, and hence homonuclear, spin–spin coupling in carbon atoms is not encountered. Furthermore, the spin quantum number of the ^{12}C is zero, thus ^{13}C and ^{12}C do not allow heteronuclear spin coupling. Finally, there are effective ways to decouple the possible interaction of ^{13}C atoms with available protons. This usually gives rise to a single line in the spectrum for a certain kind of carbon.

^{13}C NMR spectroscopy allows the measurement of all nuclei in most organic compounds ranging from small to medium-sized. In addition, quadrupole and indirect spin–spin coupling constants have also been measurable. There have been significant developments in interpreting chemical shifts using theoretical treatments allowing accurate determination of structure to render the technique to a versatile tool. The use of ^{13}C as an isotopic tracer in reactions is one important application that might be exploited. By introducing a molecule enriched in ^{13}C, identification of different forms of carbon in a complex mixture has been possible by rigorous analysis of the chemical shifts and fine structures. This method applies even to systems that are biological in nature.

3.6.1 Identification of Molecules

The carboxyl carbon is represented by the large peak on the left, while the methyl carbon is represented by the four smaller peaks on the right. The indirect spin–spin interaction with the three methyl protons in methyl carbon causes the splitting of the resonance into four components.

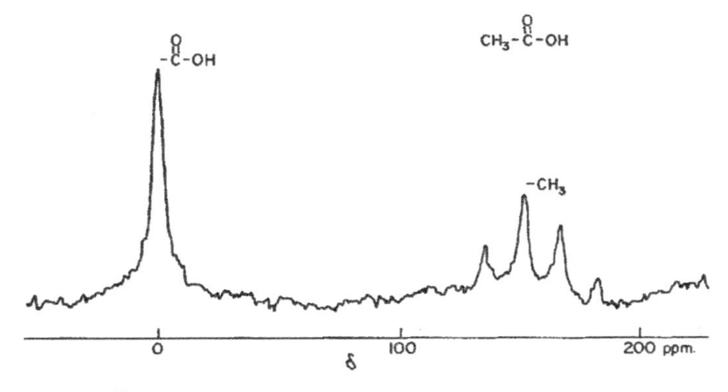

Figure 3.11 ^{13}C NMR spectrum of acetic acid.

As shown in Figure 3.11, methyl or hydroxyl protons do not exhibit any coupling with the carboxyl carbon. (Figure 3.11). Since the possibility of both carbon atoms in the molecule to be ^{13}C is extremely low due to the low abundance of ^{13}C, no ^{13}C-to-^{13}C coupling is observed.

3.6.2 Soil Organic Matter Studies

NMR spectroscopy was successfully used to characterize the structural properties of soil in particular humic substances [15, 16]. Later on, the application of FT-^{13}C NMR spectroscopy could elucidate the structure of extracted humic acids present in soil samples [17, 18].

Since then, solid-state ^{13}C NMR spectroscopy has become a key technique for studying soil structures of organic matters [19, 20]. The combined analysis of ^{13}C and ^{15}N NMR spectroscopy has been very useful in elucidating specific issues in studies of soil organic matters. ^{13}C in association with ^{15}N NMR can help characterize the structure of bulk soils or determine fractions of soil organic matter (SOM), as well as provide information on soil carbon and nitrogen cycling. In addition, the mechanism of pollutant binding processes in soils may also be elucidated.

3.6.3 Physical Separations and Structural Characterization of Organic C in Bulk Soils

The possibility to determine the structure of SOM in bulk soils as the extracts have only a portion of the total for many soils is a major advantage of the cross-polarization magic angle spinning ^{13}C NMR ((CPMAS ^{13}C NMR) technique. The magic angle spinning is crucial for solid materials for compensating the chemical shift anisotropy. The cross-polarization between the concentrated ^{1}H and the dilute ^{13}C spins enhances the signal to allow the investigation of solid materials at a natural abundance of ^{13}C [19]. CPMAS ^{13}CNMR spectra of bulk soils collected from several layers of a Haplic Phaeozem from the Halle, Germany region are shown in Figure 3.12. Table 3.12 shows the assignments of peaks for ^{13}C NMR spectra. The major signals are observed at 30, 56, 72, 105, 119, 130, 150, and 175 ppm. The signals at 72 and 105 ppm together with the shoulders around

65 and 80–90 ppm are assigned to polysaccharides. The peak at 30 ppm is due to the signal for methylenic C, which has the origin of long-chain aliphatic molecules including lipids, fatty acids, cutin acids, and other likely unidentified biopolymers. Broad resonances are marked between 30 and 55 ppm which are indicative of the presence of proteins or the linkage of peptides. In lignin, the signals for methoxyl C are observed at 56 ppm, while that for phenolic C, C-substituted aromatic C, and protonated aromatic C were noted at 150, 130, and 119 ppm, respectively. The signal at 130 ppm shows high relative intensity due to the C-substituted aromatic C and suggests a great change of the original lignin structure and/or the presence of materials with similar structures from other sources like soot and charcoal in the soil. The signal at 175 ppm is assigned to the amide and carboxyl groups in different species. ^{13}C NMR spectrum of soil or humic compounds show peaks at 0–45 ppm for alkyl C, 45–110 ppm for O-alkyl C, 110–160 ppm for aromatic C, and 160–210 ppm for carboxyl C [21].

Table 3.12 Different peaks in solid-state ^{13}C NMR spectra (referenced to TMS = 0 ppm) [22]

Chemical shift range (ppm)	Assignment
220–160	Carboxyl/carbonyl/amide carbons
160–140	Aromatic COR or CNR groups
140–110	Aromatic C–H carbons, guaiacyl C–2, C–6 in lignin, olefinic carbons
110–90	Anomeric carbons of carbohydrates, C–2, C–6 of syringyl units of lignin
90–60	Carbohydrate-derived structures (C–2 to C–5) in hexoses, C–a of some amino acids, higher alcohols
60–45	Methoxyl groups and C–6 of carbohydrate and sugars C–a ofmost amino acids
45–25	Methylene groups in aliphatic rings and chains
25–0	Terminal methyl groups

The influence of soil texture on soil organic carbon and nitrogen dynamics and storage so far received a surge of interest. Numerous efforts have been directed toward the employment of solid-state ^{13}C NMR to characterize SOM in various fractions of soil [24].

Haplic Phaeozem

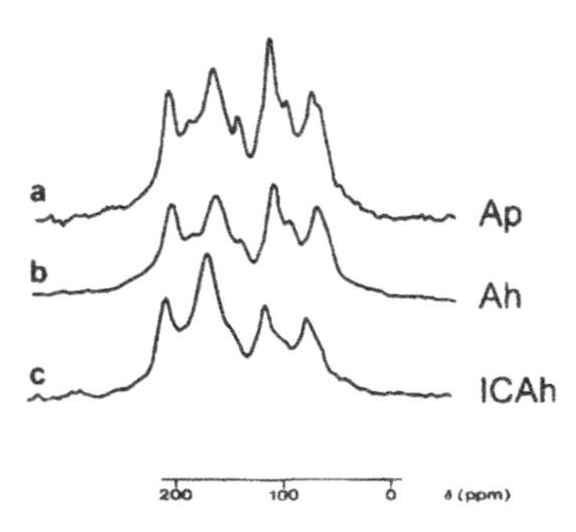

Figure 3.12 ^{13}C NMR spectra of bulk soil from several layers of a Haplic Phaeozem from Halle, Germany [23].

References

1. Mazzola, E. P., Lambert, J. B. and Ridge, Clark D., *Nuclear Magnetic Resonance Spectroscopy: An Introduction to Principles, Applications, and Experimental Methods*, 2nd Ed. (John Wiley, New Jersey), 2018.
2. Chatwal, G., R. and Sham, K. Anand, *Instrumental Method of Chemical Analysis*, 5th Ed. (Himalaya Publishing, Mumbai), 2002.
3. Data Organic Chemistry, NMR Spectroscopy: Hans Reich NMR Collection - Content (organicchemistrydata.org).
4. Parimao, P., *Pharmaceuical Analysis*, 1st Ed. (CBS, New Delhi), 1998.
5. Pavia, L., Lampman, G. M., Kriz, G. A. and Vyvyan, J. R., *Introduction to Spectroscopy*, 5th Ed. (Cengage Learning, India), 2007.
6. Kalsi P. S., *Spectroscopy of Organic Compound*, 6th Ed. (New Age International, New Delhi), 2004.
7. Sharma, Y. R., *Elementary Organic Spectroscopy*, 3rd Ed. (S. Chand, New Delhi), 1992.
8. Schneider, H. J. and Hoppen V., Carbon-13 nuclear magnetic resonance substituent-induced shieldings and conformational equilibriums in cyclohexanes, *J. Org. Chem.*, 43, 3866, 1978.
9. Kleinpeter E. and Seidl, P. R., The γ- and δ-effects in ^{13}C NMR spectroscopy in terms of nuclear chemical shielding (NCS) analysis, *J. Phys. Org. Chem.*, 18, 272, 2005.

10. Macomber, S. R., *A Complete Introduction to Modern NMR Spectroscopy* (Wiley Interscience, New York), 1998.

11. Silverstein R. M., Bassler, G. C. and, Morrill, T. C., *Spectrometric Identification of Organic Compounds*, 5th Ed. (Wiley, New York), 1991.

12. Wehrli, F. W. and Wirthlin, T., *Interpretation of Carbon-13 NMR Spectra*, 2nd Ed. (Heyden, London), 1983.

13. Kegley, S. E. and Pinhas, A. R., *Problem and Solutions in Organometallic Chemistry* (University Science Books, Mill Valley, CA), 1986.

14. Tidwel, T. T., *Ketenes*, 2nd Ed. (Wiley Interscience, New York), 1995.

15. Barton, D.H.R. and Schnitzer, M., A new experimental approach to the humic acid problem, *Nature*, vol. 198, 217–218, 1963.

16. Neyroud, J. A. and Schnitzer, M., The chemistry of high molecular weight fulvic acid fractions, *Can. J. Chem.*, 52, 4123–4132, 1972.

17. Gonzalez Vila, F. J. and Lentz, H., FT-^{13}C nuclear magnetic resonance spectra of natural humic substances, *Biochem. Biophys. Res. Comm.*, 72, 1063–1070, 1976.

18. Wilson, M. A., Pugmire, R. J., Zilm, K. W., Gob, K. M., Heng, S., and Grant, D., Cross-polarization ^{13}C NMR spectroscopy with magic angle spinning characterizes organic matter in whole soils, *Nature*, 294, 648–650, 1981.

19. Wilson, M. A., *NMR Techniques and Applications in Geochemistry and Soil Chemistry* (*Pergamon Press*, Oxford), 1987.

20. Wilson, M.A., Chapter 10: Application of nuclear magnetic resonance spectroscopy to organic matter in whole soil in humic substances in soil and crop sciences, In: MacCarthy, P., Clapp, C. E., Malcolm, R. L. and Bloom, P. R. (eds.), *Humic Substances in Soil and Crop Sciences: Selected Readings* (American Society of Agronomy, Crop Science Society of America, and Soil Science Society of America), 221–260, 1990. https://doi.org/10.2136/1990.humicsubstances.c10

21. Kogel-Knabner, I., ^{13}C and ^{15}N NMR spectroscopy as a tool in soil organic matter studies *Geoderma*, 80, 243–270, 1997.

22. Knicker, H., Hatcher, P. G. and Scaroni, A. W., A solid-state ^{15}N NMR spectroscopic investigation of the origin of nitrogen structures in coal, *Int. J. Coal Geol.*, 32, 255–278, 1996.

23. Schmidt, M. W. I., Knicker, H., Hatcher, P. G. and Kogel-Knabner, I., Impact of brown coal dust on the organic matter in particle-size fractions of a mollisol, *Org. Geochem.*, 25, 29–39, 1996.

24. Balesdent, J. and Mariotti, A., Chapter 3: Measurement of soil organic matter turnover using C^{13} natural abundance, In: Boutton, T. and Yamasaki, S. (eds.), *Mass Spectrometry of Soils* (Marcel Dekker, New York), 83–112, 1996.

Chapter 4

Electron Spin Resonance Spectroscopy

Preeti Gupta,[a] S. S. Das,[b] and N. B. Singh[c]
[a]*Department of Chemistry, DDU Gorakhpur University, Gorakhpur, India*
[b]*Department of Chemistry (ex-professor), DDU Gorakhpur University, Gorakhpur, India*
[c]*Department of Chemistry and Biochemistry, SBSR and RDC, Sharda University, Greater Noida, India*
n.b.singh@sharda.ac.in

4.1 Introduction

Electron spin resonance (ESR) also known as electron paramagnetic resonance (EPR) or electron magnetic resonance (EMR) spectroscopy is a method for the studies of paramagnetic materials. It is a branch of absorption spectroscopy in which the frequency of electromagnetic radiations in the microwave region (0.04–25 cm) is absorbed by a paramagnetic substance to produce a transition between the magnetic energy level of an electron with unpaired spins. The splitting of magnetic energy levels produced by the action of the magnetic

Spectroscopy
Edited by Preeti Gupta, S. S. Das, and N. B. Singh

ISBN 978-981-4968-32-4 (Hardcover), 978-1-003-41258-8 (eBook)
www.jennystanford.com

field occurs due to the presence of unpaired electrons contained in an ion or a molecule. The basic principle depends on the fact that atoms, ions, molecules, or molecular fragments with odd numbers of electrons exhibit characteristic magnetic properties. The electron possesses a spin and due to this spin, it has a magnetic moment. ESR spectroscopy, after its discovery in 1944 by E. K. Zavoisky, has widely been used as a very accurate and informative technique for the studies of different kinds of paramagnetic species either in solid or liquid states. The basic principle of ESR and NMR spectroscopy are quite similar but their applications and instrumentations are quite different [1–3].

In ESR spectroscopy, electromagnetic radiation having a frequency in the microwave region is absorbed by a paramagnetic substance to induce a transition between magnetic energy levels of an electron with unpaired spins till the splitting of magnetic energy is done by applying a static magnetic field.

4.2 Basic Principles and Origin of ESR Spectra

The principle of ESR depends upon the fact that an electron is a charged particle and the fundamental properties of an electron, i.e., (i) mass (ii) charge, and (iii) spin or intrinsic angular momentum are associated with the spin angular momentum and its value is given by

$$S = \sqrt{s(s+1)}\frac{h}{2\pi} \tag{4.1}$$

The spin angular momentum (S) is a vector quantity. The electron spins around its axis and its spinning motion causes it to act like a tiny bar magnet to produce a magnetic field and the magnetic field of circular current is equal to the magnetic dipole moment μ_S which is given by

$$\mu_S = -g\frac{eh}{4\pi mc}\sqrt{s(s+1)}\frac{h}{2\pi} \tag{4.2}$$

Eq. (4.2) can also be written as

$$\mu_S = -g\beta\sqrt{s(s+1)} \tag{4.3}$$

where g is the electron *g-factor* for the free electron (g = 0.0023), e is the charge, m is the mass of the electron, and β is the Bohr's magneton which is a factor for converting angular momentum into magnetic moment. The value of β is equal to $\frac{\text{eh}}{4\pi\text{mc}} = 9.27 \times 10^{-24}\,\text{JT}^{-1}$. The negative sign indicates that the magnetic moment vector acts in the opposite direction to that of the angular momentum vector. When the electron is placed in a magnetic field (strength H in tesla), the electron's magnetic dipole will precess around the axis of the field.

Similar to NMR spectroscopy, the Larmor frequency of precession ω, in this case, is also given by

$$\omega = \gamma H \tag{4.4}$$

γ is the magnetogyric ratio of the dipole, or in the other words, it is the ratio of magnetic moment to the angular momentum.

Further, the magnetic energy of interaction E on putting the electron in a magnetic field of strength, H, is given by

$$E = -\mu_s H \cos\theta \tag{4.5}$$

where θ is the angle between the axis of the electron's dipole and the direction of the magnetic field. But as per quantum mechanical restrictions, only certain values of θ are allowed. If μ_H is the components of μ_s in the direction of the field, then one can write

$$\cos\theta = \frac{\mu_H}{\mu_s} \tag{4.6}$$

Therefore, the equation for the magnetic energy of interaction E can be simplified as

$$E = -\mu_H H \tag{4.7}$$

where μ_H is defined as

$$\mu_H = -g\beta M_s \tag{4.8}$$

where M_s is spin quantum number, β = electrons Bohr's magneton. Therefore, with the help of Eqs. (4.7) and (4.8) we can write

$$E = g\beta H M_s \tag{4.9}$$

Under the influence of magnetic field H, the spin of the unpaired electron M_s can align in two different ways to create two spin states, i.e., when $M_s = +1/2$ and $M_s = -1/2$. The alignment can either be in parallel direction to the magnetic field which corresponds to the lower energy state for $M_s = -1/2$ or antiparallel to the direction to the applied magnetic field corresponding to a higher energy state when the value of $M_S = +1/2$. Therefore, corresponding to the two values of M_S the magnetic interaction energy can also have the following two values:

$$E_- = E_0 - \frac{1}{2} g\beta H \ (\text{for } M_s = -\frac{1}{2}) \tag{4.10}$$

$$E_+ = E_0 + \frac{1}{2} g\beta H \ (\text{for } M_s = +\frac{1}{2}) \tag{4.11}$$

where E_0 is the energy in the absence of the magnetic field. The negative sign (–) corresponds to the lower energy state and the positive sign (+) corresponds to the higher energy state.

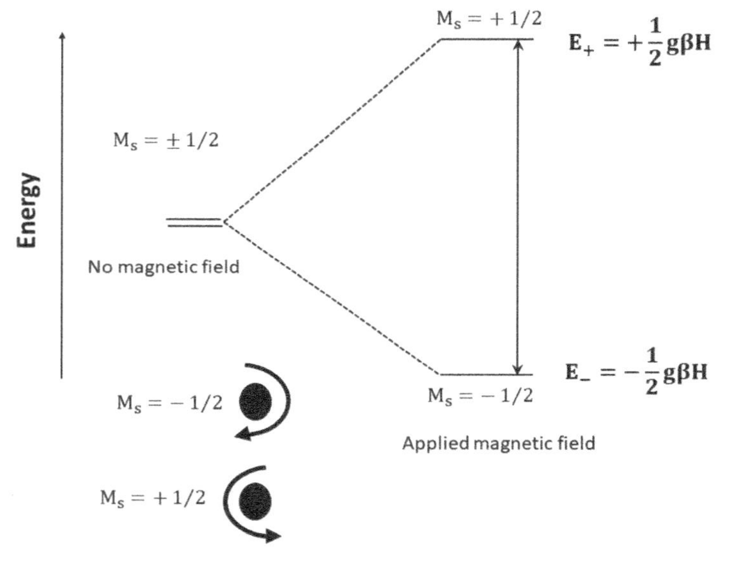

Figure 4.1 Splitting pattern of energy levels and the possible transition in presence of a magnetic field.

When the transition between two energy levels E_+ and E_- takes place, i.e., from the higher level to the lower one, then absorption of a quantum of radiation of frequency υ (in the microwave region) occurs. Thus, one can write

$$\Delta E = h\upsilon = E_+ - E_- = E_0 - \frac{1}{2} g\beta H - \left(-\frac{1}{2} g\beta H \right) \tag{4.12}$$

or, $$h\upsilon = g\beta H$$

where υ is the frequency of absorbed radiation in cycles/sec. This condition is called the resonance condition and under such conditions, the transition between these two Zeeman levels may occur which is shown in Figure 4.1. This resonance condition is the root cause for the origin of an ESR spectrum and Eq. (4.12) is the fundamental equation of ESR spectroscopy.

4.3 The Electron g-Factor

g-Factor is a unitless quantity that is a measure of the intrinsic magnetic moment of electrons. It is also known as Zeemann splitting factor, Lande splitting factor, or spectroscopic splitting factor.

Table 4.1 g-values of different compounds [4]

Compounds	g-value
$MoO(SCN)_5^{2-}$	1.935
$VO(acac)_2$	1.968
CH_3	2.0026
$C_{14}H_{10}$(anthracene) cation	2.0028
$C_{14}H_{10}$(anthracene) anion	2.0029
$Cu(acac)_2$	2.13

The g-value is defined as the proportionality constant between the frequency and the field at which the resonance occurs and it is given

$$g = \frac{h\upsilon}{\beta H} \tag{4.13}$$

It is proportional to the magnetic moment of the molecule which is being studied by ESR spectroscopy. In carbon-centered organic free radical, the value of the g-factor is always close to 2.0023 for unbounded electrons. But generally, the g-factor values of bounded unpaired electrons in a molecule are not equal to the g-value for free unbounded electrons. For organic radicals, the g-value ranges between 1.99–2.01 very close to the free electron value. Whereas, for transition metal compounds, large variations are observed in the g-values due to spin-orbit coupling and zero-field splitting. In this case, the g-values range from 1.4 to 3.00.

4.3.1 Factor Affecting g-Factor

There are the following factors that greatly influence the g-values:

(i) The magnitude of the g-value depends upon the orientation of the molecule containing an unpaired electron with respect to the magnetic field.

(ii) The g-values are not constant rather they depend upon the type of species/ molecules containing unpaired electrons.

(iii) In the gas or solution phase, due to the free movement of the molecules, several orientations are possible. As such, the g-value is averaged as the overall orientation of the molecule. But in the solid or crystal phase, the 'g' value is fixed because of the restricted movement of the molecule.

(iv) In crystals, the g-value depends on the direction of the external magnetic field with respect to the crystallographic axes *x*-, *y*-, and *z*. The three, possible g-values along *x*-, y-, and z-axes are indicated by g_x, g_y, and g_z. These three, possible g-value of a crystal can be obtained by orienting the crystal in these three perpendicular directions with respect to the external field. This phenomenon is known as anisotropy of the electron g-value [1, 5]. The following g-value notations are used for crystals that have axial symmetry.

  (a) $g_\perp$ denotes the measured g-values when a particular crystal is kept perpendicular to the direction of the external magnetic field.

  (b) $g_\parallel$ represents the measured g-values of the crystal whose axis is parallel to the direction of the external magnetic field.

(v) If θ is the angle between *the* *z*-axis and the applied magnetic field then the g-value is given by

$$g^2 = g_\parallel^2 \cos^2\theta + g_\perp^2 \sin^2\theta \tag{4.14}$$

(vi) It depends upon temperature.

4.3.2 Determination of g-Values

The g-values can be experimentally determined by using a reference sample material such as DPPH (diphenyl picryl hydrazyl free radical) and then comparing the g-values of the unknown sample with the g-values of the reference-free radical sample. DPPH has a g-value of 2.0036. Both the sample and the reference material are placed into the same double sample cavity cell and their ESR spectra are recorded with the help of a dual channel recorder. The ESR spectrum will give signals and the separation of centers of the two resonance signals will represent the magnetic field difference (ΔH) at the two positions of the sample (Figure 4.2). Thus, the g-value of the unknown sample can very easily be calculated by using the equation

$$g = g_S - \left[\frac{\Delta H}{H}\right] g_S \tag{4.15}$$

where g_S represents the g-value separation of the reference sample DPPH. ΔH indicates the separation of the centers of the two spectra. H is the strength of the applied external magnetic field [1].

Besides DPPH, tetracene positive ion having a g-value equal to 2.0026 and perylene positive ion with a g-value equal to 2.00258 can also be used as standard reference samples.

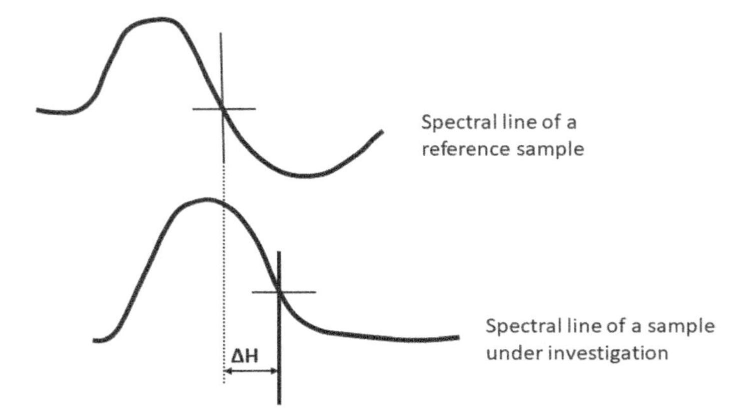

Figure 4.2 Determination of ΔH the separation of the two centers of the spectral lines of the saturation and relaxation reference and unknown sample.

4.4 Saturation and Relaxation in ESR Spectroscopy

Saturation means that no further absorption, no other resonance peaks, no more signals, and no broadening of signals take place in ESR spectra. These conditions for saturation are obtained when the number of electrons in both the energy levels (higher and lower energy states) becomes equal. In this condition, there will be no further absorption of radiation in the radiofrequency region will take place and no resonance signals would appear. This condition is known as saturation.

Under normal conditions of measurements, the number of electrons in the two energy states is not equal because the electrons in the higher energy state are regularly coming back to the lower energy state by losing energy. The loss of energy by an electron in the high energy level is known as relaxation. There are two types of radiation-less relaxation processes.

4.4.1 Spin–Lattice Relaxation Process

In this due to interaction, the energy ΔE is transferred to the neighboring atoms either in the same molecule or in some other molecules.

4.4.2 Spin–Spin Relaxation Process

In this case, the energy of the spinning electron in the upper energy state is transferred to the neighboring electron present in the lower energy state.

The fast relaxation process will prevent the saturation of electrons in the upper energy level and hence they are responsible for the appearance of broad absorption lines in the ESR spectrum. On the contrary, slow relaxation leads to the appearance of narrow absorption lines in the spectrum.

4.5 Splitting of ESR Signals: Hyperfine Interaction

The ESR spectrum signals exhibit splitting due to the interactions between the spinning electrons and adjacent splitting magnetic nuclei. This is known as hyperfine splitting or hyperfine interaction in which the ESR signals are split by the neighboring nuclei. Important information can be obtained from the analysis of hyperfine interaction lines. The identity of the radical can be fixed by the study of its hyperfine structure and thus can be used as a fingerprint for that radical. When unpaired electrons are placed in the magnetic field then the energy E gets splits into two, i.e., ($E_- = -\frac{1}{2}g\beta H$ and $E_+ = +\frac{1}{2}g\beta H$).

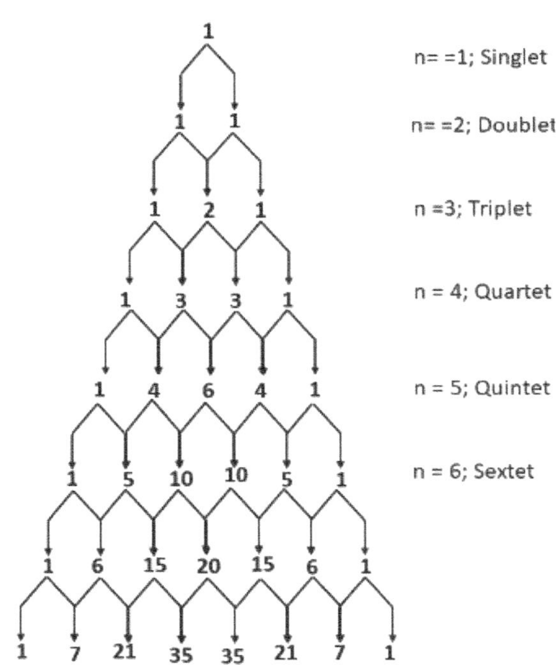

Figure 4.3 Pascal's triangle showing the coefficient of binomial expansion $(1 + x)^n$.

When a single electron interacts with one nucleus then the ESR signals will split up into $2I + 1$ line. But when there are n equivalent nuclei then the ESR signals will be split into $2NI + 1$ multiple lines (where I is the spin quantum number of the nucleus). The relative intensities of the split lines can be determined by the number of interacting nuclei. All the split lines will be of equal intensity when only one nucleus is interacting with the unpaired electron. Whereas, in the case of multiple nuclei interaction, the relative intensities of the lines will follow a binomial distribution. i.e. the relative intensities of the lines will be proportional to the coefficient of binomial expansion of $(1 + x)^n$ as shown in Figure 4.3 [1].

When the interaction of electron spinning between two energy states and the nuclear spin of the proton takes place then each energy state gets further split up into two more new energy levels corresponding to $M_S = +1/2$ and $M_S = -1/2$, where M_S is the nuclear spin angular quantum number. Therefore, one would observe four different energy levels as shown in Figure 4.4. This shows that in the ESR spectrum of one proton and one electron system (H atom), there will be two peaks corresponding to the two transitions, i.e., $M_S = -1/2$ to $M_S = -1/2$ and $M_S = +1/2$ to $M_S = +1/2$.

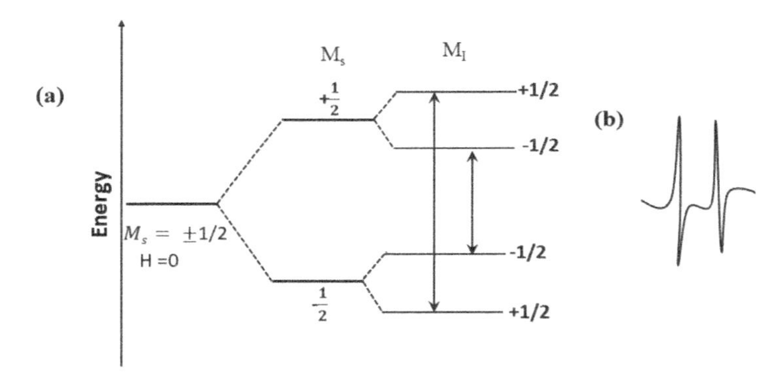

Figure 4.4 Hyperfine splitting in the ESR spectrum of H_2 atom.

Generally, the interaction energy with the neighboring nuclei is given by

$$E_+ = g\beta HM_S + AM_SM_I \tag{4.16}$$

and the selection rule for ESR spectroscopy $\Delta M_S = \pm 1, \Delta M_1 = 0$.
where A is the hyperfine coupling constant, ΔM_1 is the nuclear spin quantum number, ΔM_S is the electron spin quantum number with the help of the above energy equation. The energies of four energy levels of the hydrogen atom can be written as

$$E_{1/2\to 1/2} = \frac{1}{2}g\beta H + A\frac{1}{2}\frac{1}{2} = \frac{1}{2}g\beta H + \frac{1}{4}A \quad (M_s = \frac{1}{2} M_I = \frac{1}{2}) \tag{4.17}$$

$$E_{1/2\to -1/2} = \frac{1}{2}g\beta H - \frac{1}{4}A$$

$$E_{-1/2\to-1/2} = -\frac{1}{2}g\beta H + \frac{1}{4}A$$

$$E_{-1/2\to1/2} = -\frac{1}{2}g\beta H - \frac{1}{4}A$$

where all the four energy levels are shown in Figure 4.4.

In the case of deuterium, the value of nuclear charge $I = 1$, and when the $M_S = +1/2$, there will be three values of $M_I = 1, 0, -1$. Similarly, M_I will also have three values equal to –1, 0, and 1, when the value of $M_S = -1/2$. Therefore, due to splitting there will be six energy levels but according to the selection rule $\Delta M_S = \pm 1/2$ and $\Delta M_I = 0$, the number of allowed transitions in deuterium will be only three, and correspondingly three lines for three transitions will appear in the ESR spectrum. All these lines are of equal intensities. Figure 4.5 shows various transitions in deuterium.

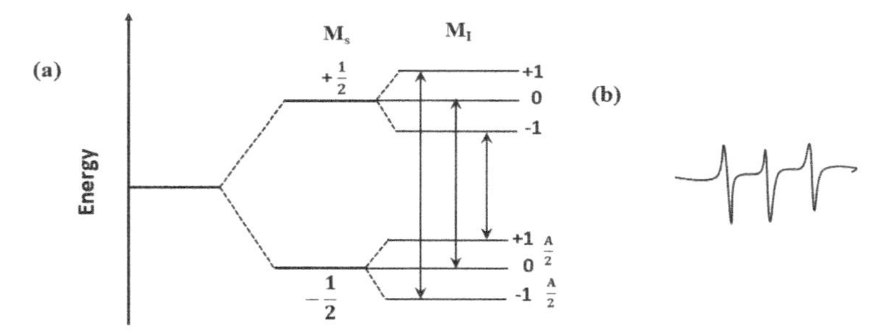

Figure 4.5 Hyperfine splitting in the ESR spectrum of a deuterium atom.

However, in the case of methyl radical, the interaction between unpaired electrons and three hydrogen nuclei will take place. Accordingly, there will be four values for M_I corresponding to M_S = +1/2 and M_I = –1/2 which are mentioned below:

$$M_I = +3/2, +1/2, -1/2, -3/2 \text{ for } M_S = ½$$

$$M_I = -3/2, -1/2, +1/2, +3/2 \text{ for } M_S = -½$$

However, the selection rule predicts only four transitions to occur so there will be four spectral lines in the ESR spectrum of methyl radical. The relative intensities of these four spectral lines will be in the ratio 1 : 3 : 3 : 1. The four, possible transition of methyl radical is shown in Figure 4.6.

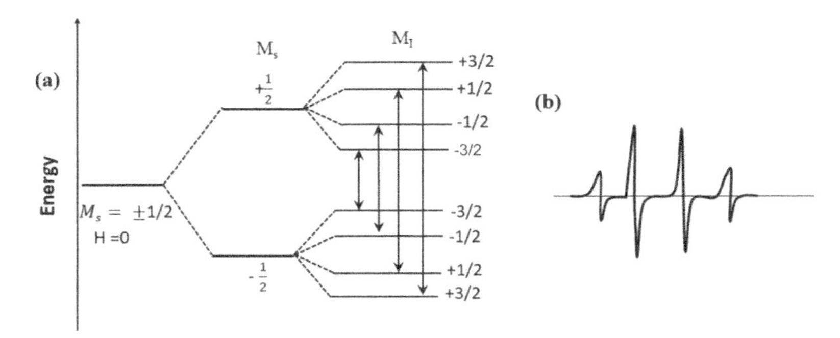

Figure 4.6 Hyperfine splitting in the ESR spectrum of methyl free radical.

The relative intensities of spectral lines are determined by the number of interacting nuclei. If only one nucleus is interacting then all spectral lines will have equal intensities however in the case of multiple nuclei, for the most common spin = ½, the intensities of the lines are determined with the help of a binomial equation.

4.6 Zero-Field Splitting: Kramer Degeneracy

In ESR spectra, the splitting of spin levels of electrons takes place even in the absence of an applied magnetic field. This phenomenon is called *zero-field splitting*. Mostly this splitting occurs in transition metal complexes having more than one unpaired electron. Because of zero-field splitting more than expected transitions are observed under the influence of an external magnetic field.

It can also be defined as that phenomenon in which the degeneracy of spin states of electrons is removed by the internal magnetic field of the unpaired electron even in the absence of an external magnetic field. It takes place in those atoms, molecules, or complexes which contain an even number of electrons.

When the system consists of an odd number of unpaired electrons, then the spin degeneracy of each level becomes double degenerate. The phenomenon is known as *Kramer's degeneracy.*

If a single electron is present in the system, then there is no complication in the ESR spectra. But when more than one electron, for example, two electrons are present in the sample under investigation then some internal splitting is observed. Let us consider a two, unpaired electron system (S = ½ + ½ = 1). In this case when of both the electron are parallel then the spin $M_S = +1$ and if the other is antiparallel then the spin will be $M_S = 0$ and when both are antiparallel then $M_S = -1$. If we draw their energy level, the spin state +1 and −1 are degenerate. In zero-field splitting, when there was no external magnetic field, one observed only one line.

Even in the absence of any external magnetic field, the above splitting takes place which is known as zero-field splitting whose spin is parallel to the external magnetic field will have lower energy and the other one which is opposite to the applied magnetic field possess higher energy. On applying the external magnetic field, then the degenerate (±1) energy state will split into two lines (+1 and −1). So, yielding three lines will be observed as shown in Figure 4.7.

According to the selection rule $\Delta M_S = 0, \pm 1, \pm 2, \pm 3, \ldots$ in such a system transition can take place from $-1 \rightarrow 0$, $0 \rightarrow +1$ and from −1 to +1 in presence of an external magnetic field. These three types of transitions will lead to the appearance of three lines in the spectra. But since the transition from $-1 \rightarrow +1(\Delta M_S = \pm 2)$ is very feeble, as such only two spectral lines will be observed in the ESR spectrum. Theoretically, three lines should appear but actually one gets only two lines. Since the transition from the level $0 \rightarrow +1$ and $-1 \rightarrow 0$ have equal energy, therefore, they exhibit degeneracy and only one ESR signal is observed. Further, since this system contains an even number of unpaired electrons (i.e., = 2), Kramer's degeneracy will not be applicable and each level will not doubly degenerate.

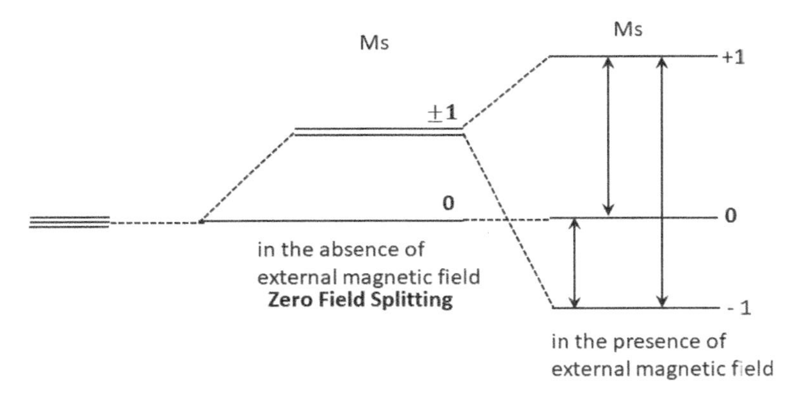

Figure 4.7 Zero-field splitting in the ESR spectrum for the system containing one unpaired electron.

The d^5 system (Mn^{2+}) contains an odd number (5) unpaired electrons. So, in this case Kramer degeneracy will be applicable [Figure 4.8(b)].

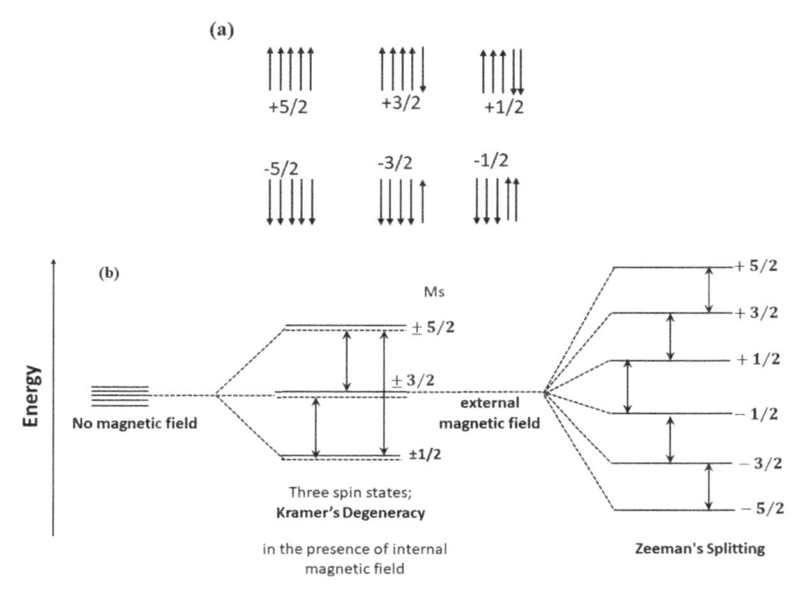

Figure 4.8 (a) Different spin arrangement in the d^5 system, (b) Kramer's degeneracy in the ESR spectrum for the system containing an odd number of electrons.

In the zero magnetic fields, all the spin levels of electrons remained degenerate. The presence of an internal magnetic field caused the splitting of spin states and it will give three doubly degenerate spin states with M_S = ±1/2, ±3/21, ±5/2. The arrangement of five spin states with the value of M_S is shown in Figure 4.8 (a) and the splitting pattern is shown in Figure 4.8 (b). Theoretically, three transitions are shown in Figure 4.8 (b) and the transition from +1/2 → +5/2 (ΔM_S= ±2) will not be observed due to its very low intensity. Whereas, the transitions from +1/2 → +3/2 and +3/2 → +5/2 will be observed even in the absence of any external magnetic field. This is known as Kramer's degeneracy which is caused by the intrinsic magnetic field of the system.

When the external magnetic field is applied to the system, the three doubly degenerate spin states will further split into six levels and five transitions will occur, i.e., $-1/2 \rightarrow +1/2$, $+1/2 \rightarrow +3/2$, $+3/2 \rightarrow +5/2$, $-3/2 \rightarrow -1/2$ and $-5/2 \rightarrow -3/2$.

4.7 Isotropic Coupling Constant

Due to hyperfine interaction with a single proton (I = 1/2) a single unpaired electron (S =1/2) splits into two energy levels: one pair for M_S = +1/2 and one pair for M_S = −1/2 (as shown in Figure 4.4). When an unpaired electron interacts with deuterium (^{2}H) for which I = 1, then each electron level splits futher into three (shown in Figure 4.5). The splitting is a field-dependent phenomenon at lower magnetic fields but at the higher fields, the maximum value of splitting is equal to A/2 where A is known as the *isotropic hyperfine coupling constant*. Similarly, for each value of M_S, the splitting occurs into $2I + 1$ energy levels when an unpaired electron interacts with a nucleus whose nuclear spin is equal to I. Mostly free radical electrons have several magnetic nuclei which can be separated into different groups of equivalent sets on the basis of the symmetry of a molecule. It is considered that these equivalent nuclei interact with free radical electrons as one nucleus.

4.7.1 Isotropic Hyperfine Splitting from Single Set of Equivalent Protons

Let us first consider one unpaired electron system which interacts with a set of two equivalent protons. In this case, the energy levels can be obtained by the successive splitting of each level (Figure 4.9). The first nucleus interacts with the two levels, i.e., M_S= +1/2 and M_S = −1/2 levels of the unpaired electron and gets split into two. The amount of splitting is A/2. The second nucleus then interacts and due to this each level further splits by A/2 into two levels. Since both the nuclei are equivalent, therefore, the hyperfine coupling constant value A will also be the same. In view of the two-fold degeneracy of nuclear spins, the relative intensities of the allowed transitions will be 1:2:1 till the conditions $kT >> A$ and $g\beta H >> A$ are obeyed by the system. The selection rule for the single nucleus system, $\Delta M_S = \pm 1$ and $\Delta M_I^t = 0$ remains the same for these systems also.

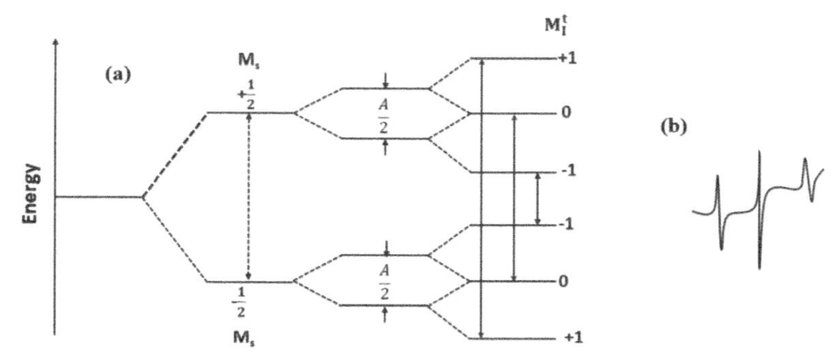

Figure 4.9 Splitting of energy levels and transition of a system with one unpaired electron and two equivalent nuclei [2].

4.7.2 Isotropic Hyperfine Splitting from Multiple Set of Equivalent Protons

When the protons of the two sets are chemically non-equivalent from each other then their splitting constant values are different. Consider a free radical system containing two non-equivalent protons 1 and 2 whose hyperfine coupling constants are designated respectively as A_1 and A_2, where $A_1 > A_2$.

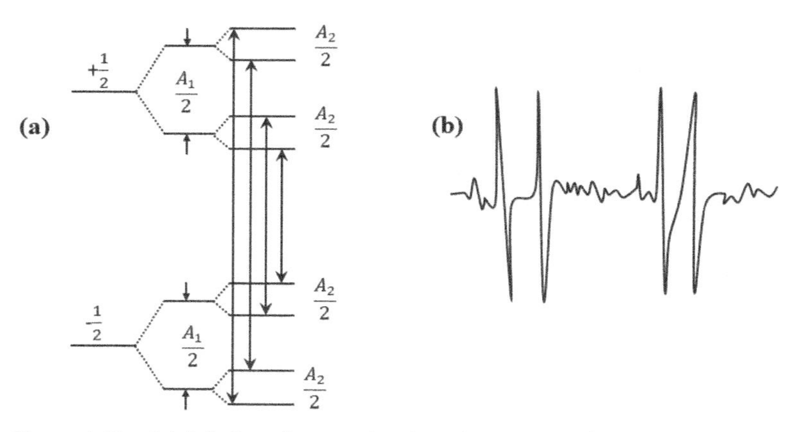

Figure 4.10 (a) Splitting of energy levels and transition of a system with one unpaired electron and two inequivalent nuclei, (b) EPR spectrum of the glycolic acid radical (HOCHCOOH) in aqueous solution at 298 K [6].

Firstly, the protons of type will interact with the electron and get split into four energy levels (two due to $M_s = +\frac{1}{2}$ and two due to $M_s = -\frac{1}{2}$), where the separation of splitting between two lines is $A_1/2$. The second set of protons 2 further splits each of these four levels into two more levels separated from each other by $A_2/2$. The energy level diagram has been shown in Figure 4.10. The selection rules for the allowed transitions are again the same for which $\Delta M_s = \pm 1$ and $\Delta M_I^t = 0 = \Delta M_I^1 + \Delta M_I^2$.

Now consider a radical containing three protons out of which two are equivalent and one is different. If the hyperfine coupling constant for each equivalent proton is A_1 and A_2 is the hyperfine coupling constant for the third proton and $A_1 \gg A_2$, then the energy level diagram for such a system can be drawn in presence of a magnetic field. The splitting takes place due to two equivalent nuclei plus other nuclei having the value $I = 1/2$. Radical CH_2OH formed by photolysis of a methyl $-H_2O_2$ is a suitable example of this case. Its ESR spectrum is shown in Figure 4.11. The smaller splitting takes place in the –OH proton while large splitting takes place due to $-CH_2$ protons.

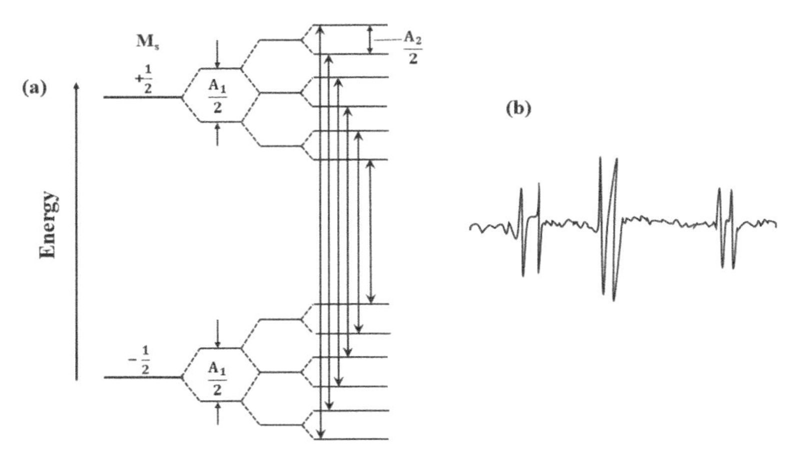

Figure 4.11 Splitting of energy levels and transition of a system with two inequivalent nuclei and one different (b) EPR spectrum of the CH_2OH radical in methanol at 299 K [7].

In general, if the molecule contains sets of m and n equivalent protons, then the number of clear visible lines in the ESR spectrum of such a molecule will be equal to $[(m + 1)(n + 1)]$. The spectrum in such cases can be studied by first considering the splitting pattern from the largest hyperfine splitting, i.e., A_n. The further splitting takes place due to m protons whose hyperfine splitting constant is A_m. In the case of naphthalene anion radical $m = n = 4$ (Figure 4.12). The number of spectral lines in this case according to the $(m + 1)$ and $(n + 1)$ formula will be equal to 25 but it should be noted that the central line of the spectrum is 36 times more intense than that of the outermost lines [8].

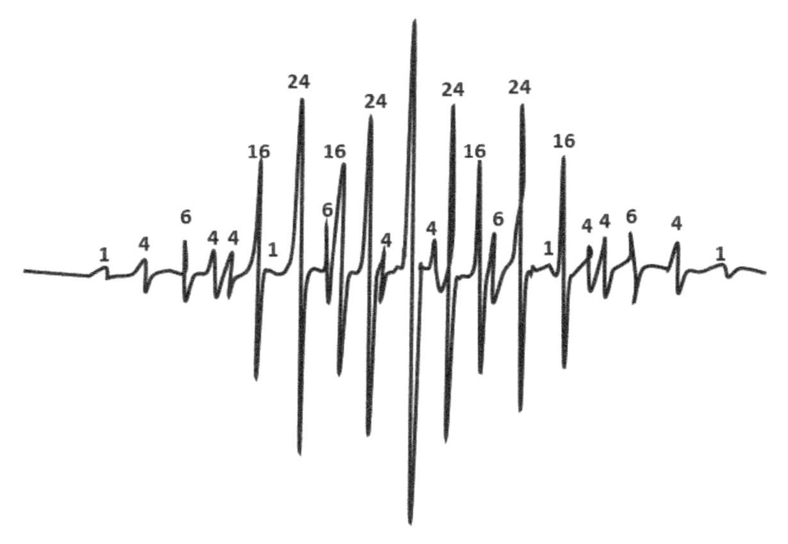

Figure 4.12 EPR spectrum of the naphthalene anion radical in dimethoxyethane (K^+ is the counter ion) at 298 K [8].

4.8 Anisotropy and Hyperfine Splitting

When the sample under investigation is in a solution state, the molecule is continuously moving. Therefore, the interactions are the same in all directions. Thus, the hyperfine splitting is called *isotropic* and is denoted by the symbol A.

But in the case of solids, depending upon the orientation of the sample crystal, the interaction is changed in different directions. So '*A*' is then known as the *anisotropic coupling constant*.

Generally, in solids, the external magnetic field is applied along *the* z-axis. So, the value of the anisotropic coupling constant along *the* z-axis is known as $A_{\parallel}$ and its value along x- and y-directions are denoted as $A_{\perp}$. The average value of anisotropic coupling constant A is then calculated as

$$A_{\text{aver}} = \frac{1}{3}[A_{\parallel} \cdot A_{\perp}] \tag{4.18}$$

In solid crystalline systems, an ESR spectrum depends on the orientation of the sample and the strength of the externally applied magnetic field. Anisotropy occurs in solids, frozen solutions, radicals obtained by irradiating crystalline substances, radicals trapped in host matrixes of solids, and point defects in single crystals. Some paramagnetic systems, which exhibit anisotropy can be grouped into three classes:

(i) Free radicals
(ii) Transition ions surrounded by ligands and
(iii) Point defects

4.8.1 Anisotropy in Crystals

The knowledge of symmetry is very important in crystals. In addition to symmetry, the local symmetry around any unpaired electron center is also important. The local symmetry could be classified into three types, i.e., cubic, uniaxial, and rhombic.

4.8.1.1 *Cubic system*

This could further be subdivided into three subsystems: (i) cuboid, (ii) octahedral, and (iii) tetrahedral. In all these three, anisotropy is absent in the ESR studies since all the three principal-axes values are equal. In the case of cubic systems, the positions of ESR signals is isotropic. The g-factor is isotropic and independent of magnetic field direction i.e. one can write

$$g_x = g_y = g_z \tag{4.19}$$

Therefore, the g-factor is a scalar quantity and the spin Hamiltonian is written in the following form

$$\Delta E = \hat{H} = g\beta\left(H_x \hat{s}_x + H_y \hat{s}_y + H_z \hat{s}_z\right) \quad (4.20)$$

An electron in a negative ion vacancy (F-center) in an alkali halide is a suitable example of this type [Figure 4.13(a) and (b)].

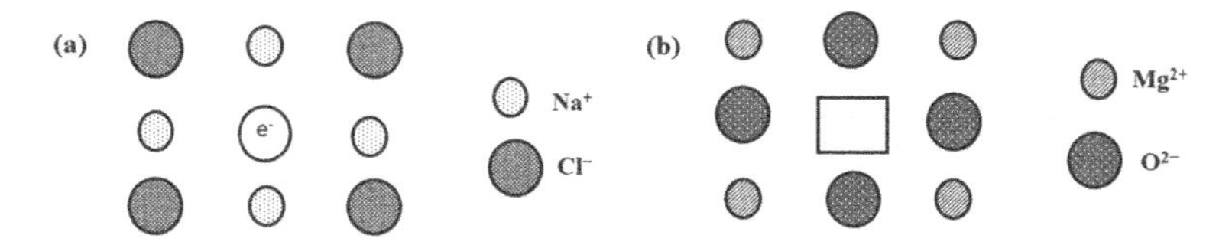

Figure 4.12 (a) F-center in NaCl crystal with cubic symmetry, (b) V-center in MgO crystal with tetragonal symmetry.

4.8.1.2 *Uniaxial symmetry*

On the other hand, the high local symmetry can be reduced from octahedral to tetragonal by introducing an imperfection along one of the axes. The V-center in MgO or CaO contains one unpaired electron [6, 9–11]. Ideally, in the crystal of MgO, the Mg^{2+} and O^{2-} ions are situated on octahedral symmetry sites. But on irradiation with X-ray at low temperature, V-center is formed. One electron is removed from one of the six oxygen ions (near the existing positive magnesium-ion vacancy) and the resulting ion, O^- is displaced away from the vacancy. The type of defect has been illustrated in Figure 4.12b. Due to this defect a fourfold uniaxial symmetry (= tetragonal symmetry) is obtained. The g-factor becomes anisotropic and depends on the direction of the magnetic field. If the field is parallel to the axis, then one can write

$$g_x = g_y = g_\perp \text{ and } g_z = g_{||}$$

Further
$$g_x = g_y \neq g_z \quad (4.21)$$

where z is the symmetry axis.

The values of $g_\perp$ and $g_{||}$ are obtained from the relation

$$g_\perp = \frac{h\upsilon}{\beta H_\perp} \text{ and } g_{||} = \frac{h\upsilon}{\beta H_{||}} \quad (4.22)$$

The angular dependence of the ESR spectrum of the V-center in MgO is represented in Figure 4.13.

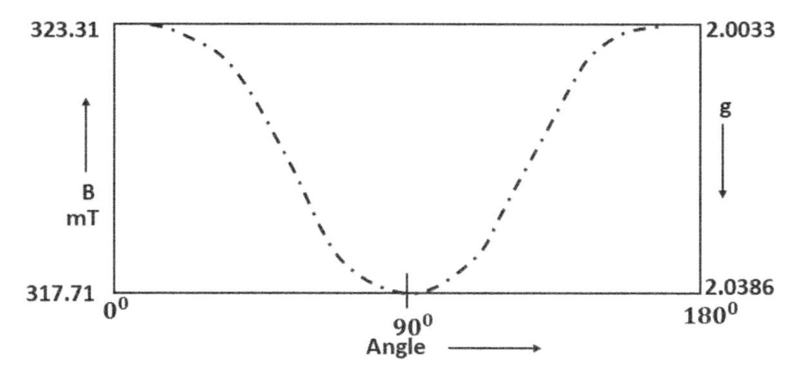

Figure 4.13 ESR spectrum presenting angular dependence of the V^- center in MgO.

The effective g-factor, in this case, is given by

$$g_{\text{eff}}^2 = g_{\perp}^2 \sin^2\theta + g_{\|}^2 \cos^2\theta \tag{4.23}$$

where θ is the angle between the magnetic field H and the symmetry axis of the defect, i.e., the z-axis. Angle θ is 0^0 when H is parallel to *the* z-axis ($H \,||\, Z$) and angle θ is 90^0 when H is perpendicular to the z-axis ($H \perp Z$). The spin Hamiltonian is then written as

$$\hat{\text{H}} = \beta\left(g_x H_x \hat{s}_x + g_y H_y \hat{s}_y + g_z H_z \hat{s}_z\right)$$

$$= \beta\left[g_{\perp}\left(g_x H_x \hat{s}_x + g_y H_y \hat{s}_y\right) + g_{\|}\left(H_z \hat{s}_z\right)\right] \tag{4.24}$$

4.8.1.3 *Rhombic symmetry*

The systems containing rhombic local symmetry are very complex. This type of system can be studied by considering the defect center (shown in Figure 4.14) generally found in those alkali halides which possess the rock salt structure (for example KBr). In this case, the defect is due to the superoxide O_2^- which is a paramagnetic diatomic molecule having a single unpaired electron and which has replaced a diamagnetic Cl^- ion [12]. In this example O–O interatomic direction is used as the z-axis of the system. The axis lies in the plane, whereas, the y-axis is directed out of the plane. The axes x and y are not equal. The symmetry of this defect in this system is rhombic instead of uniaxial due to interaction with the neighboring atoms. The g-factor is dependent on the magnetic field direction and is anisotropic in nature since $g_x \neq g_y \neq g_z$. The spin Hamiltonian in this case is represented by the following equation

$$\hat{H} = \beta\left(g_x H_x \hat{s}_x + g_y H_y \hat{s}_y + g_z H_z \hat{s}_z\right) \tag{4.25}$$

The g-factor g_x, g_y, and g_z can be evaluated from the line positions measured with the magnetic field along *the* x-, y-, and z-direction. The effective average of g for a particular orientation can be obtained with the help of the following expression:

$$g_{\text{eff}}^2 = g_x^2 \cos^2\theta_{H_x} + g_y^2 \cos^2\theta_{H_y} + g_z^2 \cos^2\theta_{H_z} \tag{4.26}$$

where θ_{H_x}, θ_{H_y}, and θ_{H_z} are the orientation angles between the magnetic field H and x-, y-, and z-axes.

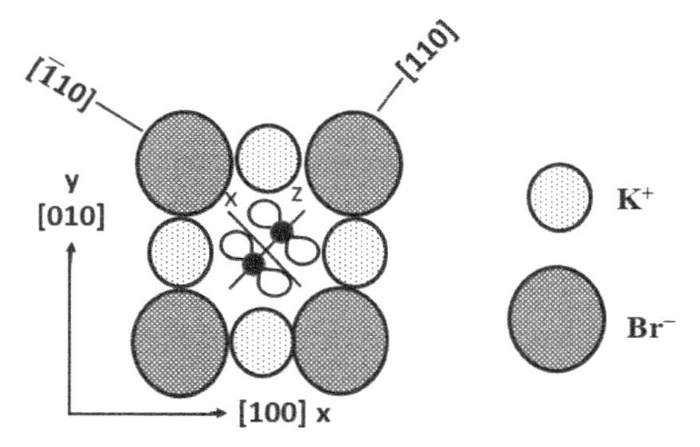

Figure 4.14 Projection onto plane xy of a unit cell for an alkali-halide crystal having the rock salt structure, showing a substitutional O^{2-} ion site.

4.9 Instrumentation

The basic components of an ESR spectrometer are (i) a source for microwave radiation, (ii) a sample cell, (iii) a system for transmitting the radiation energy to the sample cell, (iv) a magnetic system, (v) a detection system, and (vi) a recorder. The schematic diagram of ESR spectroscopy is shown in Figure 4.15 [3].

4.9.1 Source

This can be subdivided into the following different units:

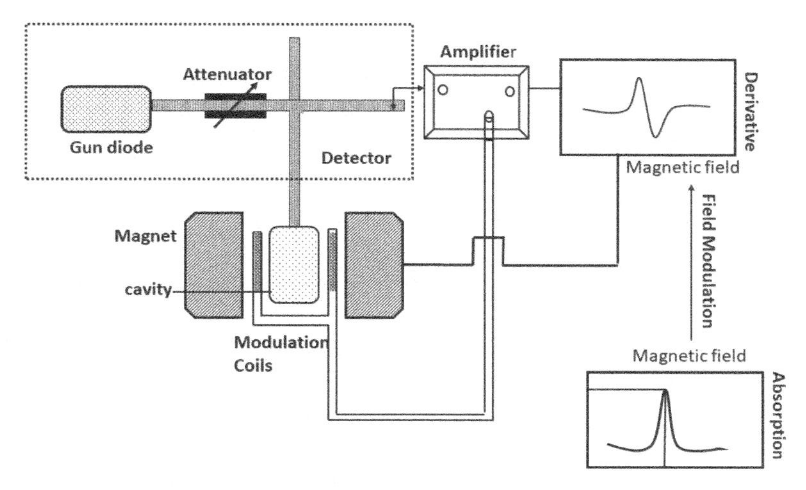

Figure 4.15 Schematic diagram of ESR spectrometer.

4.9.1.1 *Microwave oscillator*

Generally, a klystron oscillator is used to produce microwave radiation. This klystron oscillator can suitably work over a small frequency region. This could be stabilized against temperature fluctuations by immersing it in an oil bath or by air cooling. It is kept at a fixed frequency with the help of an automatic control circuit and can provide a power of about 300 mW.

4.9.1.2 *Wave guide and wave meter*

The microwaves are propagated in waveguides which is a hollow rectangular brass tubes. It propagates the microwave radiations to the sample. The wave meter is attached to know the frequency of microwave radiation produced by the klystron oscillator. It is generally calibrated in units of frequency, i.e., in MHz.

4.9.1.3 *Attenuator*

The attenuator is used to adjust the power of the microwave radiation which is going to strike the sample. The power of the microwave passing through the waveguide is continuously decreased by putting a piece of resistive materials into the waveguide. The piece is shown as variable attenuators which reduces the full power of the klystron into the suitable power for the sample.

4.9.1.4 *Isolator*

Isolators are used in ESR spectrometers to minimize the vibrations in the microwave's frequency produced by the klystron source. A strip of ferrite material is used as an isolator, which prevents the reflection of microwave radiation power back into the klystron source. It also stabilizes the frequency of the source.

4.9.2 Sample Cell/Cavity

In most ESR spectrometers, sample cells in the form of dual sample cavities are used for simultaneous observation of the sample and the reference material. Such a sample cavity may be rectangular (TE 120 cavity) or cylindrical (TE 011 cavity) in shape. The resonance sample cavity is generally called the heart of the ESR spectrometer and is made in such a way that it maximizes the applied field along the sample dimensions to cause spin resonance.

4.9.3 A Magnet System

The sample resonant cavity is placed between the two poles of an electromagnet which provides a homogenous magnetic field whose strength can be varied from 0–500 Gauss. A highly regulated power supply energizes the magnet to produce stable magnetic fields.

4.9.4 Detector

The most common detector used in ESR spectrometer is silicon crystal which converts the microwave radiation power into a direct current input.

4.9.5 Recorder

The signals coming out from the phase-sensitive detector and the sweep unit are recorded in their derivative form in order to signal-to-noise ratio.

4.9.5.1 *Working of instrument*

The substance whose ESR spectra have to be studied is kept in the sample resonant cavity. The klystron oscillator is then operated to produce monochromatic microwave radiations. After passing through the isolator, wave meter, and attenuator, the radiations enter the circulator placed in magic-T. Then, this microwave radiation reaches the detector which acts as a rectifier and converts the microwave power into a direct current. The magnetic field in which the resonant cavities containing the sample and the reference material are varied slowly to a suitable value at which the resonance may occur. The recorder records this resonance as an absorption peak. Generally, ESR spectra are recorded in the derivative form to enhance the sensitivity and resolution. For this purpose, the magnetic field is changed for several minutes. By doing so, the recorder will show the derivative of the microwave absorption spectrum against the applied magnetic field (as shown in Figure 4.16). The ESR spectrum has been obtained by plotting intensities against the strength of the magnetic field. But it is useful

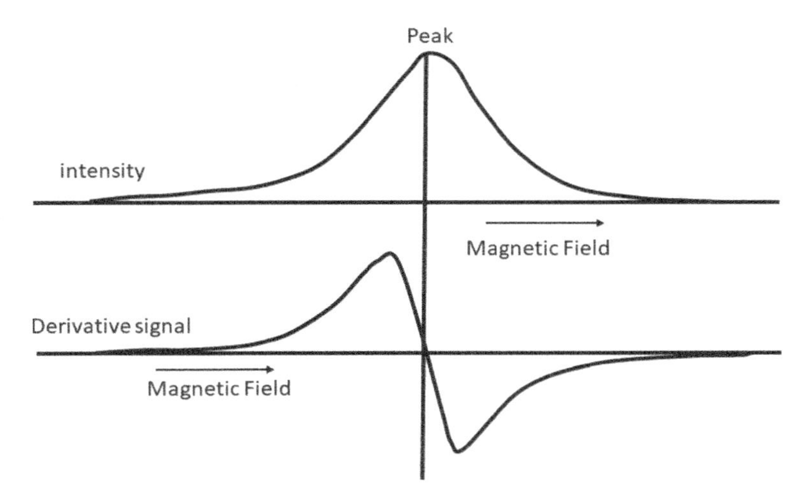

Figure 4.16 Absorption peak and derivative signal for ESR spectrum.

to represent the ESR spectrum as a derivative curve where the first derivative of the absorption curve is plotted against the strength of the magnetic field. The total area occupied by the absorption curve or the derivative curve is proportional to the number of unpaired electrons in the sample. For obtaining the number of unpaired electrons in the unknown sample, a comparison is done with the line shape of the standard reference sample whose number of unpaired electrons is known. 2,2-diphenyl-1-picrylhydrazyl (DPPH) is most widely used as standard material whose splitting factor g = 2.0036 and it contains 1.53×10^{21} unpaired electrons per gram.

4.10 Application of ESR Spectroscopy

This spectroscopy technique has a wide range of applications in various fields which are as follows:

- It is useful in the structural determination of different organic and inorganic compounds and free radicals.
- It is used to study the reaction velocities of various organic reactions and also the rate constant for the bimolecular reaction.
- ESR spectrum is a significant tool for distinguishing valance states. In the chromia-alumina catalyst, the superimposed three phases of chromium have been differentiated by this technique.
- It also provides information about the oxidation state of metal.
- It has a significant role in estimating many metal ions such as manganese ion, vanadium ion, and polynuclear hydrocarbon, Cu(II), Cr(II), Gadolinium(III), Fe(III), and Ti(III).
- ESR has been used in different biological systems such as detecting the presence of free radicals in diseased tissues, the functioning of oxidative enzymes.

References

1. Rabek, J.F., *Experimental Methods in Polymer Chemistry* (John Wiley, Bristol), 1980.
2. Weil, J. A., Bolton, J. R., *Electron Paramagnetic Resonance Elementary Theory and Practical Applications*, 2nd Ed. (John Wiley, USA), 2007.
3. Lund, A., Shiotani, M., Shimada, S., *Principles and Applications of ESR Spectroscopy* (Springer, USA), 2011.
4. Atherton, N. M., *Principles of Electron Spin Resonance* (Ellis Horwood and PTR Prentice Hall, Physical Chemistry Series, Ellis Horwood, Chichester), 1993.

5. Bersohn, M. Baird, J. C., *An Introduction to Electron Paramagnetic Resonance* (Benjamin, New York), 1966.
6. Wertz, J. E., Auzins, P., Griffiths, J. H. E. and Orton, J. W., *Faraday Discuss. Chem. Soc.*, 28, 136, 1959.
7. Livingston, R. and Zeldes, H., Paramagnetic resonance study of liquids during photolysis: hydrogen peroxide and alcohols, *J. Chem. Phys.*, 44, 1245, 1966.
8. Bolton, J. R., Fraenkel, G. K., Electron spin resonance study of paring theorem for alternant hydrocarbons: 13C splittings in anthracene positive + negative ions, *J. Chem. Phys.*, 40, 3307, 1964.
9. O'Mara, W. C., *Trapped Hole-Centers in Magnesium Oxide*, Ph.D. thesis (University of Minnesota, Minneapolis, MN, U.S.A.), 1969.
10. Delbecq, C. J., Hutchinson, E., Schoemaker, D., Yasaitis, E. L. and Yuster, P. H., ESR and optical-absorption study of the V1 center in KCl: NaCl, *Phys. Rev.*, 187, 1103, 1969.
11. Patten, F. W. and Keller, F., EPR identification of the structure of the V1 color center in KCl, *J. Phys. Rev.*, 187, 1120, 1969.
12. Zeller, H. R. and Känzig, W., *Helv. Phys. Acta*, 40, 845, 1967.

Chapter 5

Infrared Spectroscopy

Preeti Gupta,[a] S. S. Das,[b] and N. B. Singh[c]
[a]*Department of Chemistry, DDU Gorakhpur University, Gorakhpur, India*
[b]*Department of Chemistry (ex-professor), DDU Gorakhpur University, Gorakhpur, India*
[c]*Department of Chemistry and Biochemistry, SBSR and RDC, Sharda University, Greater Noida, India*
preeti17_nov@yahoo.com

5.1 Introduction

The studies of infrared absorption spectra gained importance in the year 1935 when IR absorption of certain organic molecules containing –OH groups was reported [1]. When this work was reported, it immediately opened up a way for extensive studies of chelation and intermolecular H-bond formation in organic compounds [2]. Very truly, one can say that out of all the regions of the electromagnetic spectrum, the infrared region can give the best information for elucidation of the structure of organic molecules

Spectroscopy
Edited by Preeti Gupta, S. S. Das, and N. B. Singh

ISBN 978-981-4968-32-4 (Hardcover), 978-1-003-41258-8 (eBook)
www.jennystanford.com

[1–5]. The simple experimental setup of IR spectroscopy and useful pieces of information, which one can obtain from the IR spectrum, have made this method very popular for determining the structures of several organic and inorganic compounds [6–12]. Even though IR spectroscopy is a much better, more useful, and faster method than any other analytical method, there are certain limitations to it. Molecules that have higher molecular weights and are more complex are difficult to be studied by this technique [13].

5.2 Principle of Infrared Spectroscopy

All organic compounds and many inorganic compounds absorb IR radiation. To understand the reason for the absorption of radiation of this energy, the molecules may be assumed as a system of balls of varying masses (the atoms of a molecule) and springs of varying strengths (the chemical bonds of molecules). Further, the molecules are not only in constant motion relative to each other but also the nuclei within a single molecule are continuously changing position with reference to one another undergoing vibrations and rotations. The atoms in a molecule do not remain in fixed relative positions but vibrate about some mean position due to the elasticity of chemical bonds. Thus, it is the vibration rather than rotation that makes the molecule inferred active by causing a change in the dipole moment.

For example, let us consider a carbon dioxide (CO_2) molecule in which the three atoms are arranged linearly and there is a net positive charge on the carbon and negative charges on the two oxygen atoms (Figure 5.1).

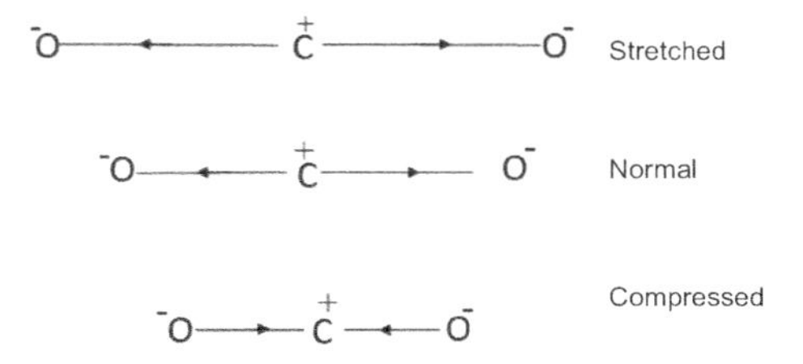

Figure 5.1 The symmetric stretching of the CO_2 molecule.

Suppose this molecule vibrates in the symmetric stretch mode of vibration in which both C–O bonds change simultaneously and the molecule is either stretched or compressed. This clearly indicates that the dipole moment remains zero throughout this motion and therefore, this particular vibration becomes *infrared inactive*. However, in the antisymmetric stretch, one bond stretches while the other is compressed and vice versa. Thus, there is a periodic change in the dipole moment. So, this vibration of the CO_2 molecule will be *infrared active*.

In addition to these two vibrations, there exist two bending modes of vibration also which are *infrared active* as in both bending vibrations the dipole moment changes periodically.

Let us first discuss vibrational motions in a diatomic molecule and then simultaneous vibrational and rotational motions in diatomic molecules will be taken up. The vibrations in polyatomic molecules shall be discussed in the last.

5.3 The Vibration in a Diatomic Molecule

Let us consider a diatomic molecule of the type, AB (HCl, HBr, NO, CO, etc.) in which the two nuclei A and B are supposed to either stretch toward each other or go away causing the bond along the nuclei either to compress or to elongate. Thus, this diatomic molecule can perform vibrations either by compression or elongation like a coiled spring in which the distance between the two atoms of the molecule is constantly changing. This could be clearly illustrated in Figure 5.2.

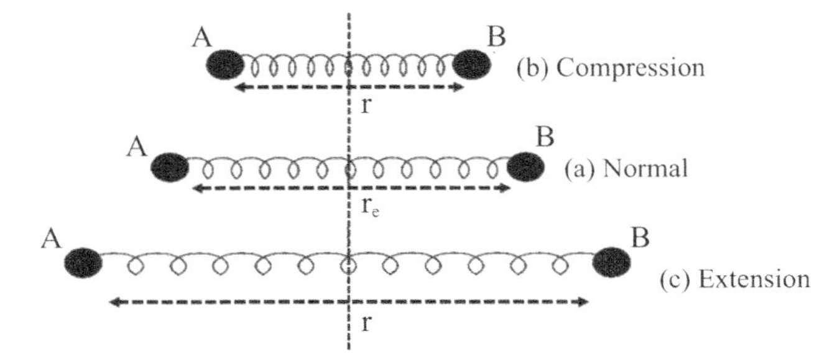

Figure 5.2 Vibrations in a diatomic molecule (a) normal position (b) compression in which the two atoms come close and (c) elongation in which the two atoms go away.

The point O (in Figure 5.2) represents the center of mass, r_e is the distance of separation between the two atoms at a normal position corresponding to minimum energy (i.e., bond length, r is the new distance of separation of the two atoms when stretched or compressed as shown Figure 5.2(b) and (c). The displacement of A with respect to B is given by $r–r_e$. Since the compression and elongation of a bond are supposed to behave like a coiled spring obeying Hooke's law, the restoring force, F can be obtained from the expression

$$F = -k(r - r_e) \tag{5.1}$$

where k is the force constant. The potential energy, V, is proportional to the square of relative displacement of the atom and is given by

$$U = \frac{1}{2}k(r - r_e)^2 \tag{5.2}$$

while the kinetic energy (K) in terms of reduced mass (μ) and velocity (v) is given by

$$K = \frac{1}{2}\mu v^2 \tag{5.3}$$

The oscillation motion of a vibrating diatomic molecule may be compared to the simple harmonic vibration of a particle of mass equal to μ and a displacement equal to $r - r_e$, i.e., Δr. In the diatomic molecule, the two atoms are bonded through a bond with a force constant k and the nuclei of these two atoms oscillate with a characteristic frequency, $\overline{\omega}_{OSC}$. This vibrational frequency depends on the mass of the system and the force constant. It could be shown that the oscillation frequency, ω_{OSC} (in the unit of Hertz) is given by the expression

$$\omega_{OSC} = \frac{1}{2\pi}\left(\frac{k}{\mu}\right)^{1/2} \tag{5.4}$$

where μ the reduced mass of the system is given by

$$\frac{m_1 m_2}{m_1 + m_2}$$

where m_1 and m_2 are the points masses of two atoms A and B situated at distances r_1 and r_2, respectively, from the center of mass at any given instance.

However, in the units of wave number the frequency is given by the following expression:

$$\bar{\omega}_{OSC} = \frac{1}{2\pi c}\left(\frac{k}{\mu}\right)^{1/2} \text{ (in cm}^{-1}\text{)} \tag{5.5}$$

where c is the velocity of light expressed in cm/s.

The vibrational energies like other molecular energies are quantized. The allowed vibrational energies for a system can be calculated with the help of the Schrodinger equation:

$$\frac{\partial^2\varphi}{\partial(\Delta r)^2} + \frac{8\pi^2\mu}{h^2}\left\{E - \frac{1}{2}\left(\Delta r^2\right)\right\} = 0 \tag{5.6}$$

Since the vibrations of a diatomic molecule, with certain approximations, could be compared to a simple harmonic oscillator, the Schrödinger wave equation [Eq. (5.6)] gives the following result for the vibrational energy, E_v:

$$E_v = \frac{h}{2\pi}\left(\frac{k}{\mu}\right)^{1/2}\left(v + \frac{1}{2}\right) \text{ or } E_v = \left(v + \frac{1}{2}\right)h\omega \tag{5.7}$$

where v is vibrational quantum number and can have values $v = 0, 1, 2, 3, ...$ On converting Eq. (5.7) in the units of wave number (in cm^{-1}), we can write

$$\varepsilon_v = \frac{E_v}{hc} = \left(v + \frac{1}{2}\right)\bar{\omega}_{OSC} \text{ in cm}^{-1} \tag{5.8}$$

The various allowed energy levels for simple harmonic vibrator molecules can thus be described by Eq. (5.8) and the allowed energy levels are shown in Figure 5.3.

5.3.1 Zero-Point Energy

By putting $v = 0$ in Eqs. (5.7) and (5.8), the lowest vibrational energy can be obtained whose values of energy in the units of Joules and cm^{-1} are given by Eq. (5.9).

$$E_0 = \frac{1}{2}h\omega_{OSC} \text{ (in Joules) and } \varepsilon = \frac{1}{2}h\bar{\omega}_{OSC} \quad \text{(in cm}^{-1}\text{)} \tag{5.9}$$

The quantity $\frac{1}{2}h\omega_{OSC}$ (in Joules) or $\frac{1}{2}h\bar{\omega}_{OSC}$ (in cm^{-1}) is known as the *zero-point energy,* which corresponds to the energy of molecules in the vibrational ground state. It depends upon the strength of the chemical bond and the atomic masses. However, it must be mentioned here that the diatomic molecules can never have a zero vibrational state since the atoms are never at rest completely relative to each other. Even at absolute zero temperature, when all other forms of motion such as rotational, and translational cease, the vibrational motion still exists.

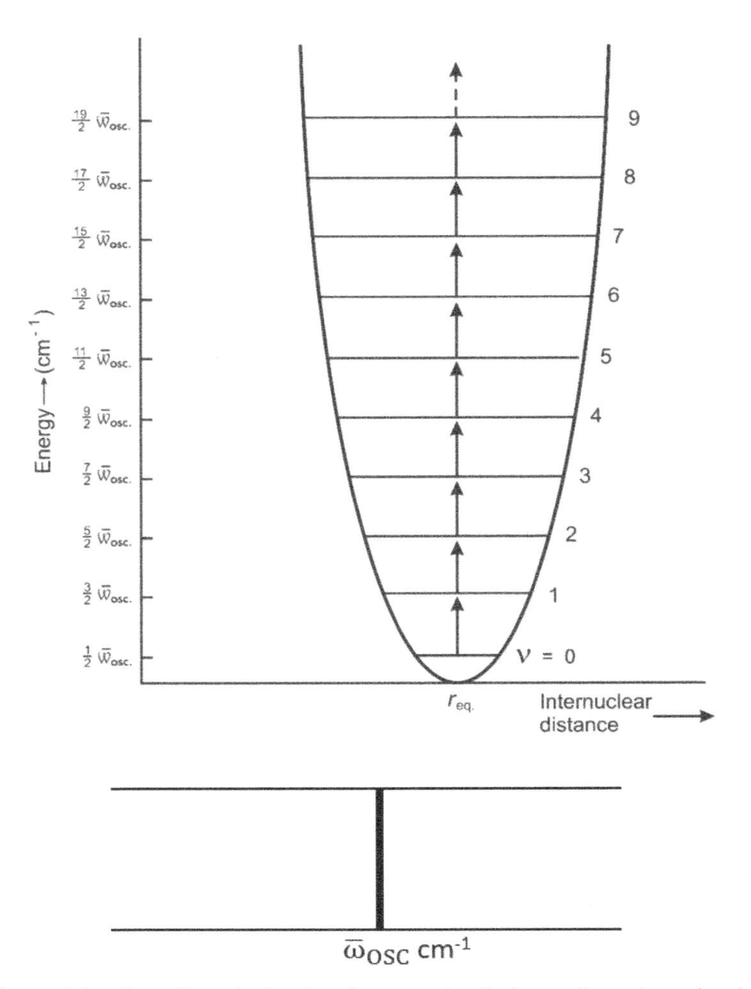

Figure 5.3 The allowed vibrational energy levels for a diatomic molecule performing simple harmonic oscillation. The sign (↑) shows transitions between them.

5.3.2 Infrared Spectra and Selection Rule

The simple selection rule for the vibrating diatomic molecule undergoing vibrational changes and performing harmonic oscillation is given as

$$\Delta v = \pm 1 \tag{5.10}$$

This suggests that each mode of vibration will give only one band. The plus sign (+) applies to absorption spectra while the minus sign (–) indicates emission spectra. Moreover, since vibration spectra are generally determined by absorption phenomenon, the selection rule becomes $\Delta v = +1$.

When the selection rule is applied to emission or absorption for any initial value of v, it could be shown then that during emission from a higher state $v + 1$ to the initial state v, or during absorption from v to $v + 1$ state, the vibrational energy is given by

$$\varepsilon_{v+1} - \varepsilon_v = \left(v + 1 + \frac{1}{2}\right)\overline{\omega}_{\mathrm{osc}} - \left(v + \frac{1}{2}\right)\overline{\omega}_{\mathrm{osc}}$$

$$\varepsilon_{(v+1)\to v} = \overline{\omega}_{\mathrm{osc}} \ (\text{in cm}^{-1}) \tag{5.11}$$

and

$$\varepsilon_{v\to(v+1)} = \left(v + \frac{1}{2}\right)\overline{\omega}_{\mathrm{osc}} - \left(v + 1 + \frac{1}{2}\right)\overline{\omega}_{\mathrm{osc}}$$

$$\varepsilon_{v\to(v+1)} = \overline{\omega}_{\mathrm{osc}} \ (\text{in cm}^{-1}) \tag{5.12}$$

And since the energy difference between two levels when expressed in cm^{-1}gives the wave number, therefore wavenumber of the spectral line other absorbed or emitted is simply given by

$$\overline{\upsilon}_{\mathrm{spectra}} = \overline{\varepsilon} = \overline{\omega}_{\mathrm{osc}} \ (\text{in cm}^{-1}) \tag{5.13}$$

These results show that in a simple harmonic model, the vibrational levels are equally spaced and transition between any two neighboring states will give rise to an equal amount of change in energy as shown in Figure 5.3. Eq. (5.13) also indicates that the vibrational spectrum must have single lines if rotational energy changes are not associated with it. All the vibrational lines obtained from any simple harmonic oscillator will have the same frequency (Figure 5.3).

If radiations in the IR region are passed through a sample of diatomic molecules and the transmitted radiations are analyzed (with the help of a monochromator and detector of spectrometer) then one obtains an infrared spectrum. This IR spectrum is obtained as a result of the absorption of radiation and its interaction with the vibrational motions of the molecules. But the spectrum is observed only when the vibration involves a change in the dipole moment of the molecule. Due to this reason, homonuclear diatomic molecules do not give any vibrational spectra in the IR region, whereas heteronuclear diatomic molecules give vibrational spectra. The position of the line in the observed spectrum gives useful information about the difference between the various energy levels while the intensity of the spectral lines provides information about the population of levels involved in the transition. The spectrum, thus, obtained is characteristic of a particular diatomic molecule.

5.3.3 Limitations and Drawbacks of Simple Harmonic Oscillator Model

1. The lowest vibrational energy states can very well be explained for diatomic molecules performing simple harmonic oscillations but in the case of higher vibrational energy states, this simple harmonic oscillator model does not give correct explanations for the experimental findings.
2. The harmonic oscillator model suggests equally spaced energy levels but experimentally this is not true. It has been found that the spacing between the lower energy levels is more and as the value of v increases, the spacing between the level decreases with the increasing value of v.
3. The phenomenon of overtones and combinations are observed in the vibrational spectra of diatomic molecules but the harmonic oscillator model is not able to explain these phenomena.
4. The prediction of zero-point energy on the basis of a simple harmonic oscillator model creates a difference between the wave mechanical and classical approaches to molecular vibrations. According to wave mechanics, the molecule must always vibrate to some extent.

This requires the framing of another suitable model which could explain these shortcomings. This new model is named *anharmonic oscillator model.*

5.4 The Diatomic Molecules as Anharmonic Oscillator

In reality, the molecules do not exactly obey the laws of simple harmonic motions. Therefore, a more complicated behavior of the bonds connecting the two atoms is expected. It will be better if diatomic molecules are considered to perform anharmonic oscillations. In spite, the bonds which may be considered to behave perfectly elastic for small compression and extensions, are not so homogenous for larger distortions and they do not obey Hooke's law. Thus, a more complicated behavior of the real bonds should be expected.

It was P.M. Morse [14] who gave a mathematical equation for the potential energy of diatomic molecules doing anharmonic oscillators. This equation is called as P.M. Morse function and is written as

$$U = D_e\left[1-\exp(-\beta\left[r_e - r\right])\right]^2$$

$$U = D_e\left[1-\exp\left(-\beta\,\Delta r\right)\right]^2 \tag{5.14}$$

where D_e is the dissociation energy of a particular diatomic molecule, β is constant for that molecule, r is internucleus distance and r_e is the internuclear distance corresponding to the minimum in the Morse energy curve, so that Δr = (r_e – r). The constant β is given by the relation

$$\beta = \overline{\omega}_{\mathrm{OSC}}\left(\frac{2\pi^2\, c\mu}{D_e \mathrm{h}}\right)^{1/2} \tag{5.15}$$

where μ, ω_{OSC}, and c are reduced mass, oscillation frequency, and velocity of light, respectively. D_e is the equilibrium dissociation energy and measured at the minimum of the potential energy curve. Further, the dissociation energy, D_o, corresponds to the energy of the lowest vibrational level and is given by

$$D_o = D_e - \frac{1}{2}\mathrm{h}\omega_{\mathrm{OSC}} \tag{5.16}$$

When Eq. (5.16) is used in the Schrödinger equation [1], then one gets a different result for allowed energy levels. This may be written as

$$\varepsilon_v = \frac{E_v}{hc} = \left(v + \frac{1}{2}\right)\overline{\omega}_e - \left(v + \frac{1}{2}\right)^2 \overline{\omega}_e x_e + \left(v + \frac{1}{2}\right)^3 \overline{\omega}_e y_e \text{ (in cm}^{-1}\text{)} \tag{5.17}$$

where $\overline{\omega}_e$ is an equilibrium oscillation frequency expressed in wave numbers and x_e, y_e, etc. are *anharmonicity constants*. But the value of the y_e is very much less than the value of x_e. So, one can easily write Eq. (5.17) as

$$\varepsilon_v = \frac{E_v}{hc} = \left(v + \frac{1}{2}\right)\overline{\omega}_e - \left(v + \frac{1}{2}\right)^2 \overline{\omega}_e x_e \text{ (in cm}^{-1}\text{)} \tag{5.18}$$

This x_e is always small and positive for the bond stretching vibration. Due to this, the vibrational levels do not have equal spacings and become closer to each other with increasing values of v. These vibrational levels are shown in Figure 5.4.

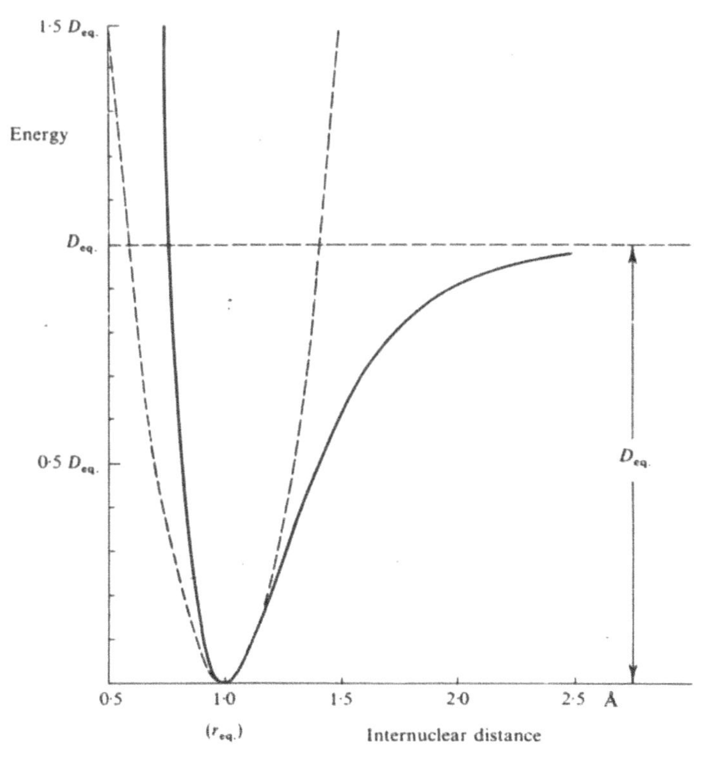

Figure 5.4 The Morse curve in which the potential energy of an anharmonic diatomic molecule is shown by solid lines and the energy curve for a harmonic oscillator is shown by dotted lines [5].

The above Eq. (5.18) can also be rearranged as

$$\varepsilon_v = \left(v + \frac{1}{2}\right)\left\{\overline{\omega}_e - \left(v + \frac{1}{2}\right)\overline{\omega}_e x_e\right\} \text{ (in cm}^{-1}\text{)} \tag{5.19}$$

On comparing the energy levels of the harmonic oscillator (Eq. 5.8) with Eq. (5.19) of an anharmonic one, it can be written:

$$\overline{\omega}_{OSC} = \overline{\omega}_e \left\{1 - x_e\left(v + \frac{1}{2}\right)\right\} \text{ (in cm}^{-1}\text{)} \tag{5.20}$$

This suggests that anharmonic oscillation type diatomic molecules also behave like harmonic oscillators but in this case the oscillation frequencies, ω_{OSC} decrease with increasing values of vibrational levels.

5.4.1 Zero-Point Energy for Anharmonic Oscillator

For the ground state when the values of v = 0, the oscillation frequency of the ground state, $\overline{\omega}_0$, can be obtained with the help of Eq. (5.20). It is given by

$$\overline{\omega}_{osc} = \overline{\omega}_0 = \overline{\omega}_e \left(1 - \frac{1}{2}x_e\right) \text{ (in cm}^{-1}\text{)} \tag{5.21}$$

and the corresponding energy for the ground state level, v = 0, can, therefore, easily be obtained by putting v = 0 in Eq. (5.19), i.e.,

$$\varepsilon_0 = \frac{1}{2}\overline{\omega}_e \left(1 - \frac{1}{2}x_e\right) \text{ (in cm}^{-1}\text{)} \tag{5.22}$$

This is defined as *zero-point energy* for a diatomic *anharmonic oscillator*. This value differs from the zero-point energy of the harmonic oscillating system by a factor $(1 - \frac{1}{2}x_e)$.

5.4.2 Infrared Spectra and the Selection Rule

The selection rule for all the transitions in an anharmonic oscillator is given as

$$\Delta v = \pm 1, \pm 2, \pm 3, \ldots \tag{5.23}$$

According to this selection rule, the long jump transitions are also possible, i.e., transitions from the ground state to state 1, state 2, or state 3, and so on are also permissible (Figure 5.5). However, generally, the spectral lines corresponding to $\Delta v = \pm 1, \pm 2, \pm 3$ possess the observable intensity. Therefore, only three transitions in an anharmonic oscillator are considered, which are being discussed below:

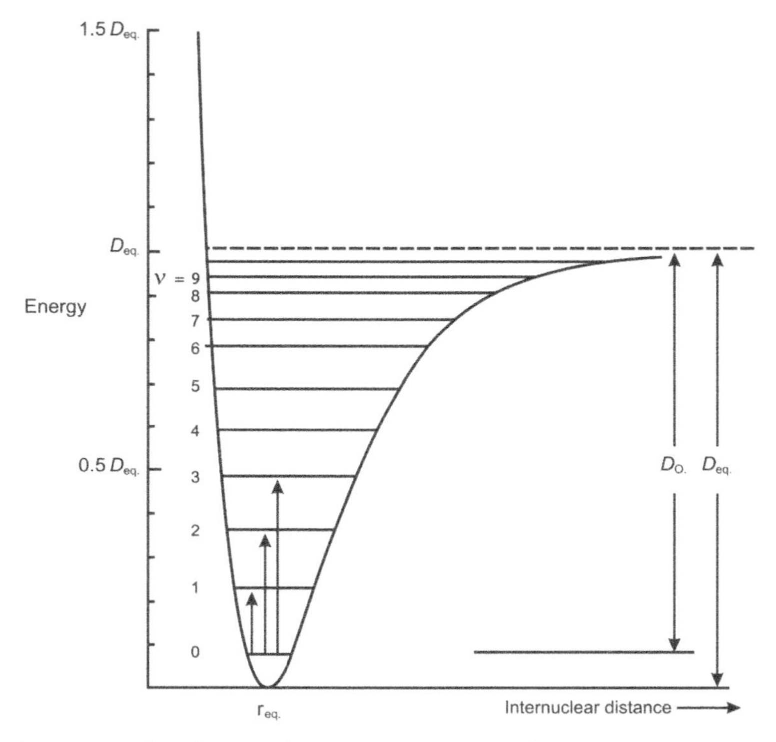

Figure 5.5 The allowed vibrational energy levels for a diatomic molecule undergoing an anharmonic oscillation [5].

1. Transition from $v = 0$ to $v = 1$, i.e., $\Delta v = \pm 1$ will yield ***fundamental*** bands with sharp spectral lines having good intensity. The energy change (ΔE) in this transition is given by

$$\Delta E = E_{v=1} - E_{v=0} = hc\overline{\omega}_e(1 - 2x_e) \text{ (in Joules)} \tag{5.24}$$

and the frequency of such radiations will be given by

$$\Delta\bar{\varepsilon}_1 = \bar{\upsilon}_1 = \frac{\Delta E}{hc} = \frac{E_{v=1} - E_{v=0}}{hc}$$

$$= \overline{\omega}_e\,(1-2x_e)\ \text{(in cm}^{-1}\text{)} \tag{5.25}$$

2. Transition from vibrational level $v = 0$ to $v = 2$, i.e., $\Delta v = \pm 2$ gives ***first overtone*** bands. The spectral lines in this case will be less intense than the fundamental bands. The energy change in this transition is given by the equation

$$\Delta E = E_{v=2} - E_{v=0} = 2hc\overline{\omega}_e(1-3x_e)\ \text{(in Joules)} \tag{5.26}$$

and the frequency is given by

$$\bar{\varepsilon}_2 = \bar{\upsilon}_2 = \frac{\Delta E}{hc} = \frac{E_{v=2} - E_{v=0}}{hc} = 2\overline{\omega}_e(1-3x_e)\ \text{(in cm}^{-1}\text{)} \tag{5.27}$$

3. The transition from the ground state vibrational level $v = 0$ to the third vibrational level $v = 3$, i.e., $\Delta v = \pm 3$ will show ***second overtone*** bands. The spectral lines of these bands have very negligible intensities. The energy change, in this case, is given as

$$\Delta E = E_{v=3} - E_{v=0} = 3hc\,\overline{\omega}_e(1-4x_e) \tag{5.28}$$

and the frequency of the second overtone is obtained by the following equation:

$$\bar{\varepsilon}_3 = \bar{\upsilon}_3 = \frac{\Delta E}{hc} = \frac{E_{v=3} - E_{v=0}}{hc} = 3\overline{\omega}_e\left(1-4x_e\right)\ \text{(in cm}^{-1}\text{)} \tag{5.29}$$

These three transitions will produce three spectral lines in the spectra of diatomic molecules undergoing anharmonic oscillations. The frequencies of the *first* and *second overtone* bands are 2 and 3 times greater than the frequency of the fundamental band. Therefore, they appear in the region of a shorter wavelength than that of the *fundamental* band.

The IR spectral studies of HCl confirm these theoretical predictions. In its spectrum, a very intense absorption at 2886 cm^{-1}, a less intense one at 5668 cm^{-1}, and a very feebly intense at 8347 cm^{-1} are observed [5]. The infrared spectrum for the HCl molecule is shown in Figure 5.6 which clearly shows one signal with strong intensity followed by two more signals of weak intensities (Figure 5.6a). Further, depending on the resolution of the equipment, the bands also show a different types of features and fine structures as shown in Figure 5.6b.

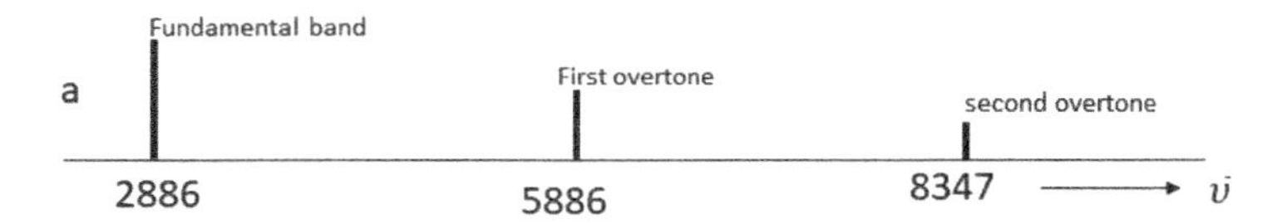

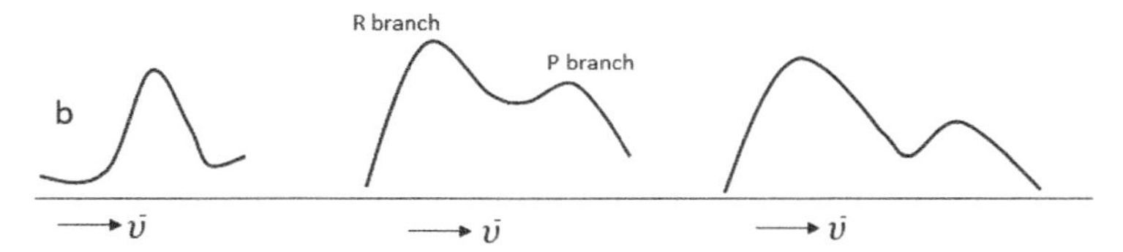

Figure 5.6 (a) The positions and intensities of the fundamental band, first overtone, and second overtone in HCl. (b) The fundamental band of HCl under low, moderate, and high-resolution spectrometer is also shown.

5.4.3 Hot Bands

Till now the transition from the vibrational level $v = 1$ to other higher vibrational levels has been ignored. It is well known that the population of molecules at the level $v = 1$ is very less. Therefore, such transitions have very low frequencies. When the temperature is increased, the higher vibrational levels, i.e., $v = 1, 2, 3, \ldots$ become more populated and the transitions from levels $v = 1 \rightarrow v = 2$, or from $v = 2 \rightarrow v = 3$, etc. will be observed and is possible to calculate the frequencies of such transitions. For a transition from first level $v = 1$ to $v = 2$, the change in energy level is given by

$$\Delta E = E_{v=2} - E_{v=1} = \text{hc}\,\bar{\omega}_e(1 - 4x_e) \text{ (in cm}^{-1}) \tag{5.30}$$

and the frequency of this radiation can be obtained with the help of the following equation

$$\bar{\upsilon}_{1\rightarrow 2} = \frac{\Delta E}{\text{hc}} = \frac{E_{v=2} - E_{v=1}}{\text{hc}} = \bar{\omega}_e(1 - 4x_e) \text{ (in cm}^{-1}) \tag{5.31}$$

$$\bar{\upsilon}_{2\rightarrow 3} = \frac{\Delta E}{\text{hc}} = \frac{E_{v=3} - E_{v=2}}{\text{hc}} = \bar{\omega}_e(1 - 6x_e) \text{ (in cm}^{-1}) \tag{5.32}$$

At high temperatures, weak absorption bands are observed near the fundamental band υ_1 at a lower frequency near the fundamental band, $\bar{\upsilon}_1$. These low-intensity bands are called *hot bands* since they are noticed only at higher temperatures.

5.5 The Vibrating Rotating Diatomic Molecule (Vibrational–Rotational Spectra)

Let us now discuss a diatomic molecule that simultaneously performs vibrations and rotations. A molecule absorbs energy and causes a jump from one vibrational state to another (within the same electronic levels). When vibrational change is accompanied by a simultaneous rotational change, then *vibrational–rotational spectra* are obtained. The spectrum consists of a band made up of a series of closely spaced lines. The fine structures of rotational bands confirm that simultaneous rotation and vibration have occurred in such molecules. As pointed out earlier, here also only those molecules possessing permanent dipole moment can produce *rotational vibrational spectra,* and such systems are known as *rotating vibrators* or *rotating oscillators.*

5.5.1 Energy Levels of Rotating Vibrators

If we consider that the diatomic molecule is performing rotations and vibrations independently so that there is no interaction between these two vibrations, then the total energy of the rotating vibrator will simply be the sum of vibrational and rotational energies and one can write:

$$E_{\text{total}} = E_{\text{vibrational}} + E_{\text{rotational}}$$

and this total energy, E_{total}, is defined by the quantum number, *v*, and *J* as $E_{v,J}$. Here, it should be made clear that the rotating vibrating diatomic molecule may either *perform harmonic* vibrations and behave as *a rigid rotor*, or as a *non-rigid rotor* executing *anharmonic* oscillations. These two cases are discussed below in two separate sections, respectively.

5.5.1.1 Case I: Diatomic molecules performing harmonic oscillations and behaving as a rigid rotator

The total energy of such a system can be obtained by adding both the rotational energy [Eq. (4.12), from Section 4.1.2] and the vibrational energy (Eq. 5.8) of a rigid rotator diatomic molecule performing harmonic oscillations. Thus, one can write

$$E_{\text{total}} = E_{\text{v,J}} = E_{\text{v}} + E_{\text{J}} = \text{hc}\,\overline{\omega}_{\text{OSC}}\left(v + \frac{1}{2}\right) + B\text{hc}\,J(J+1)\;(\text{in Joules}) \tag{5.33}$$

and in terms of wave numbers, the above energy equation can be written as

$$\varepsilon_{\text{v,J}} = \frac{E_{\text{v,J}}}{\text{hc}} = \overline{\omega}_{\text{OSC}}\left(v + \frac{1}{2}\right) + BJ(J+1)\;(\text{in cm}^{-1}) \tag{5.34}$$

where $B = \text{h}/8\pi^2 I_{\text{C}}$.

5.5.1.1.1 The IR spectra, selection rule, and P, Q, and R branches

The selection rules for such combined motions for a harmonic oscillator and rigid rotator will be

$$\Delta v = \pm 1 \text{ and } \Delta J = \pm 1 \tag{5.35}$$

These selection rules are the combination of the selection rules for vibrational and rotational motions of diatomic molecules as described earlier. According to this selection rule, a diatomic molecule cannot have $\Delta J = 0$ (except under a very special case). This means that a vibrational change must also lead to a simultaneous rotational change.

When a vibrating rotating diatomic molecule is subjected to radiations and the transmitted radiations are studied with the help of a suitable spectrometer, a spectrum is obtained. This vibrational–rotational spectrum appears as a result of different transitions between different vibrational and rotational levels of the molecule. The rotational levels corresponding to two vibrational levels $v = 0$ and $v' = 1$ are shown in Figure 5.7. In these two vibrational levels, the rotational levels corresponding to $v = 1$ are designated as J' and those in the lower vibrational level $v = 0$ are designated as J.

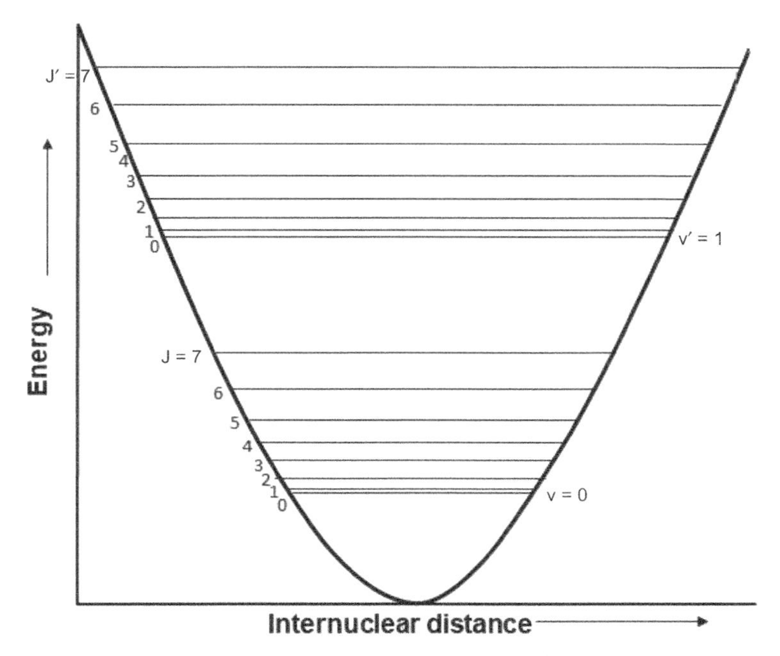

Figure 5.7 The rotational energy levels *J* and *J*′ for two different vibrational states *v* = 0 and *v*′ = 1 of a diatomic molecule [5].

Let us now calculate the change in energy for such simultaneous transitions. Suppose transition takes place between the vibrational levels *v* to *v*′ and rotational levels *J* and *J*′. Then the energy change for the harmonic oscillator and rigid rotator model can be calculated with the help of Eq. (5.33) as follows:

$$E_{\mathrm{v,J}} = E'_{\mathrm{v,J}} + E_{\mathrm{v,J}} = \mathrm{hc}\,\overline{\omega}_{\mathrm{OSC}}\left(v' - v\right) + B\mathrm{hc}\left[J'\left(J' + 1\right) - J\left(J + 1\right)\right] \quad \text{(in Joules)} \qquad (5.36)$$

and in the units of wave number as

$$\Delta\varepsilon_{\mathrm{v,J}} = \overline{\omega}_{\mathrm{OSC}}\left(v' - v\right) + B\left[J'\left(J' + 1\right) - J\left(J + 1\right)\right] \quad (\text{in cm}^{-1}) \qquad (5.37)$$

But, since the vibrating rotating diatomic molecule is behaving as a harmonic oscillator, therefore, the allowed transition will only be those for which $v' - v = \Delta v = 1$. Therefore, Eq. (5.37) can be rewritten as

$$\Delta\varepsilon_{\mathrm{v,J}} = \overline{\omega}_{\mathrm{OSC}} + B[J'\left(J' - 1\right) - J\left(J + 1\right)] \quad (\text{in cm}^{-1}) \qquad (5.38)$$

Eq. (5.38) is called the analytical expression for a vibrating rotating rigid diatomic molecule performing simple harmonic oscillations. It can also be inferred from Eq. (5.38) that the frequencies of spectral lines (in cm^{-1}) will vary according to the values of J' and J, respectively. Now two situations may then arise:

(i) If $\Delta J = +1$ then $J' = J + 1$. This condition represents ***R branch*** in the spectrum and the frequencies of these spectral lines will be given by

$$\overline{\upsilon}_{\text{spect}} = \Delta\varepsilon_{\text{v,J}} = \overline{\omega}_{\text{OSC}} + 2B[(J+1)(J+2) - J(J+1)]$$

$$\overline{\upsilon}_{\text{sp}} = \Delta\varepsilon_{\text{v,J}} = \overline{\omega}_{\text{OSC}} + 2B(J+1) \text{ (in cm}^{-1}\text{)} \tag{5.39}$$

where J = 0, 1, 2, 3, 4, ... Further, from Eq. (5.39), one may obtain the frequencies of R branch lines for different values of J.

when $J = 0$, $\overline{\upsilon}_{\text{sp}} = \overline{\omega}_{\text{OSC}} + 2B$

$J = 1$, $\overline{\upsilon}_{\text{sp}} = \overline{\omega}_{\text{OSC}} + 4B$

$J = 2$, $\overline{\upsilon}_{\text{sp}} = \overline{\omega}_{\text{OSC}} + 6B$

$J = 3$, $\overline{\upsilon}_{\text{sp}} = \overline{\omega}_{\text{OSC}} + 8B$

$J = 4$, $\overline{\upsilon}_{\text{sp}} = \overline{\omega}_{\text{OSC}} + 10B$

(ii) If $\Delta J = -1$ then $J' = J - 1$. This condition represents ***P branch*** in the spectrum and the frequency of spectral lines will be given by

$$\overline{\upsilon}_{\text{spect}} = \Delta\varepsilon_{\text{J,v}} = \overline{\omega}_{\text{OSC}} - 2BJ \text{ (in cm}^{-1}\text{)} \tag{5.40}$$

where J' = 1, 2, 3, 4 .... Further with the help of Eq. (5.40), one may obtain the frequencies of P branch lines at different values of J.

when $J = 1$, $\upsilon_{\text{sp}} = \overline{\omega}_{\text{OSC}} - 2B$

$J = 2$, $\overline{\upsilon}_{\text{sp}} = \overline{\omega}_{\text{OSC}} - 4B$

$J = 3$, $\overline{\upsilon}_{\text{sp}} = \overline{\omega}_{\text{OSC}} - 6B$

$J = 4$, $\overline{\upsilon}_{\text{sp}} = \overline{\omega}_{\text{OSC}} - 8B$

Further, when $\Delta v = \pm 1$ and $\Delta J = 0$, then the frequency of the spectral line is given as

$$\overline{\upsilon}_{\text{spect}} = \overline{\upsilon}_0 = \Delta\varepsilon_{\text{v,J}} = \overline{\omega}_{\text{OSC}} \text{ (in cm}^{-1}\text{)} \tag{5.41}$$

Since according to selection rules, $\Delta J = 0$, transitions are not allowed, so the transition with frequency for which $\overline{\upsilon}_0 = \overline{\omega}_{\text{osc}}$ is not observed in the spectra and the line at $\overline{\omega}_{\text{osc}}$ is missing. This is referred to as ***Q* branch** and is shown by the dotted line in the spectrum (Figure 5.8).

The frequency, $\overline{\omega}_{OSC}$ is known as the *band center*. Eqs. (5.39) and (5.40) suggest that the spectrum will contain equally spaced lines with a spacing of 2B on each side of the band center, $\overline{\upsilon}_0$. Therefore, the infrared spectrum of a diatomic molecule of this type (rigid rotator and harmonic oscillator) will show two types of bands (*P and R branches*) with the dip in the center (Q branch) corresponding to $J = 0$.

The R and P branches of lines are obtained in the spectrum when the values of $\Delta J = \pm 1$. However, other branches namely S and O can also be observed if the values of $\Delta J = \pm 2$. These other branches of lines are mentioned in Figure 5.8.

5.5.1.2 *Case II: Diatomic molecules performing anharmonic oscillation and behaving as non-rigid rotators*

The total energy of such a non-rigid rotating diatomic molecule performing anharmonic oscillations is given by the expression:

$$E_{v,J} = hc\,\overline{\omega}_{OSC}\left[\left(v+\frac{1}{2}\right)-x_e\left(v+\frac{1}{2}\right)^2\right]+Bhc\left[J(J+1)\right]\ \text{(Joules)} \tag{5.42}$$

or

$$E_{v,J} = hc\,\overline{\omega}_{OSC}[\left(v+\frac{1}{2}\right)(1-x_e\left(v+\frac{1}{2}\right)]+Bhc\left[J(J+1)\right]\ \text{(Joules)} \tag{5.43}$$

In the above equation the small centrifugal distortion constants D, H, etc. have been ignored for simplicity.

5.5.1.2.1 *Selection rules and P, Q, and R branches*

For anharmonic oscillator non-rigid rotator diatomic molecules, the selection rule becomes:

$$\Delta v = \pm 1, \pm 2, \pm 3, \ldots \text{ and } \Delta J = \pm 1 \tag{5.44}$$

According to these selection rules, such diatomic molecules will not have $\Delta J = 0$. This means that a vibrational change must be accompanied by a simultaneous rotational change. For a fundamental vibrational transition $\Delta v = +1$, one will obtain a series of transitions when the value of $\Delta J = +1$ and another series of transitions when

$\Delta J = -1$. These series of transitions are known as R and P branches, respectively.

The change in energy in the case of an anharmonic oscillator and non-rigid rotator diatomic molecules may be evaluated with the help of Eq. (5.42) and is written as

$$\Delta E_{v,J} = E_{v'J'} - E_{v,J} = hc\,\bar{\omega}_{OSC}\left[\left(v' + \frac{1}{2}\right) - x_e\left(v' + \frac{1}{2}\right)^2\right] + Bhc\left[J'\left(J'+1\right)\right]$$

$$-hc\bar{\omega}_{OSC}\left[\left(v + \frac{1}{2}\right) - x_e\left(v + \frac{1}{2}\right)^2\right] + Bhc\left[J(J+1)\right] \quad (5.45)$$

If only $v = 0$ to $v = 1$ transitions are considered and rotational quantum number in $v = 0$ state is designated as J while that in the $v = 1$ state as J', the energy change (in Joules) in such transitions is given by

$$\Delta E_{v,J} = E_{v'=1,J'} - E_{v=0,J} = hc\,\bar{\omega}_{OSC}\left(1 - 2x_e\right) + Bhc\left[J'\left(J'+1\right) - J(J+1)\right] \quad (5.46)$$

and in terms of wave number (in cm^{-1}), Eq. (5.46) becomes

$$\Delta\varepsilon_{v,J} = \bar{\omega}_{OSC}(1 - 2x_e) + B[J'(J'+1) - J(J+1)] \text{ (in cm}^{-1}\text{)} \quad (5.47)$$

For simplicity, if $\bar{\omega}_0 = \bar{\omega}_{osc}(1 - 2x_e)$

then $$\Delta\bar{\varepsilon}_{v,J} = \bar{\omega}_0 + B\left[J'\left(J'+1\right) - J(J+1)\right] \text{ (in cm}^{-1}\text{)} \quad (5.48)$$

When the interaction of vibration and rotation are not ignored then the Born Oppenheimer approximation is not applicable and also when the molecules go to higher rotational levels, the centrifugal force distorts the molecule perturbing its energy. Thus, with the increasing vibrational energy, the vibrational amplitude also increases, and in those cases, the value of B ($h/8\pi^2 I_C$) depends upon the value of v. The situation becomes more complex in anharmonic vibrations where the value of rotational constant, B changes more with the vibrational energy. The value of rotational constant B becomes less in the upper vibrational states in comparison to the lower states. If B_0 is the value of B in the ground state level $v = 0$ and B_1 is the value for the upper state $v = 1$, then one can write $B_0 > B_1$. For such a transition from $v = 0 \rightarrow v = 1$, the change in the energy level equation (5.48) becomes

$$\Delta\bar{\varepsilon}_{v,J} = \bar{\omega}_0 + [B_1J'(J'+1) - B_0J(J+1)] \qquad (5.49)$$

According to the selection rule $\Delta J = +1$; then in this model also two cases will arise

(i) When $\Delta J = +1$, i.e., $J' = J+1$. This condition shows ***R branch*** lines in the spectrum and the frequencies of the spectral lines, $\bar{\upsilon}_R$ will be obtained by putting the value of J' in Eq. (5.49) as

$$\Delta\bar{\varepsilon}_{v,J} = \bar{\upsilon}_R = \bar{\omega}_0 + 2B_1 + (3B_1 - B_0)J + (B_1 - B_0)J^2 \qquad (5.50)$$

where $J = 0, 1, 2, 3, 4, \ldots$

Further from Eq. (5.50), one is able to write that

when $J = 0$, $\bar{\upsilon}_R = \bar{\omega}_0 + 2B_1$

$J = 1$, $\bar{\upsilon}_R = \bar{\omega}_0 + 6B_1 - 2B_0$

$J = 2$, $\bar{\upsilon}_R = \bar{\omega}_0 + 12B_1 - 6B_0$

$J = 3$, $\bar{\upsilon}_R = \bar{\omega}_0 + 20B_1 - 12B_0$

(ii) $\Delta J = -1$; $J' = J-1$. This condition yields ***P branch*** lines in the spectrum and the frequencies of these spectral lines, $\bar{\upsilon}_P$ will be obtained by putting the value of $J' = J-1$ in Eq. (5.49) as

$$\Delta\bar{\varepsilon}_{v,J} = \bar{\upsilon}_P = \bar{\omega}_0 + (B_1 - B_0)J^2 - (B_1 + B_0)J \qquad (5.51)$$

where $J = 1, 2, 3, 4, \ldots$

when $J = 1$, $\bar{\upsilon}_P = \bar{\omega}_0 - 2B_0$

$J = 2$, $\bar{\upsilon}_P = \bar{\omega}_0 + 2B_1 - 6B_0$

$J = 3$, $\bar{\upsilon}_P = \bar{\omega}_0 + 6B_1 - 12B_0$

The expressions of the R and P branches [Eqs. (5.50) and (5.51)] suggest that the spacings of the lines appearing in the spectrum are not equal. In the spectrum, the appearance of lines of the P branch takes place at lower frequencies and the lines of the R branch appear nearer to each other toward the higher frequency side as shown in Figure 5.9. The energy levels and various transitions in the case of diatomic molecules performing anharmonic oscillations and behaving as non-rigid rotators are shown in Figure 5.9.

The above expression gives the frequencies of the lines in the *fundamental band* of the vibration–rotation spectrum. Fundamental band is one for which the vibrational transitions take place from $v = 0$ to $v' = 1$ by absorbing energy and from $v' = 1$ to $v = 0$ by emitting the energy. However, the spectrum also contains a number of other fainter bands in comparison to the fundamental band. These fainters bands due to the transition for which $v - v' = 2, 3$, etc. Further, the transition may also occur, since the molecular motion cannot be considered completely harmonic. These fainter bands are known as harmonic/overtone bands.

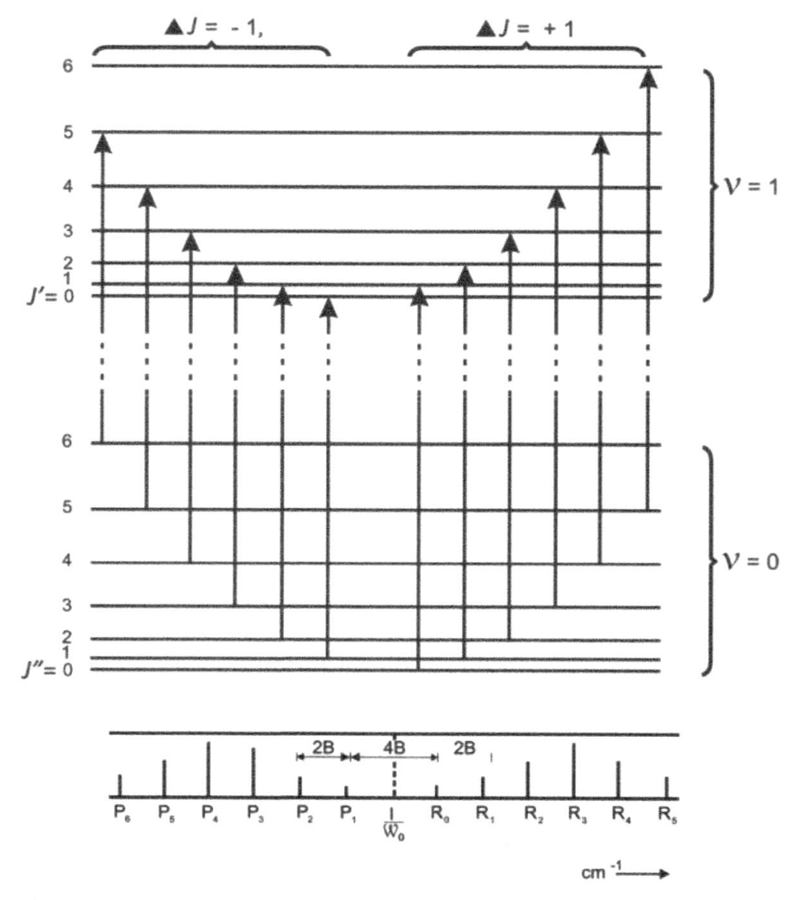

Figure 5.8 Energy levels and some transitions exhibiting vibrational–rotational harmonic motions of a rigid rotator diatomic molecule performing harmonic oscillations.

It can be clearly inferred from Eq. (5.46) that the frequency separation (in wave numbers) of rotational lines in a vibration–rotation band is equal to $h/4\pi^2 Ic$ or $2B$ cm^{-1}. However, when $J = 0$, the corresponding spectral line is missing, but the value of $\bar{\omega}_{OSC}$

can be determined from the frequencies of the lines present in the spectrum (as shown in Figure 5.9).

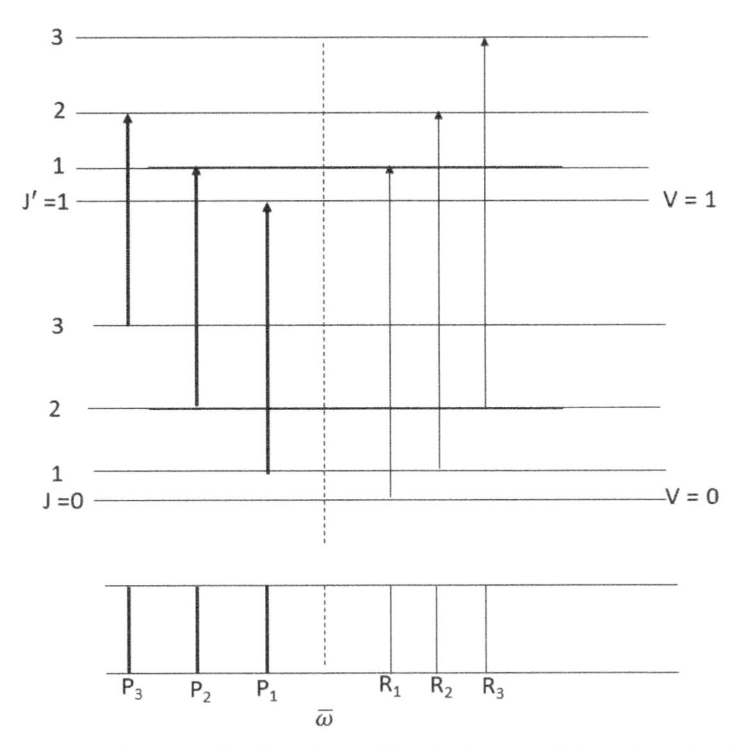

Figure 5.9 The energy levels and transitions in the case of diatomic molecules performing anharmonic oscillations and behaving as a non-rigid rotator.

5.6 Vibration and Rotation Spectra of Polyatomic Molecule

The vibration–rotation spectra of diatomic have been discussed in Section 5.5. In the present section, infrared spectra of the polyatomic molecule which are more complex than the diatomic ones shall be discussed. The number of vibrational degrees of freedom for linear and non-linear polyatomic molecules must be understood. Since polyatomic molecules are more complex than diatomic molecules. Therefore, it will be essential to describe

1. Fundamental vibrations and their symmetry
2. Influence of rotation on the spectra

5.6.1 Number of Fundamental Vibrations and Their Symmetry in a Polyatomic Molecule

In a polyatomic molecule containing N atoms, the total degrees of freedom will be $3 \times N = 3N$. A molecule can perform translational, rotational, and vibrational motions. The translational movement requires 3 degrees of freedom out of $3N$. Similarly, rotation of non-linear molecules also requires 3 degrees of freedom. Thus, the molecule is left with only $3N$–6 degrees of freedom. As such it can be said that the total number of vibrational degrees of freedom for a non-linear N-atomic molecule will be equal to $[3N - (3 + 3)] = 3N - 6$.

This means that $3N$–6 possible fundamental modes are possible for the absorption of light in the infrared region. On theoretical basis, it can be shown that molecules, such as H_2O ($N = 3$), NH_3 ($N = 4$), CH_4 ($N = 5$) and C_6H_6 ($N = 12$) should have, $3 \times 3 - 6 = 3$, $3 \times 4 - 6 = 6$, $3 \times 5 - 6 = 9$ and $3 \times 12 - 6 = 30$ possible fundamental absorption bands, respectively. But absorption in the IR region occurs only when there is a change in the dipole moment of the molecule during vibration. So certain absorption bands will not be IR active.

But when the molecule is linear, there will be no rotation about the bond axis and then the possible degree of rotational freedom will be equal to 2 only. So, it can be concluded that in a linear molecule there are $3N$–5 degrees of vibrational freedom. For example, in the case of linear triatomic molecules CO_2 ($N = 3$), there are $3 \times 3 - 5 = 4$ modes of vibrations as shown in Figure 5.10 [5].

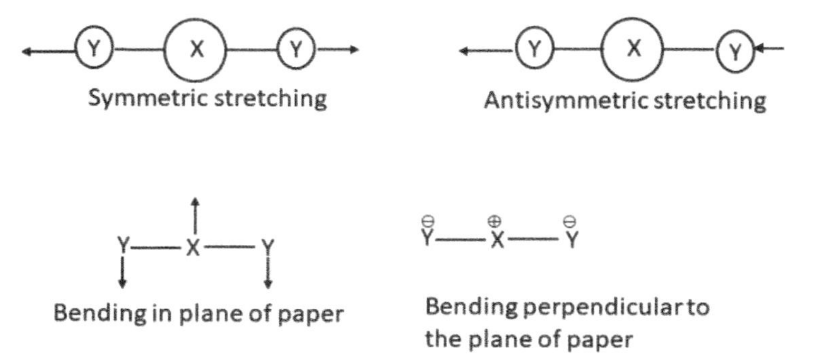

Figure 5.10 Normal modes of CO_2 molecules.

5.6.1.1 *Normal modes of vibrations in polyatomic molecules*

A molecule can vibrate only in certain normal modes. The normal modes of vibration for a polyatomic molecule are of two types:

(i) Stretching vibrations and (ii) bending vibrations

The number of stretching vibrations in both linear and non-linear N-atomic molecules is equal to $(N-1)$, whereas the number of bending in the case of linear molecules, is equal to $(2N - 4)$, and in non-linear molecules, it is equal to $(2N - 5)$, respectively [7].

Case I: Linear polyatomic molecules

Let us consider a linear triatomic molecule of the general type XY_2 (such as CO_2). In this molecule there will be $3N - 5 = 3 \times 3 - 5 = 4$ modes of vibrations, out of which $N - 1$, i.e., $3 - 1 = 2$ will be stretching vibrations and $2N - 4 = 2 \times 3 - 4 = 2$ bending vibrations. For this molecule XY_2, the normal mode of stretching and bending vibrations are shown in Figure 5.11.

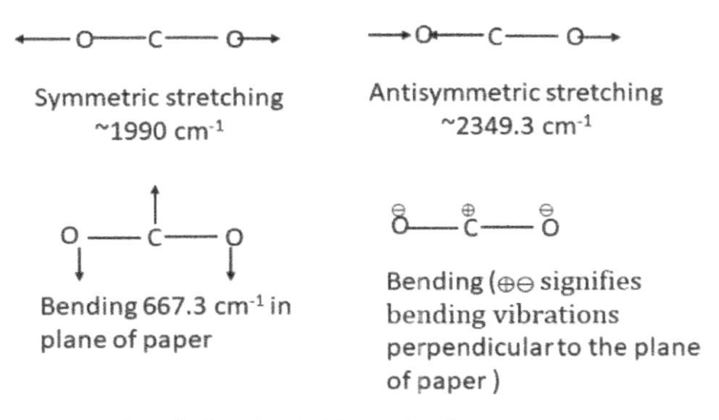

Figure 5.11 Modes of vibration in CO_2 molecules.

In symmetric stretching (υ_1), since both the atoms Y either stretch or compress, hence produces no change in the dipole moment (μ remains zero). Therefore, this vibration is infrared *inactive*. Whereas in asymmetric stretching, one end is stretched and on the other end there is compression. This produces a fluctuating dipole moment which makes this vibration an IR active. Further, there are two bending vibrations, one *in the plane* of the paper and the other *out of the plane* of the paper. These two bending vibrations differ only in the direction but have the same frequency and are equivalent in all respects except direction. Both bending vibrations are degenerate (because of having the same frequency). These two bending vibrations are infrared active due to changes in dipole moments. The symmetrical (υ_1) and antisymmetric (υ_3) vibrations are called parallel (||) vibrations and the degenerate bending vibration (υ_2) is termed a perpendicular (⊥) vibration.

The vibrations are called parallel (||) or perpendicular (⊥) according to the change in the dipole moment. If the change in dipole moment is parallel to the principle axis of the molecule then the vibrations are called parallel (||) and if the change is perpendicular to the principle axis of the molecule then the vibration is termed perpendicular (⊥).

Case II: Non-linear polyatomic molecules

In non-linear molecule of the general type XY_2 (such as H_2O or SO_2) for which $N = 3$, there must be ($3N - 6 = 3 \times 3 - 6 = 3$) vibrational modes out of which $N - 1$, i.e., $3 - 1 = 2$ will be stretching vibrations and ($2N - 5 = 2 \times 3 - 5 = 1$) bending vibration. The three normal modes of vibrations are shown in Figure 5.12.

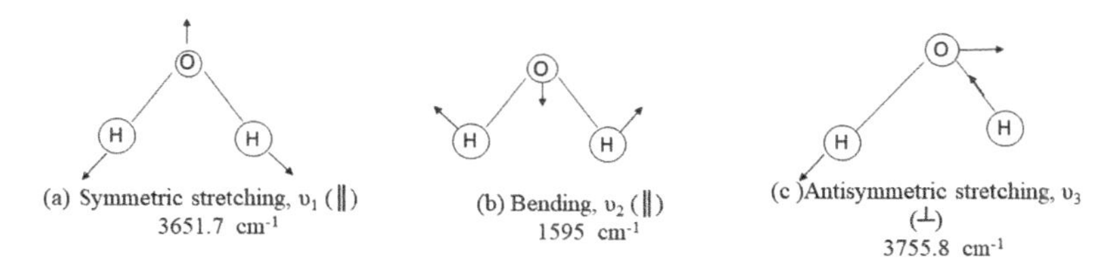

Figure 5.12 Modes of vibration in H_2O molecules.

5.6.2 The Influence of Rotation on the Spectra of Polyatomic Molecules

The spectra of polyatomic molecules are influenced because of rotations during their vibrational motions. In polyatomic molecules, the selection rules for rotational transition depend on the type of vibration which the molecule is undergoing. And it has been shown in this section that the vibrations of polyatomic molecules are of two types, parallel (||) and perpendicular (⊥) depending on the change in dipole moment either parallel or perpendicular with respect to the principle axis of rotation symmetry. Therefore, the selection rule for various types of molecules and IR spectra shall be discussed keeping in view these types of vibrations (||) or (⊥).

For a diatomic molecule, the selection rule valid for the simultaneous rotation and vibration was

$$\Delta v = \pm 1, \pm 2, \pm 3, \ldots \text{ and } \Delta J = \pm 1$$

This gives rise to a spectrum in which equally spaced line series occur on each side of a center designated as a band center.

Now in the case of vibrations of complex molecules, the selection rule changes, and for various types, this has been illustrated in the following manner.

5.6.2.1 *Linear molecules with parallel vibrations*

The selection rule for linear polyatomic molecules undergoing parallel vibrations is mentioned below

$$\Delta J = \pm 1, \Delta v = \pm 1 \text{ for simple harmonic motion}$$

$$\Delta J = \pm 1, \Delta v = \pm 1, \pm 2, \pm 3, \ldots \text{ for anharmonic motion}$$

This selection rule is similar to that discussed for the diatomic molecules. Therefore, the spectra, in this case, would also be similar to the diatomic molecule. It will consist of P and R branches with equally spaced lines on each side of the band center. No line will appear on the band center. For example, let us consider a linear HCN molecule whose spectrum is given in figure [5]. In this spectrum, the P and R branches are clearly shown with a band center appearing around 3310 cm^{-1} frequency. When the molecule become larger than the HCN molecule the value of rotational constant B becomes so small that the separate lies in the P and R branches cannot be resolved. A typical spectrum showing PR contour is then observed as shown in Figure 5.13. Non-linear molecules do not show such a P and R band. Therefore, by simply observing a P and R band in the spectrum, one can easily predict whether a molecule is linear or non-linear.

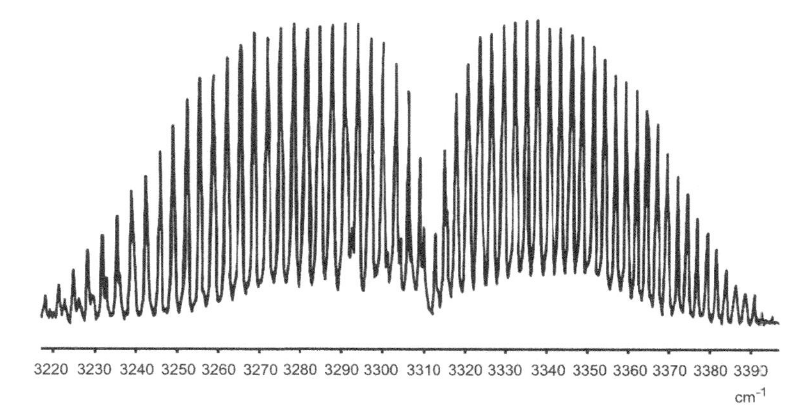

Figure 5.13 A typical PR contour in which all rotational fine structure is lost (low resolution).

5.6.2.2 *Linear molecules with perpendicular vibrations*

The selection rule applicable for vibrational–rotational spectra of linear polyatomic molecule exhibiting perpendicular vibration is written as

$$\Delta v = \pm 1 \text{ and } \Delta J = 0, \pm 1 \text{ (for simple harmonic motion)}$$

The condition that $\Delta J = 0$ clearly indicates that a vibrational change takes place without simultaneous rotational transition. A $\Delta J = 0$ corresponds to the Q branch. Therefore, the spectrum of such molecules also contains a Q branch along with P and R branches. The resultant Q line at the band center is generally very intense. One such spectrum is shown in Figure 5.8 with energy levels and transitions. This clearly shows that lines of the Q branch are superimposed upon each other at the band center and hence, an intense line is obtained.

The lines corresponding to the Q branch (transition with $\Delta J = 0$) may be obtained by using the equation

$$\varepsilon_{\text{total}} = \Delta\varepsilon_{\text{J,v}} = \left(v + \frac{1}{2}\right)\bar{\omega}_e - \left(v + \frac{1}{2}\right)^2 x_e \bar{\omega}_e + BJ(J+1), \text{ so that}$$

$$\Delta\varepsilon = \Delta e_{\text{v+1,J}} - e_{\text{v,J}},$$

and therefore, for the transition $v = 0 \rightarrow v = 1$, we write

$$\Delta\varepsilon = \left[\frac{3}{2}\bar{\omega}_e - \frac{9}{4}x_e\bar{\omega}_e + BJ(J+1)\right] - \left\{\frac{1}{2}\bar{\omega}_e - \frac{1}{4}x_e\bar{\omega}_e + BJ(J+1)\right\},$$

$$= \overline{\omega}_e - 2x_e\overline{\omega}_e$$

$$= \overline{\omega}_0 \text{ cm}^{-1} \text{ for all values of } J$$

Thus, the lines in the Q branch will be superimposed upon each other at the band center occurring at $\overline{\omega}_0$ making it very intense.

In the anharmonic vibrations, the values of the rotational constant B are different in the upper and lower vibrational state. In anharmonic vibration, an increase in vibrational energy will cause an increase in the average bond length. The rotational constant B then varies with the values of v. In the upper vibrational state, the value of B is less than that in the lower. The value of B in any vibration level can be obtained from the equation

$$B_v = B_e - \alpha(v + \frac{1}{2})$$

where B_e is the equilibrium value of B, α is a small positive constant.

According to this fact, it is possible to write $\Delta\varepsilon$ as

$$\Delta\varepsilon = e_{v+1,J} - \Delta e_{v,J},$$

$$= \frac{3}{2}\overline{\omega}_e - \frac{9}{4}x_e + B'J(J+1) - \left\{\frac{1}{2}\overline{\omega}_e - \frac{1}{2}x_e\overline{\omega}_e\ B''J(J+1)\right\}$$

$$= \overline{\omega}_0 J(J+1)(B' - B'')$$

where B' and B'' are the values of B in the upper $(v + 1)$ and lower v vibrational levels. The nature of the spectra then depends upon the difference between B values in upper and lower levels. The difference between B′ and B'' is very small and the lines cannot be resolved. In this situation, the Q branch is obtained as a broad-spectrum band.

5.6.2.3 *Symmetric top molecules with parallel vibrations*

The selection rule for vibration–rotational spectra of a symmetric top polyatomic molecule with parallel vibrations is written as

$$\Delta v = +1, \Delta J = 0, \pm 1, \Delta k = 0, \text{ where } K = [J, (J-1), \ldots 0 \ldots -J)$$

When $\Delta J = 0$, in that case, the frequencies of the spectra will be independent of K. In that situation, the spectrum of the symmetric top will be similar to the spectrum of linear polyatomic molecules showing perpendicular vibration. So, in the spectrum, the P, Q, and R branches will appear showing an intense Q band in the center.

The spacing between these two branches will be equal to $2B$. For example, molecules like the CH_3I and CH_3Br, etc. are symmetric top molecules. In such molecules, vibration–rotation energy level

$$e_{v,J} = e_{vib} - e_{rot},$$

$$\Delta\varepsilon = \left(v + \frac{1}{2}\right)\overline{\omega}_e - \left(v + \frac{1}{2}\right)^2 x_e \overline{\omega}_e + BJ(J+1) + (A-B)K^2 \ (\text{cm}^{-1})$$

where

$$\varepsilon_{vib} = \left(v + \frac{1}{2}\right)\overline{\omega}_e - (v + \frac{1}{2})^2 x_e \overline{\omega}_e \ (\text{cm}^{-1})$$

$v = 0, 1, 2, 3, ...$

and $$\varepsilon_{rot} = BJ(J + 1) + (A - B)K^2$$

where $J = 0, 1, 2, ...$ and $K = [J, (J-1), ..., 0, ..., -J]$.

5.6.2.4 *Symmetric top molecules with perpendicular vibrations*

The selection rules for symmetric top polyatomic molecule performing perpendicular vibration

$$\Delta v = \pm 1, \Delta J = 0, \pm 1, \text{and } \Delta k = \pm 1$$

The vibrational–rotational energy equation ( ) for parallel vibrations remains the same in this case also. Thus, when

(i) $\Delta J = +1$, $\Delta k = \pm 1$, one will obtain R branch spectral lines for which

$$\Delta\varepsilon = \overline{\upsilon}_{spectrscopic} = \overline{\omega}_0 + 2B(J+1) + (A-B)(1 \pm 2k) \ \text{cm}^{-1}$$

(ii) $\Delta J = -1$, $\Delta k = \pm 1$, P branch spectral lines are obtained and

$$\Delta\varepsilon = \overline{\upsilon}_{spectrscopic} = \overline{\omega}_0 - 2B(J+1) + (A-B)(1 \pm 2k) \ \text{cm}^{-1}$$

(iii) $\Delta J = 0$, $\Delta k = \pm 1$, Q branch spectral lines are found with

$$\Delta\varepsilon = \overline{\upsilon}_{spectrscopic} = \overline{\omega}_0 + (A-B)(1 \pm 2k) \ \text{cm}^{-1}$$

In these types of molecules, very complex spectra are obtained. Many sets of P and R branch lines will be obtained since for each J value there are many allowed values of k [$K = J, J-1, ..., 0, ..., -(J-1), -J$]. The Q branch is also complicated and contains a series of lines on both sides of the band center $\overline{\omega}_0$. These lines will be separated from each other by a distance of $2(A - B)$.

The IR spectra of various other types of polyatomic molecules like asymmetric will not be discussed here due to their extremely complex nature and complicated treatment.

5.7 Factor Affecting the Infrared Absorption

The vibrational frequency of a particular bond is not always the same. It differs from its observed value due to certain factors. These factors change the force constant K, and hence the vibrational frequency gets changed. Many factors are responsible for shifting the precise frequency of molecular vibration and also the effect of one factor cannot be separated from the other. Some of the factors responsible for shifting the vibrational frequencies from their normal values are discussed below.

5.7.1 Effect of Force Constant, Bond Strength, and Isotopic Substitution

The energy needed to stretch a bond depends on the bond strength and the masses of the atoms involved in the bonding. If the bond is strong, more energy will be required to stretch it and since the frequency of stretching vibration is inversely proportional to the mass of the atoms, therefore heavier atoms exhibit a vibration at lower stretching frequencies.

The vibrational frequency of a particular bond can be obtained with the help of the following Eq. 5.5 (Section 5.3).

$$\overline{\nu} = \frac{1}{2\pi c}\sqrt{\frac{K}{\mu}} \tag{5.52}$$

where μ is reduced mass ($\mu = \frac{m_1 . m_2}{m_1 + m_2}$), m_1 and m_2 are the masses of the two atoms, c is the velocity of light, and K, the force constant of the bond is a measure of the bond strength. This equation indicates that the value of the stretching frequency of the bond increases with the increasing bond strength and also by decreasing the reduced mass of the atoms involved in the bonding.

We know that the triple bonds are strongest, the double bonds are less strong while the single bonds are the weakest between the same two atoms. Thus, their vibrational frequencies are in the order of $\upsilon_{\text{triple}} > \upsilon_{\text{double}} > \upsilon_{\text{single}}$.

Their corresponding values are given below

$$\underset{2150\ \text{cm}^{-1}}{C \equiv C} > \underset{1650\ \text{cm}^{-1}}{C = C} > \underset{1200\ \text{cm}^{-1}}{C - C}$$

Increasing υ with decreasing K

Further, when the atom attached to a particular carbon atom increases in mass then the reduced mass also increases, and the frequency corresponding vibrational frequency decreases.

$$\underset{3000\ \text{cm}^{-1}}{C - H} > \underset{1200\ \text{cm}^{-1}}{C - C} > \underset{1100\ \text{cm}^{-1}}{C - O} > \underset{750\ \text{cm}^{-1}}{C - Cl} > \underset{600\ \text{cm}^{-1}}{C - Br} > \underset{500\ \text{cm}^{-1}}{C - I}$$

Increasing μ with decreasing υ

The change in the mass of atoms in any compound can be done by isotopic substitution. The substitution of a particular atom for its heavier isotope reduces the vibrational frequency of a bond. The isotopic substitution brings about a change in the reduced mass of a functional group. Since the vibrational frequency in wave numbers is related to the masses of the atoms, therefore, a change in reduced mass will shift the wave number of absorptions. For example, let us substitute deuterium, D in place of hydrogen in the C–H bond and calculate the wave number for the absorption of the C–D bond. According to Eq. (5.5), the vibrational frequency of a bond increases with the increase in the bond strength and with the decrease in the reduced mass of the system.

With the help of Eq. (5.4), one may write for C–D and C–H bonds

$$\frac{\bar{\nu}_{\text{C-D}}}{\bar{\nu}_{\text{C-H}}} = \left(\frac{\mu_{\text{C-D}}}{\mu_{\text{C-H}}}\right)^{1/2} = \left\{\frac{m_{\text{H}}.(m_{\text{C}} + m_{\text{D}})}{m_{\text{D}}.(m_{\text{C}} + m_{\text{H}})}\right\} \tag{5.53}$$

where m_{C}, m_{H}, and m_{D} are atomic masses of carbon, hydrogen, and deuterium and N_A is the Avagadro number. Therefore,

$$\frac{\bar{\nu}_{\text{C-D}}}{\bar{\nu}_{\text{C-H}}} = \left(\frac{14}{26}\right)^{1/2} = 0.7338$$

$$\bar{\nu}_{C-D} = 0.7338\,\bar{\nu}_{C-H}$$

The vibrational frequency of the C–H bond is 2900 cm^{-1} therefore

$$\bar{\nu}_{C-D} = 0.7338 * 2900 = 2128\ cm^{-1}$$

This clearly shows that the vibrational frequency of the C-D bond decreases. This effect is of great importance in IR spectral analysis.

Similarly, for O–H, the stretching vibrational frequency is observed at 3620 cm^{-1} and when H is replaced by deuterium D then the observed frequency of O–D is obtained at 2630 cm^{-1} and when calculated with the help of Eq. (5.4) $\bar{\nu}_{O-H}$ in the above manner then $\bar{\nu}_{O-H}$ comes to 2634 cm^{-1} which matches with the theoretical value.

5.7.2 Electronic Effects

When an electron-withdrawing or donating group is attached to a particular molecule then it affects the IR absorption frequency. This effect is mainly due to inductive and mesomeric effects.

5.7.2.1 *Mesomeric effect*

The presence of the pi (π) bonds in the molecule are responsible for this effect. The electron-withdrawing or electron-donating groups are represented as –R or as +R, respectively. For example, some of the +R and –R groups are mentioned below (Example 1).

The presence of electron-withdrawing substituents (–R) to a particular group causes either a decrease in the bond order or enhances the bond length which leads to a decrease in the IR frequency of that particular group. On the other hand, the presence of electron-donating substituents (+R) to any particular group, increases the bond order, and hence the bond length decreases. This then causes an increase in the IR frequency. It must also be noted here that for any particular substituent, the inductive and resonance effect must be considered simultaneously. The one which dominates will decide the IR frequency value. This effect can be illustrated with the following examples:

Example 1: Let us consider *acetone*, *but-3-ene-2-one*, and *acetophenone* whose structures are shown below and have been labeled as (a), (b), and (c).

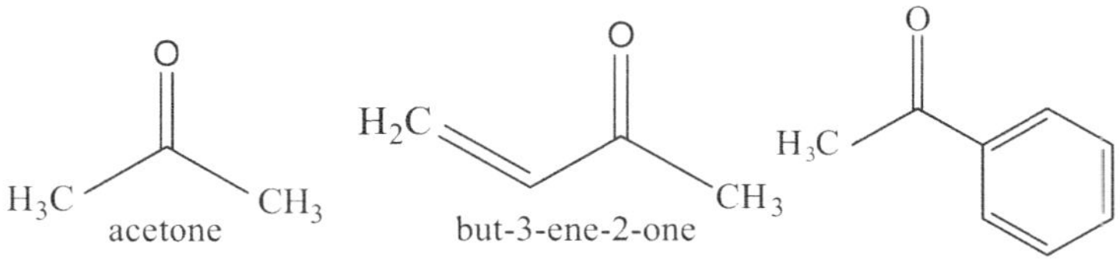

H$_3$C acetone CH$_3$ (a) H$_2$C but-3-ene-2-one CH$_3$ (b) H$_3$C (c)

The IR carbonyl stretching frequency, $\upsilon_{C=O}$ (str) appears at 1720 cm^{-1} in acetone (a) and at 1700 cm^{-1} both in but-3-ene-2-one (b) and acetophenone (c). When a carbonyl group and a carbon-carbon double bond are conjugated, both stretching frequencies are lowered. In compound (b), C=O is conjugated with α, β unsaturation whereas in acetophenone it is conjugated with an aromatic ring. In such cases, there is a contribution of type $^-$O-C=C-C$^+$. The conjugation increases the double bond character of the bond joining the C=O to the ring. Thus, The lowering of vibrational frequency by about 20 cm^{-1} in (b) and (c) as compared to (a) may be explained by considering the +R nature of the vinyl group (–CH=CH$_2$) in (b) and phenyl group (–C$_6$H$_5$) in (c). These groups will withdraw the pi electron of the C=O bond which leads to a lowering in bond order (i.e. increase in the bond length). This will weaken the bond and consequently, the stretching frequency will decrease. One may also think that C=O frequency shifts due to the mesomeric or resonance effect. Any substituent that increases the mesomeric shift will decrease the bond order of C=O and in turn, causes the lowering in C=O stretching frequency.

Another example of diene (d) and a conjugated diene (e) can be taken to explain this effect.

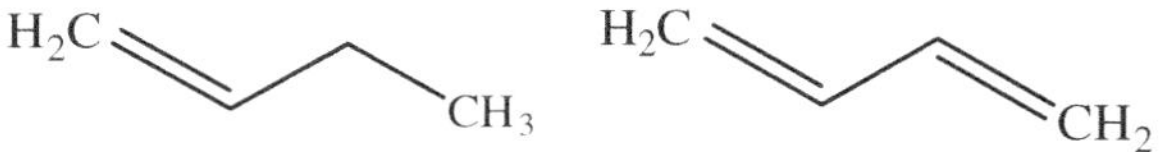

H$_2$C CH$_3$ (d) H$_2$C CH$_2$ (e)

The terminal C=C stretching frequency appears at 1650 cm^{-1} in diene (d) while in conjugated diene (e) it is at 1610 cm^{-1}. This means there is a lowering of frequency by 40 cm^{-1} in conjugated diene (e). This may be explained on the basis of conjugation in the resonance effect. The conjugated diene possesses the following three resonance structures.

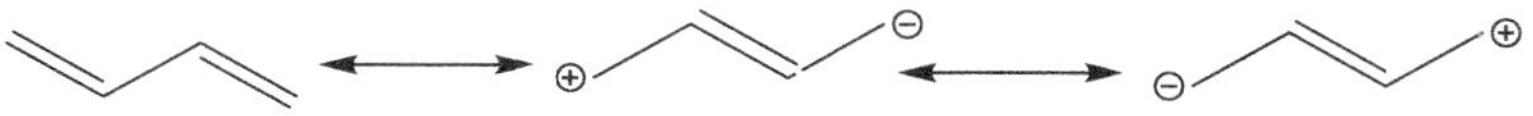

Due to resonance, the terminal bonds have a bond order value between one and two (i.e. it decreases from the normal value of two). Therefore, the frequency of C=C decreases.

5.7.2.2 *Inductive effect*

The inductive effect occurs through sigma electrons or bonds in a molecule. The symbol *+I* is given to those groups which enhance the electron density at the neighboring or bonds in a molecule. The symbol *+I* is given to those groups which donate (enhance) electron density to the neighboring atoms in a saturated carbon chain while the symbol *–I* is given to those groups which reduce (withdraw) electron density from the neighboring atoms. It is very difficult to consider the inductive effect separately from the mesomeric effect. The role of the inductive effect on the IR stretching frequency can be illustrated with the help of the following examples:

Examples 1: The aldehyde generally shows higher $\upsilon_{C=O}$ stretching frequency as compared to ketones. In ketones, an additional alkyl group is present. Alkyl groups, by the virtue of their *+I* nature, cause a decrease in C=O bond strength which eventually results in a decrease of $\upsilon_{C=O}$ frequency in ketone.

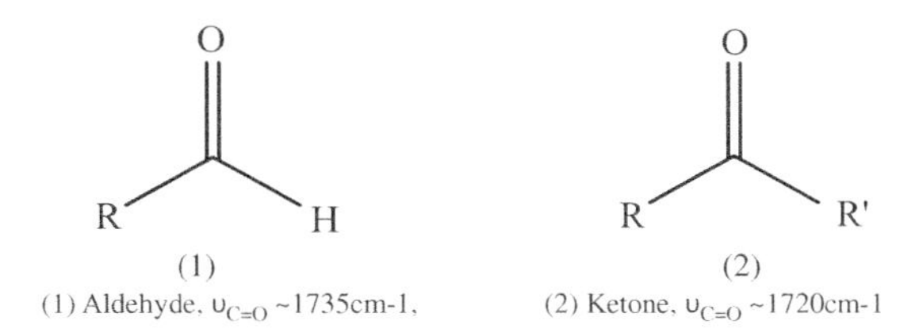

(1) Aldehyde, $\upsilon_{C=O}$ ~1735cm-1, (2) Ketone, $\upsilon_{C=O}$ ~1720cm-1

Example 2: The infrared carbonyl stretching frequency in ketone appears around 1720 cm^{-1} while those of ester and thioester their frequency appears at around 1740 cm^{-1} and 1690 cm^{-1}, respectively.

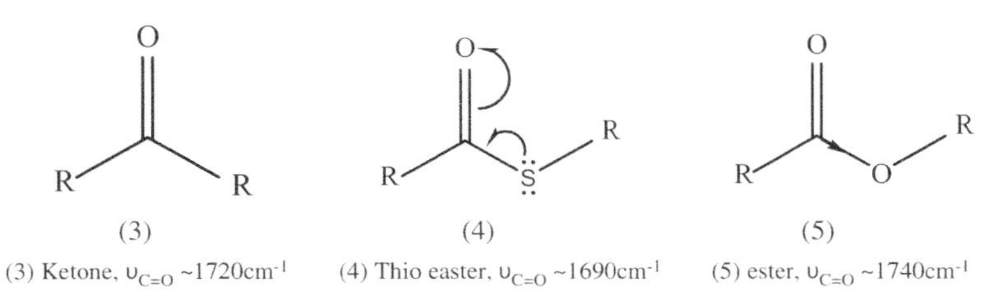

(3) Ketone, $\upsilon_{C=O}$ ~1720cm^{-1} (4) Thio easter, $\upsilon_{C=O}$ ~1690cm^{-1} (5) ester, $\upsilon_{C=O}$ ~1740cm^{-1}

The difference in stretching frequencies in these organic compounds is explained by considering the inductive effect and resonance effect together. In this thioester (4), sulfur atom has more +*R* (electron donating) than –*I* (electron withdrawing) nature. Due to this, the double bond character of the carbonyl group is lost which leads to a lowering of $\upsilon_{C=O}$ as compared to $\upsilon_{C=O}$ ketone (2). On the other hand, in ester (5), the –*I* nature of the oxygen atom is more than its +*R* nature. This causes the shortening of the C=O bond as compared to ketone (3) and hence an increase in carbonyl stretching frequency is observed as compared to ketone (3).

Example 3: The $\nu_{C=O}$ stretching frequencies in ketone, amide, and acyl chloride.

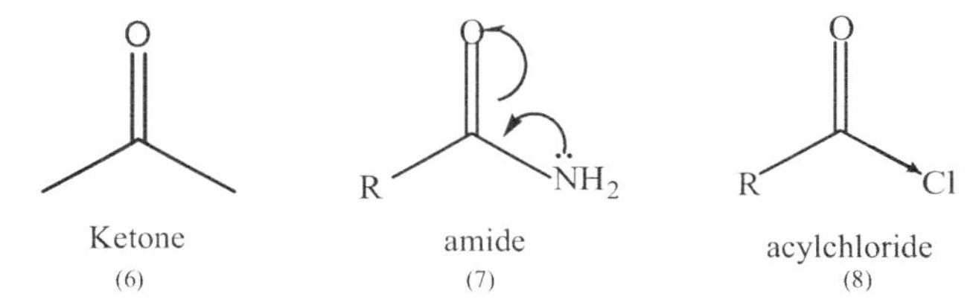

Ketone (6) amide (7) acylchloride (8)

In amide (7), the +*R* effect of the –NH_2 group produces a lengthening of the C=O bond leading to lower $\nu_{C=O}$ as compared to the corresponding ketone. In acyl chloride (8), because of the high electronegativity of Cl, the –*I* effect is more than the +*R* effect. This causes shortening of the C=O bond leading to higher stretching frequency as compared to a ketone.

Example 4: The role of *I* and *R* effects on stretching frequencies can further be illustrated by considering $\upsilon_{C=O}$ values in alkyl esters and phenyl esters.

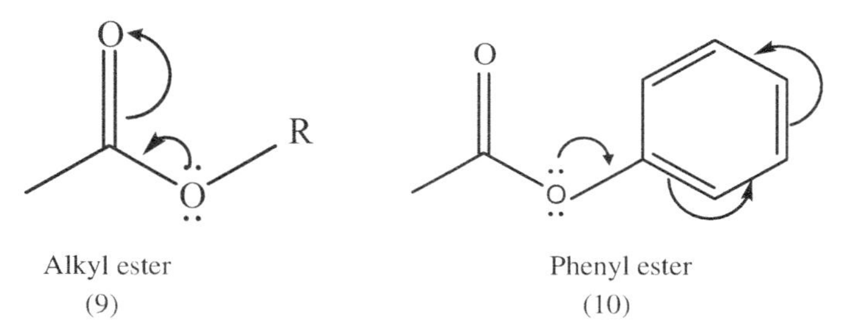

Alkyl ester (9) Phenyl ester (10)

In the alkyl ester (9), the +*R* effect of oxygen is more than its –*I* effect. The nonbonding electrons of oxygen increase the +*R* conjugation. This causes the lengthening or weakening of the C=O bond and therefore, the $\upsilon_{C=O}$ value is low as compared to phenyl ester. However, in phenyl ester, the nonbonding electrons of oxygen are partly drawn into the phenyl ring and therefore, the tendency of the lone pair of electrons of oxygen to conjugate with C=O is reduced. Thus, the $-I$ effect of oxygen dominates over its $+R$ effect. Consequently, the $\upsilon_{C=O}$ value shift toward the higher side.

5.7.3 Bond Angles Effect

Bond angles in different compounds also have some effect on the IR stretching frequencies. Among different ketones, strained cyclobutanones show the highest $\upsilon_{C=O}$. This observation can be linked to bond angle strain. In these compounds, C–C–C bond angles are reduced below the normal 120° which leads to more s character decrease. Thus, $\upsilon_{C=O}$ in this compound has a low value.

As one goes from $sp^3 \rightarrow sp^2 \rightarrow sp$ hybridization in alkane → alkene → alkyne, the s character of the C–H bond increase. The bond lengths become shorter and thus $\upsilon_{C\text{-}H}$ stretching frequencies.

$$\upsilon_{C\text{-}H}\,(\text{alkane}) > \upsilon_{C\text{-}H}\,(\text{alkene}) > \upsilon_{C\text{-}H}\,(\text{alkyne})$$

Cyclopropane shows high C–H stretching frequencies. In this case, the C–C–C bond angle is reduced considerably below the normal angle of 109.5° which increases the s character in the C–H bonds leading to a higher frequency.

5.7.4 Effect of Hydrogen Bonding

The shape and position of an infrared absorption band get changed due to the presence of hydrogen bonding. The stronger the hydrogen bonding greater shifts of the vibrational frequencies are seen from their normal values. Two types of hydrogen bonding, i.e., intermolecular and intramolecular can be identified with the help of infrared spectroscopic studies.

In intermolecular hydrogen bonding, the association of two or more molecules either of the same or a different nature takes place. For example, the existence of dimer in alcohol, carboxylic acid, etc. On the other hand, intramolecular hydrogen bonds are formed when proton donor and acceptor groups are present in a single molecule.

The IR spectra and the patterns of intermolecular hydrogen-bonded compounds depend upon the concentration of the solution and the temperature at which the measurement is being done. Generally, intermolecular hydrogen bonds are formed in a more concentrated solution in which the possibility of intermolecular hydrogen bonding is more. Therefore, a sharp band for –OH stretching in a concentrated solution of alcohol is obtained at nearly 3550 cm^{-1}. However, in a dilute solution of aliphatic alcohol, two bands are observed at 3650 cm^{-1} and 3350 cm^{-1}, respectively. The sharp band appearing at 3650 is due to the free –OH group while the broad at 3550 cm^{-1} is due to the H-bonded –OH group. It means that at a lower concentration, the possibility of intermolecular hydrogen bonding becomes less and the band appearing at 3550 cm^{-1} in concentrated solution shifts to 3650 cm^{-1}.

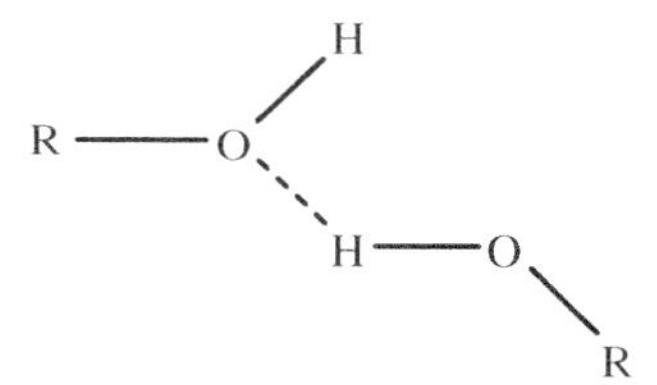

The polymeric association of O–H bonds in alcohols.

The effect of temperature also plays an important role in intermolecular hydrogen bonding. At higher temperatures the molecular association is prevented and therefore in the case of ethanol, the intensity of the sharp band appearing at 3650 increases.

The infrared spectra of benzoic acid show a broad band ranging from 2500 cm^{-1} to 3500 cm^{-1} due to hydrogen-bonded O–H stretching. It should be noted that strong hydrogen bonds exist in the stable dimeric association of carboxylic acids in the condensed phases.

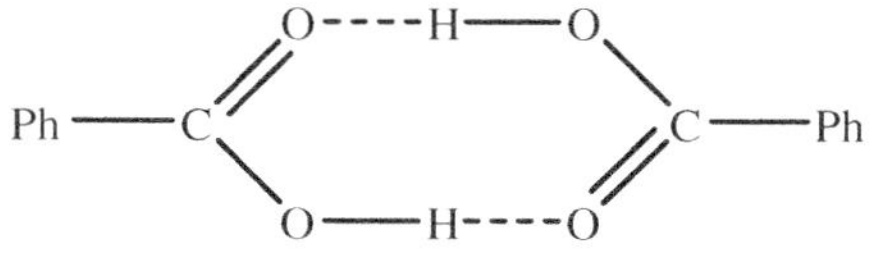

The polymeric association also occurs in carboxylic acids but the dimmers are more predominant. The free –OH stretching vibration can be seen in only extremely diluted solutions of carboxylic acids in hexane. The proportion of monomer to dimer increases in solvents such as benzene. However, in dioxane dimer is not formed because acid hydrogen itself forms bonds with the solvent.

The molecular association effects are minimum in very dilute solutions or vapor phase, but in solid state or concentrated solution, the molecular association effect is quite significant. For example, let us examine the IR spectrum of acetic acid in pure liquid form and vapor phase, $\upsilon_{C=O}$ stretching is seen at 1718 cm^{-1} in the spectrum of acetic acid in pure liquid acetic acid which is due to dimer, whereas two carbonyl bands appearing at 1733 cm^{-1} and 1786 cm^{-1} in vapor phase of acetic acid can be explained because of the presence of dimer and monomer respectively. It is therefore essential to determine the IR spectra of a sample under different conditions of physical state, temperature, and concentration.

The free –OH stretching of the carboxylic acid causes absorption to occur near 3550 cm^{-1}. However, the bonded O–H in the dimeric form show absorption in the region 2700–2500 cm^{-1} Further, when a carbonyl group C=O is involved in intramolecular hydrogen bonding, then it has a lower stretching frequency. For example, in acetophenone, the carbonyl absorption occurs at 1700 cm^{-1} but in 2-hydroxyacetophenone, having intramolecular hydrogen bonding, the carbonyl absorption occurs at 1635 cm^{-1} instead of at 1700 cm^{-1}. Thus, it is showing a decrease of 65 cm^{-1} due to intramolecular hydrogen bonding.

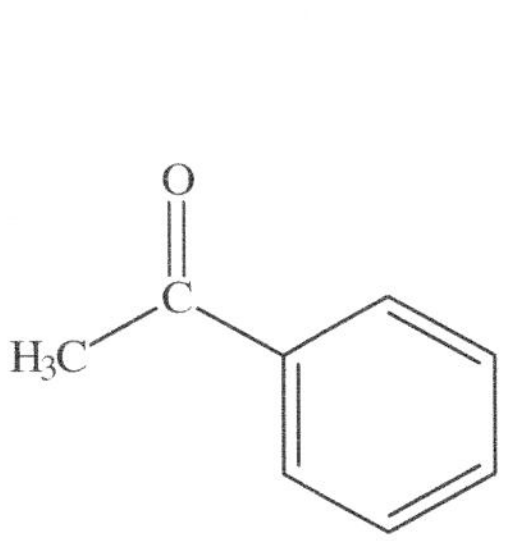

(1700 cm^{-1}) acetophenone

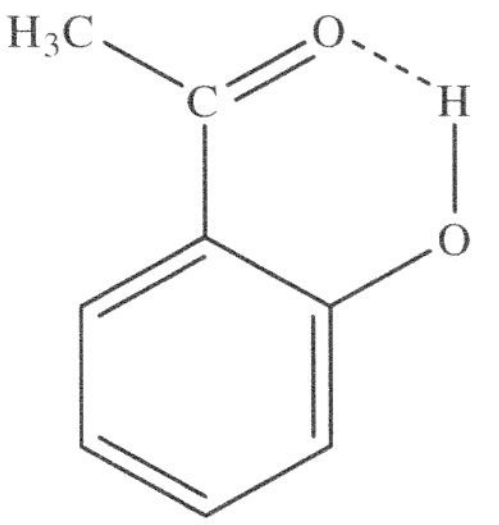

(1635 cm^{-1}) 2-hydroxyacetophenone

The IR spectral frequencies of keto and enol forms can be understood in terms of the effect of hydrogen bonding. For example, ethyl acetoacetate shows both keto and enol forms as such it gives absorption frequencies characteristics of both forms.

keto form
(1724 cm^{-1} and 1748 cm^{-1})

enol form
(1650 cm^{-1})

The alkene and aromatic π-bonds can behave as Lewis bases as such they may from hydrogen bonds to acidic hydrogens. For example, the O–H stretching frequency in phenols is lowered by 40–100 cm^{-1} when recorded in benzene solution in comparison to that recorded in CCl_4 solution.

5.7.4.1 *Intramolecular hydrogen bonding*

When acidic hydrogen such as –OH and –NH is sufficiently close to a basic center such as O=C<, O< or –N< in the same molecule, the formation of intramolecular hydrogen bonds takes place. Intramolecular hydrogen bonding is possible in systems where five- or six-membered ring formation is possible as shown in the figure below.

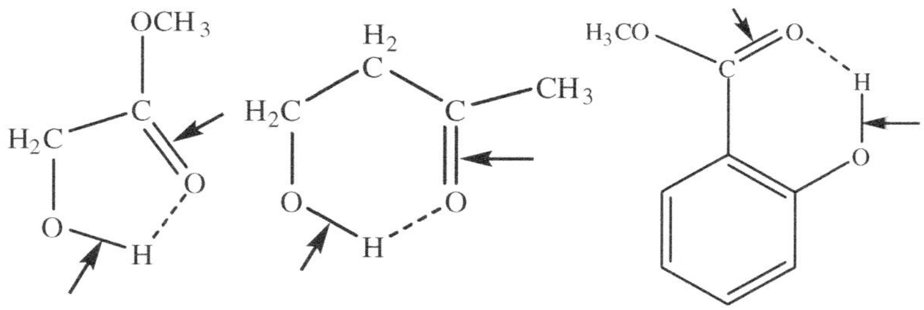

$\upsilon_{C=O}$=1748 cm^{-1}
$\upsilon_{O=H}$=3554 cm^{-1}

$\upsilon_{C=O}$=1707 cm^{-1}
$\upsilon_{O=H}$=3430 cm^{-1}

$\upsilon_{C=O}$=1684 cm^{-1}
$\upsilon_{O=H}$=3210 cm^{-1}

The frequency shifts for intramolecular hydrogen bonds compared to non-hydrogen bonded compounds.

The stretching frequency for the -OH group for an intramolecular bonded hydroxyl is generally at a lower value are broader than that for the free hydroxyl. The stretching frequency of C=O also decreases. The stretching frequencies for intramolecular hydrogen bonding are not affected by concentration change or by temperature change.

Mostly, non-associating solvents like CS_2, CCl_4, and $CHCl_3$, are used in IR spectroscopic studies. The frequencies of –OH and –NH compounds to a considerable extent. This is due to the interaction between the fundamental group of solvent and the sample.

5.8 Application and Analysis by Infrared Spectra

IR spectroscopy is a very important technique for determining the molecular structure of organic. It may also be applied to determine force constants, examine the purity of compounds, and provide information regarding the existence of hydrogen bonds in a molecular system.

5.8.1 Structural Analysis

The infrared spectrum of a complex molecule exhibits a large number of normal vibrations. Each normal mode involves some displacement of nearly all the atoms in the molecule. These normal modes may be divided into two classes of vibrations. One is known as 'skeletal vibration' in which all the atoms undergo the same displacement while the other is called 'characteristic group vibration' in which only a small portion (a small group of atoms) of the molecule shows more vigorous displacements than those of the remainder.

Linear or branched chain structures in the molecule give rise to skeleton frequencies. These fall in the range 1400–700 cm^{-1}. The following groups such as

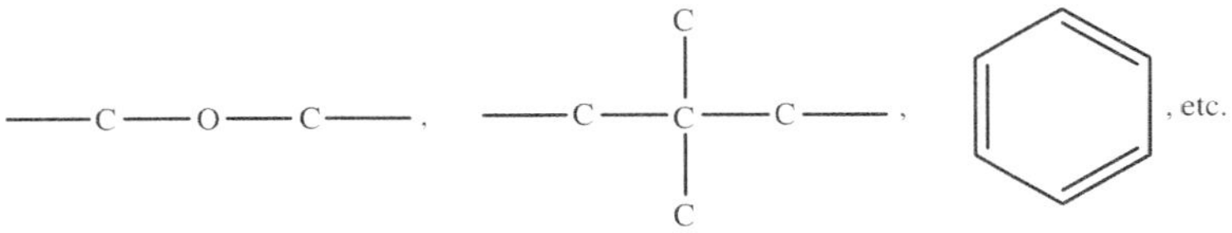

—C—O—C—, —C—C(C)(C)—C—, (benzene ring), etc.

show several skeletal modes of vibrations and therefore, there are several absorption bands in the infrared spectrum. Changing a substituent either in the chain or in the ring show a marked change in the pattern of the absorption bands. As such these bands are known as fingerprint bonds because a molecule could be recognized simply by the appearance of this part of the spectrum.

On the contrary, group frequencies do not depend on the structure of the molecule as a whole. These fall in the regions well above and well below that of the skeleton modes. Typical group frequencies have been given in Table 5.1–5.7 [3, 9–10].

Table 5.1 Characteristic Infrared bands of aliphatic hydrocarbons

Class/Group	Frequency ranges and intensities (cm^{-1})	Assignment
Alkanes	2960 (vs) 2870 (vs) 2930 (vs) 2850 (vs) 1470 (s) 1465 (s) 1380 (m-s) 1305 (m-s) 1300(m-s) 720 340–230 (s)	Methyl symmetric C-H stretching Methyl Asymmetric C-H stretching Methylene Asymmetric C-H stretching Methylene Asymmetric C-H stretching MethylAsymmetric C-H bending Methylene Scissoring Methyl symmetrical C-H bending Methylene wagging Methylene twisting Methylene rocking -C-C-C- bending
Alkenes (=CH_2)	3100–3000 (m) 1680–1600 (m-s) 1450–1200 (vs) 1000–600	=C-H Stretching C=C stretching =C-H in-plane bending =C-H out-of-plane bending
Cis-alkanes R’CH=CHR	590–290 (m)	Skeleton deformation
trans-alkanes R’CH=CHR	500–200 (m)	Skeleton deformation
Allenes	2000–1950 (s) 1100–1050 (vs)	-C=C=C- antisymmetric streching Symmetric streching
Alkynes	3300–3250 (m-s) 2260–2100 (w-s) 680–590 (s)	Terminal =C-H stretching C≡C Stretching conjugation H—C≡C bending

Table 5.2 Characteristic Infrared bands of oxygen-containing functional group

Class/Group	Frequency ranges and intensities (cm^{-1})	Assignment
Alcohols (R–OH)		
Primary alcohol $-CH_2OH$	3650–3600 (s) 1050–1020 (s)	O–H stretching C– OH stretching
Secondary alcohol –CHROH	3630–3600 (s) 11200–1080 (s)	C–O stretching C–OH stretching
teriary alcohol $-CR_2OH$	3620–3600 (s) 1160–1120 (s)	O–H stretching C–OH stretching
General –OH	3350–3250 (s)	O–H stretching (broad band)
Phenol Ph–OH	3550–3500 (s) 720–600 (s, br) 450–360 (w)	O–H stretching O–H out –of –plane bending C–OH bending
Carboxylic acids R–COOH	3300–2500 1700 1430 1240 930	O–H stretching C=O stretching C–O–H in–plane bending C–O stretching C–O–H out–of–plane bending
Aliphatic acid halides R—Cl(=O)X	~1800 (s) 970–920 (m) 450–420 (s)	C=O stretching C–C stretching Cl–C–O in plane bending
Aromatic Acid Halides Ph—Cl(=O)X	1790–1760 (s) 900–850	C=O stretching C–C stretching
Ethers –C–O–C–	1150–1080 (vs) 1230–1200 (vs) 1660–1610 1080–1020 (s) 1280–1220 (s)	C–O–C asymmetric stretching in alkyl ether C–O–C symmetric stretching in vinyl ether C=C symmetric stretching in vinyl ether C–O–C symmetric stretching in aryl alkyl ether C–O–C asymmetric stretching in aryl alkyl ether
Esters R—C(=O)OR'	1760–1720 (vs) 1730–1700 (vs) 1300–1100 1310–1250	Aliphatic C=O stretching Aromatic C=O stretching Aromatic C–O stretching Aliphatic C–O stretching

(*Continued*)

Table 5.2 (*Continued*)

Class/Group	Frequency ranges and intensities (cm^{-1})	Assignment
Anhydrides —CO O —CO	1850–1800 (variable) 1780–1710 (m–s) 1300–1100 (vs)	C=O antisymmetric stretching C=O symmetric stretching C–O–C stretching
Aldehydes andKetones	2880–2720 (m) 1740–1700 (vs) 1450–1320 (s)	Aliphatic Aldehyde C–H stretching Aliphatic Aldehyde C=O stretching H–C=O bending in aliphatic aldehydes
R–CHO & RCOR	700–630 (s) 570–520 (s) 1730–1700 1720–1680 1700–1680	C–C–CHO bending C–C=O bending C=O stretching in aliphatic ketone C=O stretching in aromatic ketone C=O stretching in aromatic ketone

Table 5.3 Characteristic Infrared bands of nitrogen-containing functional group

Class/Group	Frequency ranges and intensities (cm^{-1})	Assignment
Primary Amide R–$CONH_2$	3190–3170 3360–3340 1650–1620 (m) 1680–1660 (vs) 1420–1400 (m–s)	Primary amide NH_2 symmetric stretching Primary amide NH_2 asymmetric stretching Primary amide NH_2 bending (in solid) Primary amide C=O stretching C–N stretching
Secondary amide R–CONHR	3300–3250 1680–1700 750–760 1560–1530 (vs)	Secondary amide N–H stretching Secondary amide C=O stretching (solid and dil solution) Secondary amide N–H wagging (out of plane) Secondary amide N–H bending, C–N stretching
Tertiary amide R–$CONR_2$	1680–1630 (vs)	C=O stretching

Class/Group	Frequency ranges and intensities (cm^{-1})	Assignment
Amines	1250–1000 (m–w)	C–N stretching in primary, secondary and tertiary amine
	1350–1200 (s)	Aromatic C–N stretching

Table 5.4 Characteristic infrared bands of different nitrogen-containing compounds

Class/Group	Frequency ranges and intensities (cm^{-1})	Assignment
NO_2 group	1300–1270	Nitrate NO_2 symmetric stretching
	1660–1620	Nitrate NO_2 asymmetric Stretching
	710–690	Nitrate NO_2 bending
	870–840	Nitrate N–O stretching
	1680–1650	Nitrite N=O stretching
Aliphatic Nitro Compound R–NO_2	1390–1360 (vs)	Aliphatic nitro compound NO_2 symmetric stretching
	1570–1530 (vs)	Aliphatic nitro compound NO_2 asymmetric stretching
	650–600 (s)	Aliphatic bending of –NO_2
Aromatic Nitro Compound Ph–NO_2	1370–1320 (vs)	Aromatic nitro compound NO_2 symmetric stretching
	1500–1460 (vs)	Aromatic nitro compound asymmetric stretching
	580–520 (m)	Aromatic (–NO_2) bending
Aliphatic Nitrile —C≡N	2260–2240 (w)	Aliphatic nitrile C≡N stretching
	2240–2220 (m)	Aromatic nitrile C≡N stretching
Isonitrile	2180–2110	Aliphatic isonitrile –N≡C stretching
	2130–2100	Aromatic isonitrile –N=C stretching

(*Continued*)

Table 5.4 (*Continued*)

Class/Group	Frequency ranges and intensities (cm^{-1})	Assignment
Oxime =NOH	3600–3590 (vs) 3260–3240 (vs) 1690–1620 965–930	–OH stretching (in dilute solution) –OH stretching (in solid) –C=N–OH stretching Oxime N–O stretching
Azide $N{\equiv}N^{+}{\equiv}N^{-}$	2160–2130	Azide N≡N stretching

Table 5.5 Characteristic Infrared bands of halogen-containing functional group

Class/Group	Frequency ranges and intensities (cm^{-1})	Assignment
Fluro Alkyl $-CF_2$, $-CF_3$	1400–1000 (vs)	C–F stretching
Chloro Alkyl –C–Cl	850–550 (m)	C–Cl stretching
Bromo Alkyl –C–Br	650–500 (m)	C–Br stretching

Table 5.6 Characteristic Infrared bands of phosphorus-containing functional group

Class/Group	Frequency ranges and intensities (cm^{-1})	Assignment
Phosphate $(RO)_3P{=}O$ (R=alkyl)	1280–1250 (vs) 1050–960 (vs)	P=O stretching P–O–C stretching
Phosphate $(RO)_3P{=}O$ (R=aryl)	1320–1290 (vs) 1250–1180 (vs)	P=O stretching P–O–C stretching

Class/Group	Frequency ranges and intensities (cm^{-1})	Assignment
Phosphine $-PH_2$, $-PH$	2420–2280 (m) 1100–1050 (w–m)	P–H stretching P–H bending
Other phosphorus containing compounds	500–200 1050–700 600–300 500–200 850–500	P–S stretching P–F stretching P–Cl stretching P–Br stretching P=S stretching

Table 5.7 Characteristic Infrared bands of sulfur-containing functional group

Class/Group	Frequency ranges and intensities (cm^{-1})	Assignment
Sulphides -C-S	710–570 (m)	C-S stretching
sulphones $-SO_2-$	1180–1120 (vs) 1360–1290 (vs) 610–540 (m-s)	SO_2 symmetric stretching SO_2 asymmetric stretching SO_2 scissoring bending
Sulphonic acid $-SO_2OH$	1250–1150 (vs, br)	S=O stretching
Sulphoxide S=O	1060–1020 (s, br)	S=O stretching
Thiols -SH	~2500 (w) 700–550 (w)	S-H stretching C-S stretching

These frequencies and their spectra are highly characteristic of the group so they may be used in structural analysis. For example, the $-CH_3$ group gives four absorption peaks; one due to symmetric C–H stretching absorption between 2850 and 2890 cm^{-1}, secondly a symmetric deformation, and thirdly a symmetric deformation (i.e., the opening and closing of the H–C–H umbrella) at about 1375 cm^{-1} and fourthly a symmetric deformation at ~1470 cm^{-1}. Similarly, the >C=O group shows an intense absorption peak between 1600 and 1750 cm^{-1} –SH group shows a very sharp absorption at ~2600 cm^{-1}. Likewise, the isolated multiple bonds >C=C< or >C≡C< have highly characteristic frequencies. And these characteristic frequencies in the IR spectra are helpful in structure elucidation. For example, the IR spectrum of thioacetic acid, CH_3COSH, or CH_3CSOH. The answer is given very clearly from the infrared spectrum as shown in Figure 5.14. The sharp peaks at ~1730 cm^{-1} and ~2600 cm^{-1}

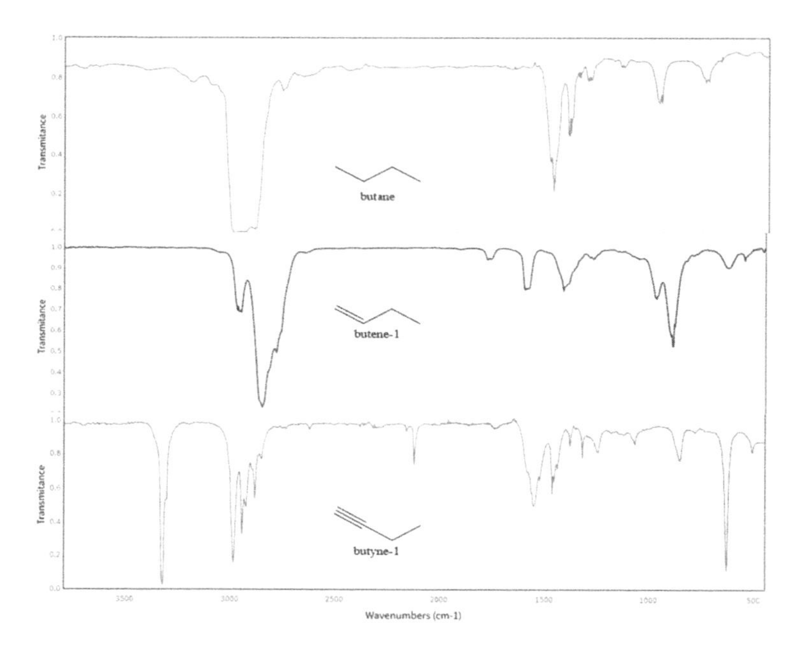

Figure 5.14 FTIR spectra of hydrocarbons: butane, butene-1, and butyne-1.

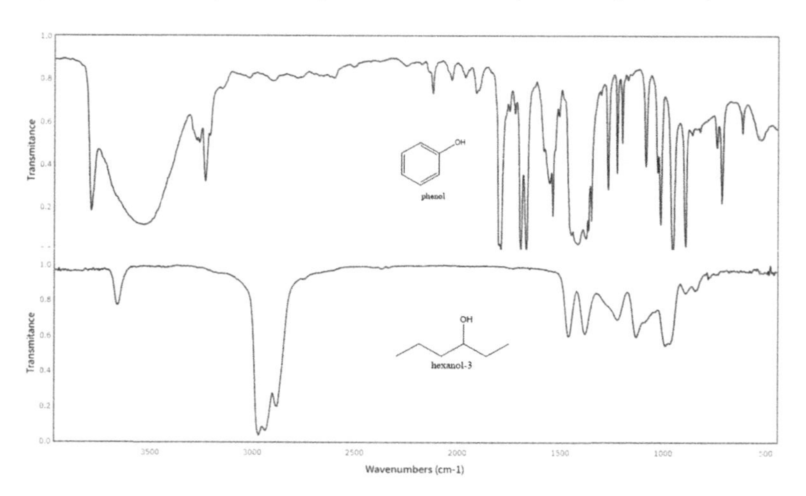

Figure 5.15 FTIR spectra of alcohol: hexanol-3 and phenol.

show the presence of >C=O and –SH groups rather than –CS and –OH groups suggesting the molecule to be CH3COSH. Further, there is no absorption at frequency 1100 cm^{-1} which indicates the absence of >C=S group (The characteristic frequency of >C=S group is 1100 cm^{-1}).

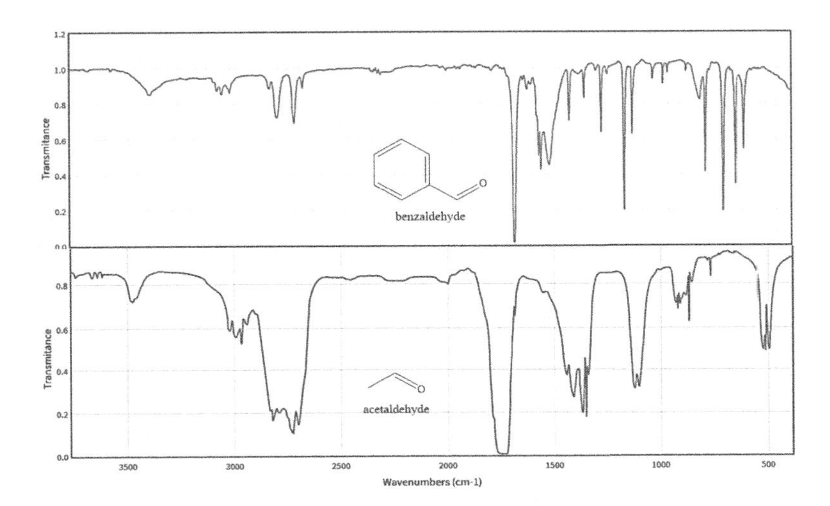

Figure 5.16 FTIR spectra of aldehydes: acetaldehyde and benzaldehyde.

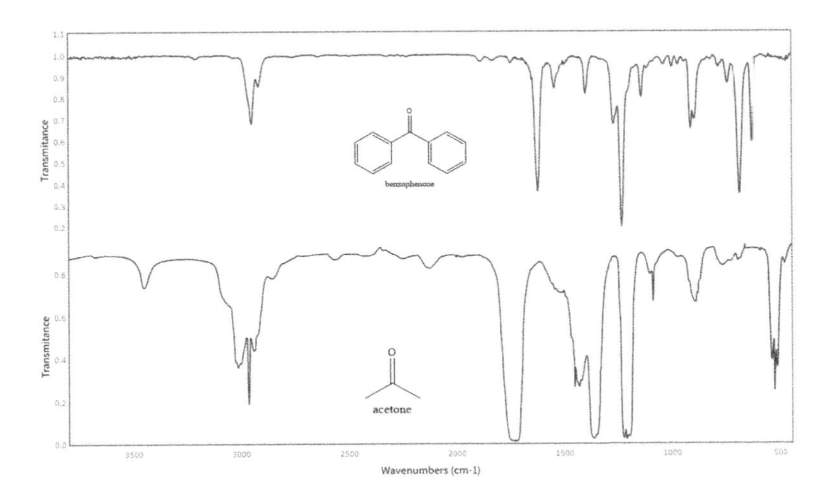

Figure 5.17 FTIR spectra of ketones: acetone and benzophenone.

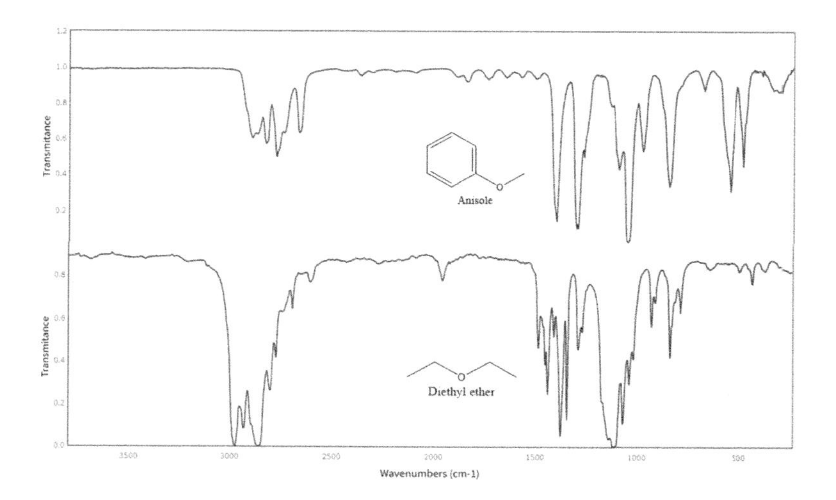

Figure 5.18 FTIR spectra of ether: diethyl ether and anisole.

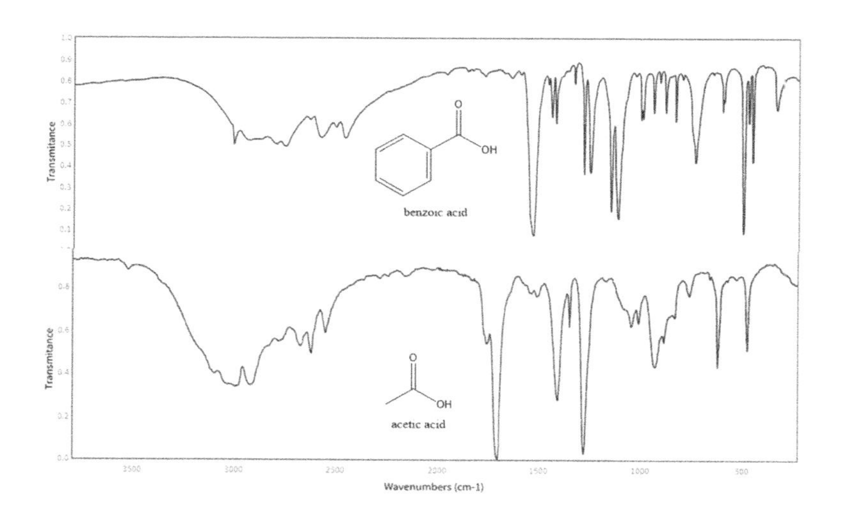

Figure 5.19 FTIR spectra of carboxylic acids: acetic acid and benzoic acid.

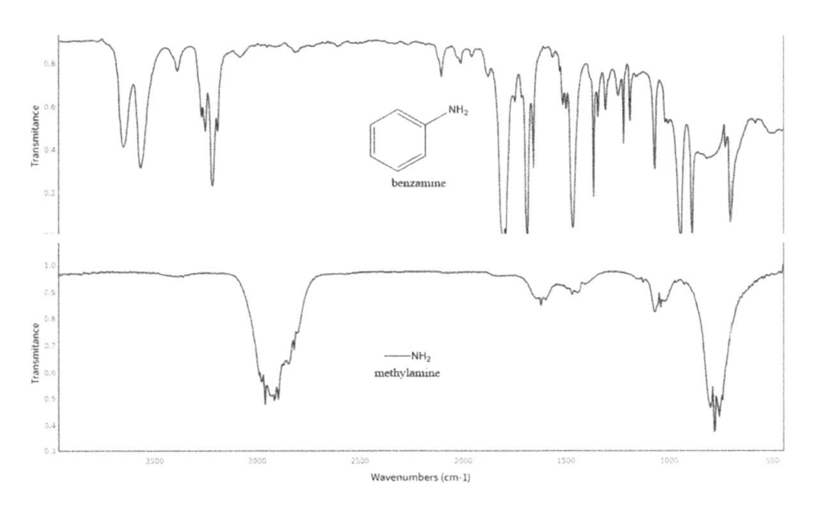

Figure 5.20 FTIR spectra of amides.

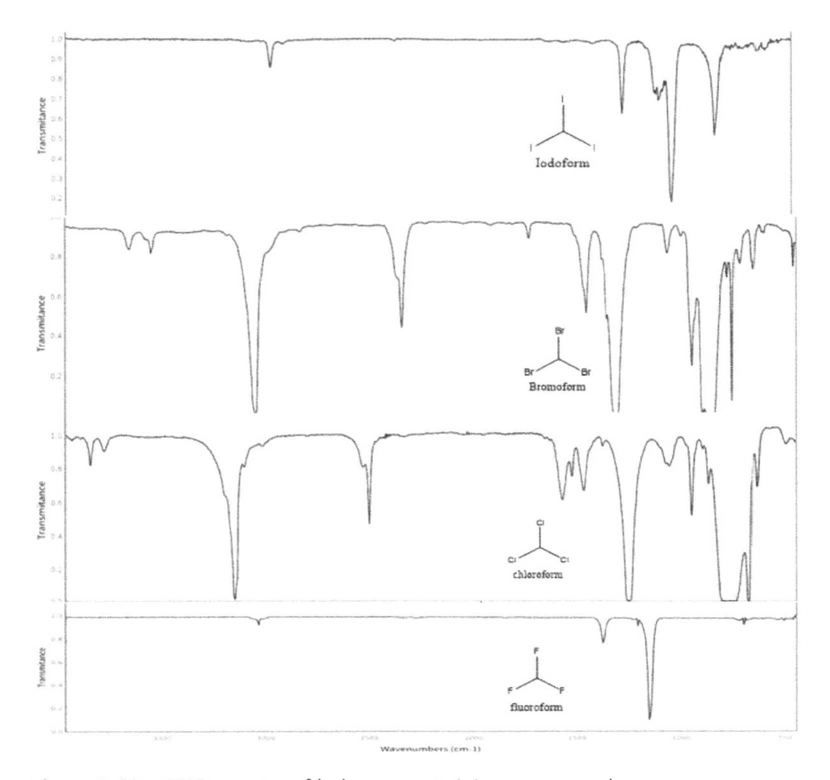

Figure 5.21 FTIR spectra of halogen-containing compounds.

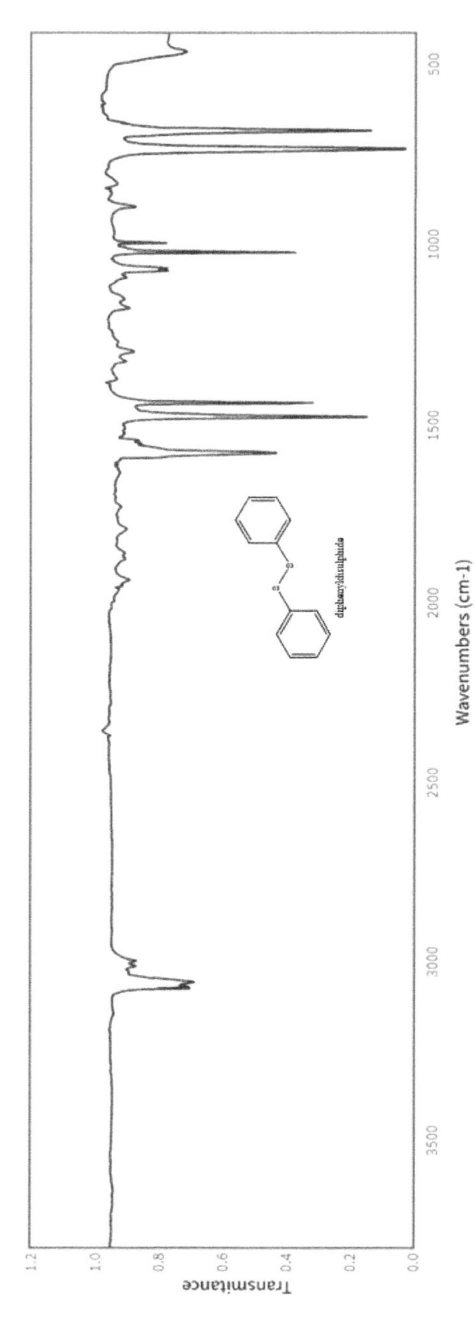

Figure 5.22 FTIR spectra of diphenyl disulfide.

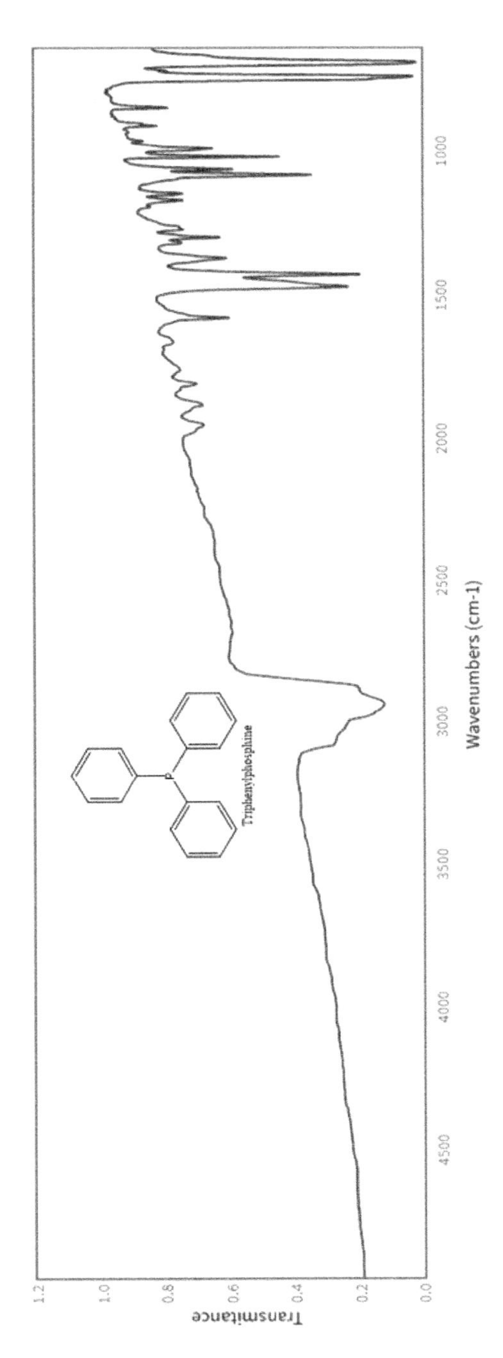

Figure 5.23 FTIR spectra of triphenylphosphine.

5.9 The Infrared Spectrometer: Instrumentation and Technique

There are two types of infrared spectrometers that are based on how the infrared frequencies are used in the measurement. In one type the infrared beam is separated in its frequencies by depression technique with the help of a grating monochromator. These spectrophotometers are known as dispersive type spectrometers. In the other type of spectrophotometers, the infrared frequencies interact to produce an interference pattern on interaction. This is then analyzed mathematically with the help of Fourier transforms, to find out the individual frequencies and their intensities. These spectrometers are known as Fourier transform infrared (FTIR) instruments.

Both types of instruments need a proper source of infrared light, a monochromator, and a detector to measure the intensity of the light.

5.9.1 Infrared Light Sources

Two common IR sources are (i) Nernst glower which is sintered mixture of the oxides of Zr, Th, Ce, Y, Er, etc., and (ii) globar filament made up of silicon carbide. The sources are in form of rods or filaments which are electrically heated nearly up to a temperature of 1800 °C.

5.9.2 Monochromator

As a monochromator, prisms were used initially to diffract the light, but in the present time, all modern instruments use diffraction gratings. These diffraction gratings may be of two types: (i) ruled gratings and (ii) holographic gratings.

In ruled grating measurements, several microscopically close parallel triangular grooves (20 to 400 grooves can be engraved per mm) are engraved on a reflective surface. Light reflected from the grating is diffracted. Interference takes place at certain angles and then the light of specific wavelengths is observed at specific angles of reflection causing constructive interference.

In holographic gratings, on optically flat glass, a photo resisting material is deposited. It is then irradiated with two beams of laser light producing interference fringes in the photosensitive part of the photoresist material. These irradiated parts of photoresist materials are washed in such a way that a grooved pattern is obtained. A reflecting material is coated on these grooves. Using this method, approximately 5000 grooves/mm is prepared. The purity of the wavelength of the monochromatic light obtained by this method is very large.

5.9.3 Detectors

Generally, thermophile detectors are used in dispersive-type instruments. In these detectors, several thermocouples are connected in series in order to add all their outputs. By connecting them in series, the sensitivity of detectors is increased.

Whereas the FTIR instrument uses either (i) thermal detector based on pyroelectric material or (ii) solid-state semiconductors which are based on photovoltaic or photoconductive principles. These solid-state semiconductors have better sensitivity and accuracy.

Extremely sensitive photovoltaic detectors use a diode-type device doped with silicon or indium antimonide or indium gallium arsenide. On the other hand, highly sensitive detectors known as mercury cadmium telluride (MCT) detectors are based on the photoconductive properties of HgCd telluride. These detectors work at the temperature of liquid nitrogen.

In FTIR instruments, thermal detectors based on pyroelectric materials or solid-state semiconductors based on photovoltaic or photoconductive principles are used for better sensitivity and accuracy. Gaseous samples are contained in glass cells with rock salt windows at both ends. Solid samples are finely ground with potassium bromide and then the mixture is pressed into a transparent disk under very high pressure. This disk is then placed directly in the infrared beam using some suitable holder.

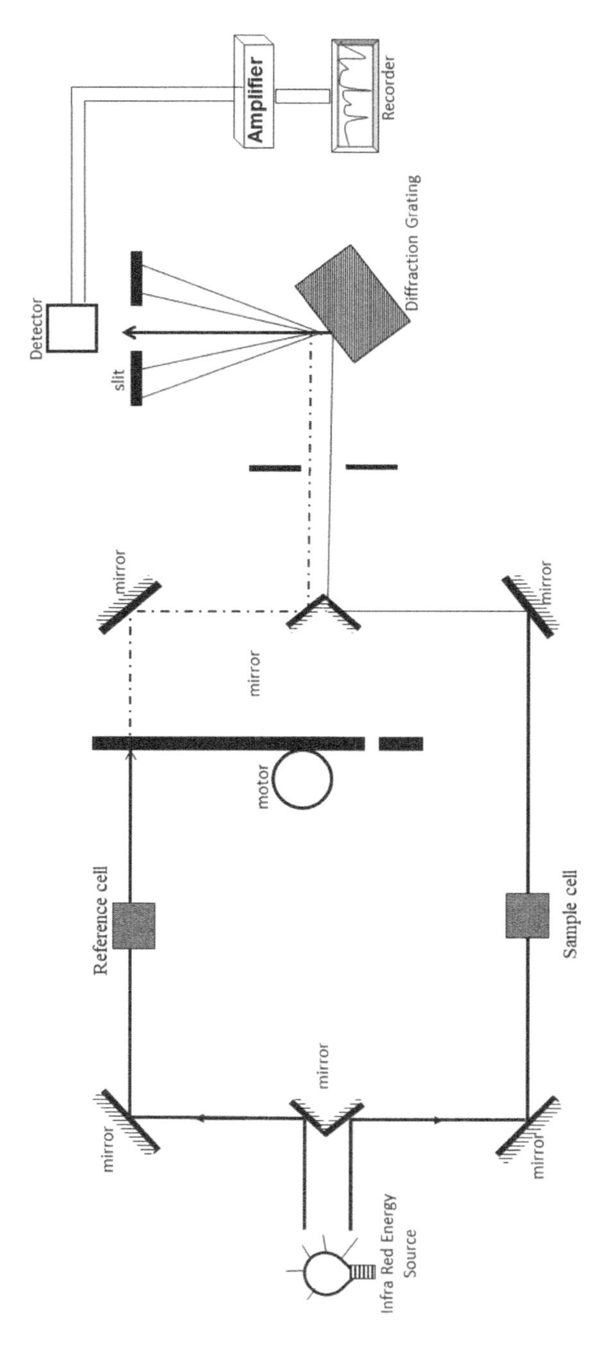

Figure 5.24 Schematic diagram of infrared spectroscopy.

5.9.4 Dispersive IR Spectrometer

A schematic diagram of a typical dispersive infrared spectrometer is shown in Figure 5.24 in which various components of the instruments have been illustrated. A beam of infrared radiation is produced from a hot wire with the help of mirrors. It is divided into two parallel radiation beams of equal intensities. The sample is placed in one beam while the reference is kept in the other beam. Both the beams then go into the monochromator which disperses the IR light and each beam is then separated into individual frequencies in the form of a continuous spectrum. The radiations are then sent into the thermocouple detector, which detects the ratio between the intensities of the two beams, i.e., reference and sample. The signals are then amplified and the spectrum of the sample is then recorded in a chart with the help of proper recording systems.

5.10 Drawback of Double Beam Dispersive Spectrometer

The double beam dispersive spectrometers have several drawbacks such as sensitivity, speed, and accuracy of wavelength. Most of the light is lost in focusing slits. Because of the slow speed of recording this instrument cannot be used for fast processes. Besides these, useful information about a molecule that may be contained in the infrared region is spectra below 400 cm^{-1} to about 20 cm^{-1}. This region is inaccessible by conventional instruments due to weak sources and insensitive detectors. All these difficulties are overcome in Fourier transform infrared spectroscopy which works on an entirely different principle based on the Michelson interferometer. So, this method is also known as interferometric infrared spectroscopy. The operation of the interferometer can be understood with the help of figure.

5.11 Fourier Transform Infrared Spectrometer

FTIR is based on the concept of interference of radiation between two beams to produce an interferogram. The interferogram is obtained as a signal because of the change in the path lengths between the two beams. The domains of distances and frequencies are then interconverted by a mathematical technique of Fourier transform [16].

5.11.1 Basic Principle and Operation of FTIR Fourier Transform Infrared Spectroscopy

The basic principle of this spectrometer is the interaction of IR frequencies to produce an interference pattern (interferogram) which is then analyzed mathematically with the help of the Fourier transform to obtain the individual frequencies and the intensities of the signals. The most common interferometer used for FTIR spectrometric analysis is known as the Michalis interferometer.

In this spectrometer, the radiation coming out from the infrared source is passed through an interferometer to reach the sample. After passing from the sample, it then reaches a detector, the signals are amplified and the high-frequency signals are eliminated with the help of a filter. The data is received in the recorder and digitalized with the help of an analog to digital converter and then transferred to the computer for Fourier transformation.

A schematic diagram of a typical dispersive spectrophotometer is shown in Figure 5.25. The transparent material of the beam splitter is coated in such a way that it is able to reflect 50% of the incident radiation to one of the mirrors M_1 while the remaining 50% of the radiation is transmitted to the other mirror M_2. Thus, half the beam of radiation going to the mirror M_1 and half to mirror M_2 are reflected from the mirror along the same path and returns to the beam splitter S. But in the beam splitter, the two beams recombine into a single beam and emerge from the interferometer at the right angles to the incident beam in form of a transmitted beam. The transmitter is pass through the sample and then to detector D.

The recombined transmitted beam leaving the splitter S may either constructive or destructive interference depending on the relative path length S to mirror M_1 and S to mirror M_2. This constructive interference will yield an intense beam when the path lengths are the same or differ by an integral number of path lengths. On the other hand, if the path differs by half integral of path length, two beams will interfere destructively at the beam splitter (B). The optical path difference is produced with the help of moving mirror M_2 and the detector is then able to detect the resultant interference pattern.

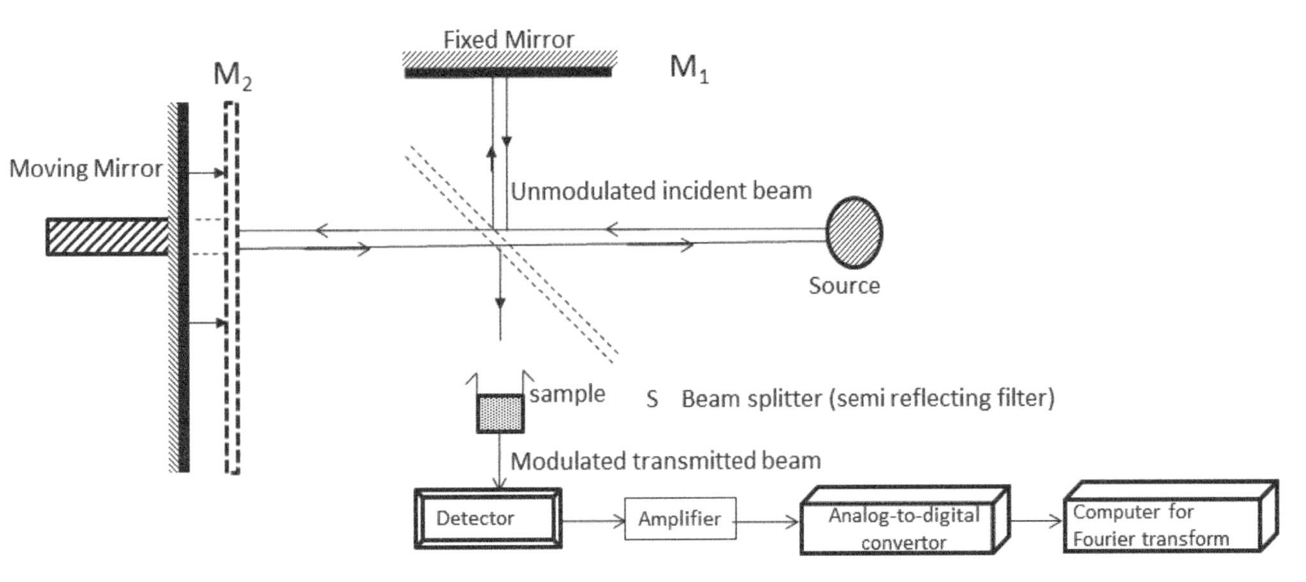

Figure 5.25 Schematic diagram of Fourier transform IR spectroscopy.

The moving mirror M2 plays an important role in this type of interferometer. The detector records alternate light and dark images as the mirror M_2 is moved slowly up and down (i.e. away or toward the beam splitter S). However, for a polychromatic light source, a complicated interferogram is obtained and the individual frequencies of light can be calculated with the help of the Fourier transform.

5.11.2 Advantages of FTIR

There are several important advantages of FTIR which uses an interferometer measurement that contains a dispersive grating monochromator. The advantages may be summarized as follows:

1. With the help of the FTIR instrument, the complete spectrum of the sample is obtained at once over the entire frequency range.
2. The slit-free optical design of the FTIR instrument is advantageous in the sense that the total output of the source can be passed through the sample continuously. This leads to a substantial increase in the energy of the radiation reaching the detector and hence a high signal and an improved signal-to-noise ratio are obtained. This advantage is also known as the Jacquinot advantage.
3. FTIR instruments have a strong and constant resolving power over the entire spectrum. In the dispersive instrument, the resolving power is especially poor at the end of the spectrum. In FTIR instruments the mirror positions and the frequency can be measured with great precision and accuracy. This is called Conne's advantage.
4. The total scanning time in the FTIR is very less than the time needed to obtain a dispersive spectrum of the same sensitivity or resolution. In the dispersive spectrometer, each scanning is done consecutively whereas in FTIR scanning of the entire frequency range is done continuously. As A result, the total scan time is reduced drastically and the signal-to-noise ratio is improved. This is called Fellgett's advantage.

References

1. Brown, J. M., *Molecular Spectroscopy*, 1st Ed. (Oxford University Press, New York), 1998.
2. Pavia, D. L., Lampman, G. M., Kriz, G. S. and Vyvyan, J. R. *Introduction to Spectroscopy*, 4th Ed. (Cengage Learning, Wasington DC), 2009.
3. Silverstein R. M. and Webster, F. X., *Spectrometric Identification of Organic Compounds*, 6th Ed. (Wiley, New York), 1997.
4. Calthup, N. B., Daly, L. H. and Wiberley, S. E., *Introduction to Infrared and Raman Spectroscopy* (Academic Press, New York and London), 1990.
5. Banwell, C. N. and McCash, E. M., *Fundamentals of Molecular Spectroscopy*, 4th Ed. (Tata McGraw-Hill Publishing Company, UK), 1994.
6. Kemp, W., *Organic Spectroscopy*, 3rd Ed. (Palgrave Publishers), 1991.
7. Reddy, K.V., *Symmetry and Spectroscopy of Molecules* (New Age International Publication, India), 2009.
8. Singh, N. B., Gajbhiye, N. S. and Das, S. S., *Comprehensive Physical Chemistry* (New age International, India), 2014.
9. Stuart, B., *Infrared Spectroscopy: Fundamentals and Applications* (John Wiley).
10. Lambert, J. B., Gronert, S., Shurvell, H. F., Lightner, D. and Cooks, R. G., *Organic Structure Spectroscopy*, 2nd Ed. (Pearson Prentice Hall), 2013.
11. Harwood, L. M. and Claridge, T. D. W., *Introduction to Organic Spectroscopy* (Oxford University Press), 1996.
12. Gunzler, H. and Gremlich, H.-U., *IR Spectroscopy: An Introduction* (Wiley-VCH, Weinheim, Germany), 2002.
13. Colthup, N. B., Daly, L. H. and Wiberly, S. E., *Introduction to Infrared and Raman Spectroscopy*, 3rd Ed. (Academic Press, New York and London), 1990.
14. Morse, P. M. Diatomic molecules according to the wave mechanics. II. Vibrational levels. *Phys. Rev.*, 34, 57–64, 1929.
15. Dyer, J. R., *Application of Absorption Spectroscopy of Organic Compounds*, 1st Indian Ed. (Prentice Hall, India), 1984.
16. Rabek, J. F., *Experimental Methods in Polymer Chemistry* (John Wiley, Bristol), 1980.
17. Griffiths, P. R., Haseth de, J. A., *Fourier Transform Infrared Spectrometry* (Wiley, New York), 1986.

Chapter 6

Microwave Spectroscopy

Preeti Gupta,[a] **S. S. Das,**[b] **and N. B. Singh**[c]
[a]*Department of Chemistry, DDU Gorakhpur University, Gorakhpur, India*
[b]*Department of Chemistry (ex-professor), DDU Gorakhpur University, Gorakhpur, India*
[c]*Department of Chemistry and Biochemistry, SBSR and RDC, Sharda University, Greater Noida, India*
ssdas3@gmail.com

6.1 Introduction

Microwave spectroscopy is the study of the pure rotational motions of molecules. It is concerned with transitions between molecule's various rotational energy levels. And if the molecule possesses a permanent dipole moment, a rotational spectrum is obtained. This very useful technique may provide accurate values of rotational constants and molecular parameters such as dipole moment, nuclear spin, etc. [1]. However, this technique has some limitations because of the following two reasons: (i) the first limitation is that the molecule must possess a permanent dipole moment in

Spectroscopy
Edited by Preeti Gupta, S. S. Das, and N. B. Singh

ISBN 978-981-4968-32-4 (Hardcover), 978-1-003-41258-8 (eBook)
www.jennystanford.com

order to become microwave active. An electric field in the rotating molecule is thus generated which then interacts with the electrical components of microwave radiations and rotational transitions are obtained. (ii) The second reason is that the sample should be in the gas phase. Because in the liquid and solid phases, the rotations are hindered due to strong intermolecular interactions. In the gas phase, the rotating molecules possess sufficient rotational energies to go from the lower rotational levels to the higher ones. (iii) Another limitation is that the rotational spectroscopy must be carried out at very low pressures (0.1 to 10 nm^{-2}) to obtain better resolutions and absorption intensities. At higher pressures, both in gas and liquid phases, the frequency of collision in the molecule is very high, and very broad spectral lines are obtained in which the lines for individual transitions are difficult to identify. Thus, the desired structural information of the molecule under investigation may not be obtained. Since the molecules are fixed at their lattices in the solid phase, therefore, rotational motion is not possible and no structural information can be obtained [2–4].

For microwave spectroscopic investigations, the molecules may be classified into five groups on the basis of their shapes and moments of inertia around three principal axes of rotation perpendicular to each other. A molecule generally possesses three moments of inertia, i.e., I_A, I_B, and I_C according to the three axes of rotations as shown in Figure 6.1.

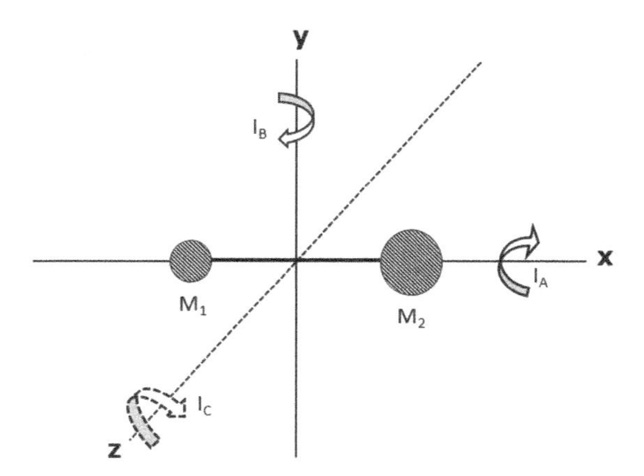

Figure 6.1 Linear molecule as rotor; I_A, I_B, and I_c are the moments of inertia due to rotations about the *x*-axis, *y*-axis, and *z*-axis, respectively.

1. Linear molecules as rotors: When all the atoms are arranged in a straight line such as H–Cl, carbon oxysulphide (O=C=S), acetylene, and O=C=O, etc., the moment of inertia $I_A = 0$ and $I_B = I_C$.

2. Spherical top molecules as rotors: In such molecules, all the three moments of inertia are equal, i.e., $I_A = I_B = I_C$. Examples are CH_4, SiH_4, SF_6, etc.

3. Symmetrical top molecules as rotors: These molecules have two moments of inertia that are equal to each other and one is different, i.e., $I_A = I_B \neq I_C$. For example, NH_3, CH_3CN, CH_3F, CH_3Cl, etc, are symmetrical top ones.

4. Planar and symmetrical top molecules as rotors: In this type, the moment of inertia $I_A = 2I_B = 2I_C$. For example, BCl_3 is one suitable example of this category.

5. Asymmetric top molecules as rotors: These molecules have three different moments of inertia, $I_A \neq I_B \neq I_C$. Water, CH_3OH, vinyl chloride, formaldehyde, etc. belong to this type.

6.2 Rotational Spectra of Linear Diatomic Molecule

As mentioned in Chapter 7 (Section 7.3), rotational energy and all forms of molecular energy are quantized. The permitted rotational energy values can be calculated by solving the Schrodinger wave equation for the system represented by that molecule.

A linear diatomic molecule such as HCl contains a hydrogen atom with a net positive charge and the chlorine atom has a net negative charge. Considering the rotation of the HCl molecule having a permanent electrical dipole moment, it is found that these positive and negative charges change periodically with time. As such the component dipole moment in a given direction fluctuates regularly. When radiation falls on the rotating HCl molecule, an interaction takes place, and energy may either be absorbed or emitted and a spectrum is thus obtained purely due to rotation.

6.2.1 The Rigid Linear Diatomic Molecule

In a diatomic molecule, when the bond distance between the atoms is not changed during rotation, then the molecule is known as a rigid rotor. In such molecules, the vibrational motions do not take place during rotation.

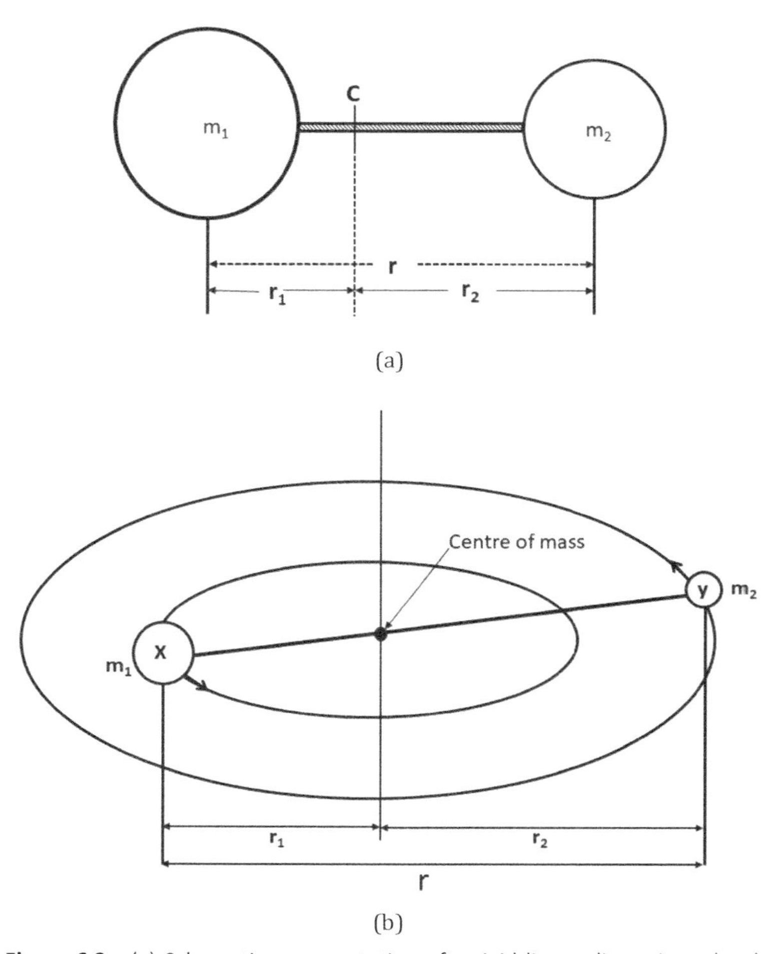

Figure 6.2 (a) Schematic representation of a rigid linear diatomic molecule having two atoms of masses m_1 and m_2 and bond length $r = r_1 + r_2$. (b) Schematic representation of rotation of a rigid diatomic molecule, when such a molecule rotates, it rotates about an axis that passes through its center of mass shown in (a) by a point C.

Such a molecule rotates about an axis that passes through its center of mass, C as shown in Figure 6.2. The two masses m_1 and m_2 are joined together by a rigid weightless bond whose length is given by:

$$r = r_1 + r_2. \tag{6.1}$$

The molecule rotates end-over-end about the center of mass C, satisfying the condition

$$m_1 r_1 = m_2 r_2 \tag{6.2}$$

Also, the moment of inertia at the point of the center of mass is given by

$$I = m_1 r_1^2 + m_2 r_2^2 \tag{6.3}$$

The values of the moment of inertia I in terms of the atomic masses m_1 and m_2 and the bond length r are given by

$$I = \frac{m_1 m_2}{m_1 + m_2} r^2 = \mu r^2 \tag{6.4}$$

where μ is known as the reduced mass of the system and can be obtained by the following equation:

$$\mu = \frac{m_1 m_2}{m_1 + m_2} \tag{6.5}$$

Thus, the rotational kinetic energy E of this system in terms of the moment of inertia I and angular velocity ω is written as

$$E = \frac{1}{2} I \omega^2 \tag{6.6}$$

This expression for E [Eq. (6.8)] may be expressed in terms of angular momentum as

$$E = \frac{1}{2} \frac{L^2}{I} \tag{6.7}$$

where L is defined in terms of rotational quantum numbers J as

$$L = \sqrt{J(J+1)} \cdot \frac{h}{2\pi} \tag{6.8}$$

where $J = 0, 1, 2, 3, \ldots$

With the help of the Schrodinger wave equation, the allowed rotational energies E_J for such a rigid diatomic rotating molecule is given by

$$E_J = \frac{h^2}{8\pi^2 I} J(J+1) \text{ (in joules)} \tag{6.9}$$

where J is the rotational quantum number and can have values $J = 0$, 1, 2, 3, ... is Planck's constant, and I is the moment of inertia of the diatomic molecule.

Eq. (6.9) provides allowed energies of various rotational levels in joules. It is desired to know the differences between these energies in the units of wave number $(\bar{\upsilon} = \Delta E / hc)$ cm^{-1} of the radiation emitted or absorbed. Normally spectra are described in terms of wave number, $\bar{\upsilon}$, the rotational energy is the equation in terms of wave number can be written with the help of relation $c = \bar{\upsilon}\lambda = \frac{\upsilon}{\bar{\upsilon}}$ as

$$\bar{\upsilon} \text{ or } \varepsilon_J = \frac{E_J}{hc} = \frac{h}{8\pi^2 IC} J(J+1)\,\text{cm}^{-1} \quad (6.10)$$

in which the energies of different energy levels are expressed in terms of wave number $\bar{\upsilon}$ and c is the velocity of light expressed in cmS^{-1}units.

The equation (6.7) may further be simplified as

$$\varepsilon_J = BJ(J + 1) \text{ (where } J = 0, 1, 2 \ldots) \quad (6.11)$$

where B denotes the rotational constant and is given by

$$B = \frac{h}{8\pi^2 IC}\,\text{cm}^{-1} \quad (6.12)$$

6.2.2 Description of Rotational Spectra of Rigid Rotating Diatomic Molecule

It has been shown in Eqs. (6.9) and (6.10) that in terms of frequency, the allowed rotational energies are given by

$$E_J = \frac{h^2}{8\pi^2 I} J(J+1)\,(\text{in joules})$$

and in the units of wave number, $\bar{\upsilon}$ as

$$\bar{\upsilon} \text{ or } \varepsilon_{J\upsilon} \text{ or } \varepsilon_J = \frac{E_J}{hc} = \frac{h}{8\pi^2 IC} J(J+1)\,\text{cm}^{-1}$$

Therefore, when $J = 1$, $\varepsilon_J = 2B\ cm^{-1}$, this represents the lowest angular momentum of the rotating molecule.

$J = 2$, $\varepsilon_J = 6B\ cm^{-1}$, $J = 3$, $\varepsilon_J = 8B\ cm^{-1}$, and so on.

In Eq. (6.11), when $J = 0$, then $\varepsilon_J = 0$, which means that the molecule is not rotating. When $J = 1$, then the value of rotational energy ε_J = 2B. This represents the lowest angular momentum of the rotating molecule. The values of ε_J can then be calculated for increasing values of J. The corresponding allowed energy levels can be shown in Figure 6.3.

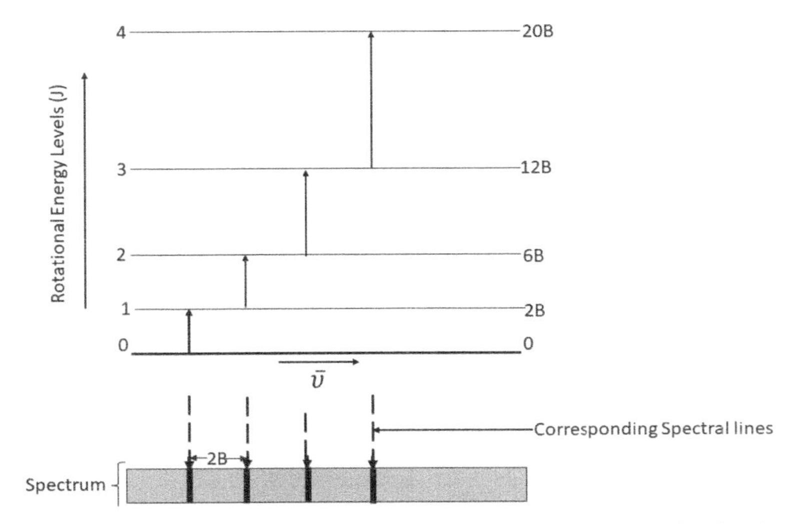

Figure 6.3 The allowed transitions between different rotational levels. The spectral lines arising from these transitions are also shown in this figure.

However, the difference between the energy levels provides useful information about the frequency of the wave number of the radiation which is absorbed or emitted when a molecule goes from one level to another. Thus, when the molecule absorbs energy in the initial rotational level J and goes to the final rotational level J′, then the difference of energy, ΔE, between two rotational levels is given by using Eq. (6.13) as:

$$\Delta E = E_{J'} - E_J = \frac{h}{8\pi^2 I} J'(J'+1) - J(J+1) \text{ (in joules)} \qquad (6.13)$$

But the selection rule suggests that only those transitions are allowed to occur for which $\Delta J = \pm 1$. This, in case of absorption of radiation

$$J' = J + 1$$

On putting the value of J' in Eq. (6.13), one can write

$$\Delta E = \upsilon = \frac{h^2}{8\pi^2 I}[2(J+1)] \quad \text{(in joules)} \tag{6.14}$$

Since $$B = \frac{h^2}{8\pi^2 IC},$$

therefore, one can write

$$\Delta E = \upsilon_{J-J'} = 2BC(J+1)] \quad \text{(in joules)} \tag{6.15}$$

and the change in energy in terms of wave number (cm^{-1}) is given by

$$\varepsilon_J = \bar{\upsilon}_{J-J'} = \frac{\upsilon_{J-J'}}{c} = 2B(J+1)\ \text{cm}^{-1} \tag{6.16}$$

6.3 Various Transitions in the Rotational Spectrum

This $\bar{\upsilon}_{J-J'}$ is the wave number of the spectral line obtained as a result of the transition between two rotational levels of the molecules. Therefore,

(i) When transition from $J = 0$ to $J' = 1$ takes place then

$$\bar{\upsilon}_{J=0\to J'=1} = 2B$$

(ii) When the transition from $J = 1$ to $J' = 2$ takes place, then

$$\bar{\upsilon}_{J=1\to J'=2} = 4B$$

(iii) When the transition from $J = 2$ to $J' = 3$ takes place, then

$$\bar{\upsilon}_{J=2\to J'=3} = 6B$$

and so on.

These suggest that the difference between the two rotational energy levels $J = 0$ to $J' = 1$, $J = 2$, and $J' = 3$, etc. will be $2B$, $4B$, $6B$, $8B$, $10B$, and so on and these transitions will produce an absorption spectrum whose lines will appear at $2B$, $4B$, $6B$, $8B$ and so on (in cm^{-1}) (Figure 6.3).

6.4 The Selection Rule

For a rigid rotating diatomic molecule, the selection rule is

$$\Delta J = \pm 1 \tag{6.17}$$

The selection rule for rotational spectroscopy suggests which rotational transitions are allowed and which are forbidden. The overall selection rule for rotational spectroscopy points out that the molecule under investigation must possess a permanent dipole moment (only polar molecule) in order to yield a rotational spectrum. Further, the selection rule as obtained with the help of quantum mechanical treatment suggested that the possible change in the angular quantum momentum J is restricted to only ±1 unit. This means that the transition

$J = 0 \rightarrow J = 1, J = 1 \rightarrow J = 2, J = 2 \rightarrow J = 3$, or vice versa and so on are only allowed. On the other hand, transition from $J = 0 \rightarrow J = 2, J = 1 \rightarrow J = 3$, or $J = 2 \rightarrow J = 4$, etc. are not allowed.

The whole rotational spectrum is given by equation (6.16) only when the molecule is asymmetric (hetero nucleus). Thus, the molecules like HCl, CO, etc. show a rotational spectrum while the (homonuclear) symmetries molecules such as N_2, H_2, O_2, Cl_2, etc. do not give any spectrum. The intensity of the rotational lines is governed by the population of the initial or ground state. The transitions for which $\Delta J = +1$ corresponds to the absorption of radiation while when $\Delta J = -1$, it suggests emission of radiation.

6.5 Intensities of Rotational Spectral Lines

After obtaining the positions of lines in the microwave spectrum, it is desirable to find out which spectral lines are more intense. The intensity depends upon the population of molecules in the initial states. Strong and intense spectral lines will be obtained when the initial state populations of molecules are greater.

The relative intensities of the spectral lines obtained by the equation

$$\bar{\upsilon}_{J \rightarrow J+1} = 2B(J+1) \text{ cm}^{-1} \quad (6.18)$$

depends upon the relative probabilities of transition between various energy levels. The probability for all the transitions for which $\Delta J = \pm 1$. This suggests that all spectral lines should have equal intensities. But this is not true, the intensities of the spectral lies are not equal. This inequality in the intensities of the spectral lines may be explained in the following manner. The total number of molecules initially present in the $J = 0$ state will be different from the molecule present in $J = 1$ or $J = 2$ states. For example, in a normal gaseous sample, in the beginning, there will be a different number of molecules at each level. As such, the different total number of molecules will participate in the transition between different levels, and hence the intensities of the spectral lines will therefore be different. This difference is directly proportional to the initial number of molecules present at each level. The larger the number of molecules in that particular state, the greater will be the probabilities of their transitions to the next higher levels. Thus, the intensity of the spectral lines will be large.

The population of molecules depends upon two factors, i.e., the Boltzmann distribution and the degeneracy of the state which governs the population of the energy levels. Besides this, the temperature also plays an important role. For a large number of molecules at thermal equilibrium, the Boltzmann expression as mentioned below gives the ratio of a number of molecules n' in the upper-level J', to that n in the lower state J.

$$\frac{n'}{n} = \frac{g'}{g} e^{-\frac{E}{kT}} \tag{6.19}$$

where g' and g are degeneracies of the energy states.
If N_J and N_0 are the numbers in the rotational energy levels J and the ground energy levels, respectively, then their ratio is given by

$$\frac{N_J}{N_0} = \frac{g_J}{g_0} = e^{-E_J/kT} = g_J e^{-E_J/kT} \quad (\text{since } g_0 = 1) \tag{6.20}$$

Further, the rotational energy levels are degenerate. The degeneracy of rotational level J is given by

$$g_J = (2J + 1) \tag{6.21}$$

(degeneracy means that for a given value of J, there are $(2J + 1)$ energy levels that have the same energy). Thus

$$\frac{N_J}{N_0} = (2J+1)e^{-E_J/kT} \tag{6.22}$$

On putting the value of $E_J = \frac{h^2}{8\pi^2 I} J(J+1)$ in Eq. (6.22), we get

$$\frac{N_J}{N_0} = (2J+1)_J \exp\left[-\frac{h^2}{8\pi^2 I} J(J+1)\right] \quad (6.23)$$

Further, since $B = \frac{h^2}{8\pi^2 I}$ cm^{-1}, Eq. (6.23) may be simplified as

$$\frac{N_J}{N_0} = (2J+1)\exp\left[-Bhc\ J(J+1)/kT\right] \quad (6.24)$$

Eq. (6.24) indicates that (i) as the value of *J* increases, the number of degenerate levels (2*J* + 1) increases rapidly, and (ii) the molecular population in each level decreases exponentially with the increasing value of *J*. Therefore, if N_J/N_0 is plotted against *J*, a curve as shown in Figure 6.4 is obtained.

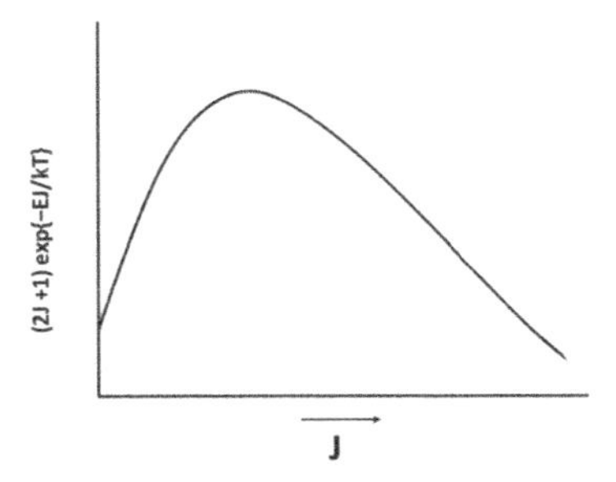

Figure 6.4 Plot between relative populations of rotational energy levels against *J*.

Thus, the relative population in the *J* = 1 state can be calculated from Eq. (6.24) by taking a typical value of B = 2 cm^{-1} and the temperature T = 300 K

$$\frac{N_J}{N_0} = \exp\left\{\frac{2\times 6.63\times 10^{-34}\times 3\times 10^{10}\times 1(2)}{1.38\times 10^{-23}\times 300}\right\} = \exp\{-0.019\} \approx 0.98$$

This calculation suggests that the equilibrium of the number of molecules in the *J* = 1 state is almost equal to the number of molecules in the *J* = 0 state. Thus, in view of the above expression, one may write the total relative population at the energy E_J as

$$\text{Population} \qquad \propto (2J+1)e^{E_J/kT} \qquad (6.25)$$

and it may also be inferred that although the molecular population in each level decreases exponentially (Eq. 6.32), the number of available degenerate levels increases rapidly with J.

A plot of $(2J + 1)\exp\{-E_J/kT\}$ versus J has been shown in Figure 6.4. The curve indicates that the population rises to a maximum and then decreases.

The maximum value of the population can be obtained by differentiating Eq. (6.24) and then setting it equal to zero. When done so, the nearest integral value of J corresponding to maximum population and therefore, the bond with maximum intensity is given by

$$J_{\max} = \sqrt{\frac{kT}{2hcB}} - \frac{1}{2} \qquad (6.26)$$

Thus, it may be concluded that since the line intensities are directly proportional to the populations of the rotational levels, the transitions between levels having very low or very high J values will have small intensities whereas a bond with maximum intensity is observed at or near the J value given by Eq. (6.26).

6.6 Isotopic Substitution

In a molecule, when a particular atom is placed by its isotope, the substitute molecule chemically remains the same but its total mass becomes different. Because of this, there will be change. When the moment of inertia increases, the value of B tends to decrease. Further, the value of B is related to the spacing between the spectral lines in the rotational spectrum. Lower is the value of B smaller is the separation between the rotational energy levels and as such in the spectral lines.

For example, in a CO molecule, if ^{12}C is replaced by ^{13}C, the mass of $^{13}C\,^{16}O$ will be more than the $^{12}C\,^{16}O$ molecule. Therefore, $^{13}C\,^{16}O$ will have a high value of I and a low value of B (the new values of B are represented by B'). Thus, the spacing between the spectral lines (rotational spectra) of $^{13}C\,^{16}O$ molecules will be less than those of $^{12}C\,^{16}O$ (Figure 6.5).

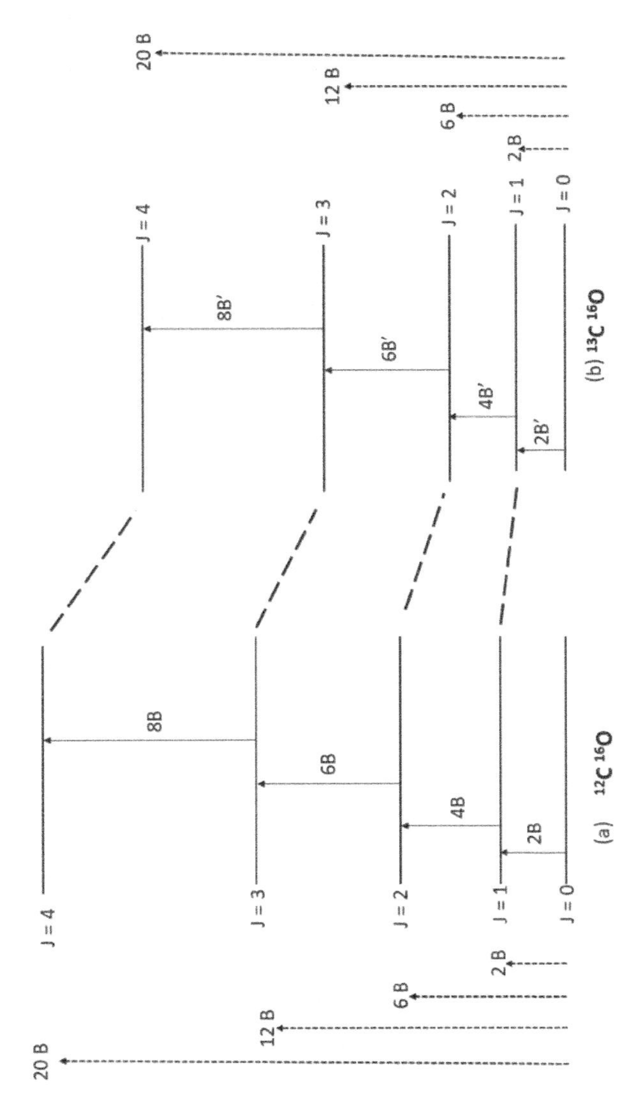

Figure 6.5 Decrease in rotational energy levels spacings by the substitution of ^{12}C by ^{13}C.

6.7 Techniques and Instrumentation

The conventional microwave spectrometer essentially contains a light source and monochromator, waveguide (beam direction), sample and sample space, detector, and spectrum recorder. A schematic diagram of the microwave spectrometer is shown in Figure 6.6 and a brief description is given below.

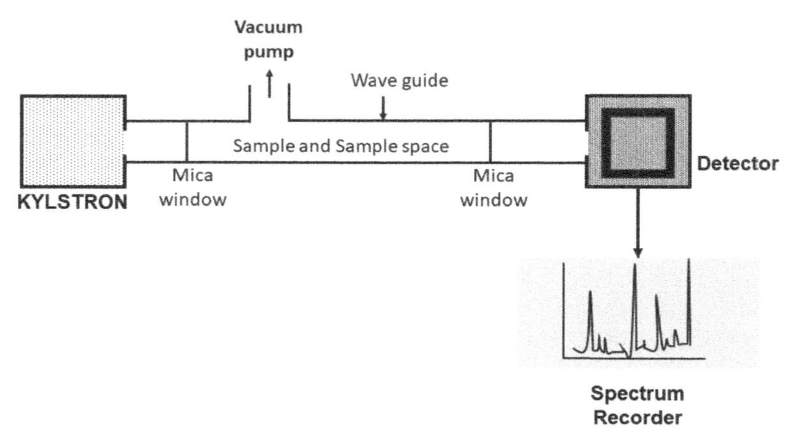

Figure 6.6 Schematic diagram of microwave spectrometer.

6.7.1 The Source and Monochromator

The main source for obtaining the radiations in the microwave region is the reflex Klystron oscillator. The main function of using a reflex Klystron oscillator is that it emits radiation of a very narrow frequency range so that it acts as its monochromator.

6.7.2 Waveguide

Since the radiation emitted by the Klystron source are difficult to handle with mirrors and lenses, therefore, hollow tubes of copper or silver having rectangular cross-section are used as waveguides. Inside this hollow tube, the radiation is confined and then the tube is bent or tapered in such a way that the direction of the radiation beam can be properly focused. The waveguides are evacuated to avoid absorption of radiation due to the presence of air in the tube.

6.7.3 Sample and Sample Space

Generally, gaseous samples are used. The sample is placed in a small portion of the evacuated waveguide which is closed at both the ends with help of thin mica windows. Small round holes are done in the tube for evacuating the tube and introducing the gaseous sample. The pressure nearly equal to 0.01 mm/Hg is required to yield the absorption spectrum.

6.7.4 Detector

A superheterodyne radio receiver is used as a detector which may be tuned to the appropriate high frequency. Also, a quartz crystal can be used as a detector which is more sensitive and easier to use. This quartz crystal is mounted on a tungsten whisker cartridge.

6.7.5 Spectrum Recorder

The signals coming out of the detector in then amplified with the help of a spectrum recorder which is generally either a cathode ray oscilloscope/oscillograph or a pen- and ink-recorder. The emitted signals produce an electrical signal which gets amplified here and then displayed as a spectrum on the oscilloscope screen or it can be recorded on chart paper with the help of a recorder.

6.8 Applications of Microwave Spectroscopy

- Microwave Spectroscopy is a specific and highly sensitive analytical tool for the analysis of gaseous samples. The microwave spectrum of a substance is very rich in lines and these lines are very sharp. Therefore, the positions of the lines can be measured with great accuracy and may be compared with the tabulated data of previously examined substances in a sample. Thus, the presence of a substance can simply be established by a comparison of the data. Further, since the intensity of a spectrum is directly dependent on the amount of a substance present in the sample, therefore the mixtures can be easily analyzed by microwave technique.

- This technique is unable to detect the presence of particular molecular grouping in a sample like –OH or –CH_3, etc., but it can distinguish the presence of isotopes in a sample.
- It can also detect different conformational isomers in a sample provided they have a different moment of inertia.
- One of the most important areas in which microwave analysis is being used is in the chemical examination of interstellar space.
- Microwave spectroscopy has successfully been used to detect simple stable molecules in space like water, ammonia, and formaldehyde.
- Microwave spectroscopy has been used to monitor and control various industrial processes.
- A microwave oven is a kitchen appliance that also employs microwave radiation primarily to cook or heat food.

References

1. Drago, R. S., *Physical Method in Inorganic Chemistry* (Reinhold Publishing), 1965.
2. Brown, J. M., *Molecular Spectroscopy* (Oxford University Press, Oxford), 2003.
3. Banwell, C. N. and McCash, E. M., *Fundamentals of Molecular Spectroscopy*, 4th Ed. (Tata McGraw-Hill, UK), 1994.
4. Singh, N. B., Gajbhiye, N. S. and Das, S. S., *Comprehensive Physical Chemistry* (New age International, India), 2014.

Chapter 7

Raman Spectroscopy

Preeti Gupta,[a] S. S. Das,[b] and N. B. Singh[c]
[a]*Department of Chemistry, DDU Gorakhpur University, Gorakhpur, India*
[b]*Department of Chemistry (ex-professor), DDU Gorakhpur University, Gorakhpur, India*
[c]*Department of Chemistry and Biochemistry, SBSR and RDC, Sharda University, Greater Noida, India*
n.b.singh@sharda.ac.in

7.1 Introduction

Raman spectroscopy is an important technique that is mostly used in the field of chemistry and material science for the identification of the molecules. It also gives information about vibrational transitions in a molecule. Therefore, Raman spectroscopy is now being used as one of the most popular tools to study the vibrational structures of molecules in association with infrared (IR) spectra. With the discovery of the Raman effect, the molecular structures can be determined. The spectra of various molecules obtained through

Spectroscopy
Edited by Preeti Gupta, S. S. Das, and N. B. Singh

ISBN 978-981-4968-32-4 (Hardcover), 978-1-003-41258-8 (eBook)
www.jennystanford.com

this spectroscopy can be used as fingerprints in the structural determination of unknown compounds. The Indian physicist C. V. Raman and his coworker K. S. Krishnan discovered this phenomenon in 1928. Although this effect was predicted earlier in the year 1923 by Prof. Smekel, it was experimentally observed by Raman [1–4]. Raman spectroscopy differs from vibrational and rotational spectroscopy in the sense that it takes into account the scattering of the radiations by the sample instead of absorption processes.

What is Raman effect? While performing some experiments on the scattering of lights by different liquids, Raman and Krishnan observed that when a beam of monochromatic light of a particular frequency interacts with the molecules of a clear dust-free transparent liquid, a small portion of the incident beam (0.1% to 1.0%) is scattered. This scattering lasts for a very brief period of time (10^{-14} s or even less) and during this period, the molecule goes into a higher excited energy level known as 'the virtual state' (virtual state is that whose energy cannot be precisely defined because it is controlled by the uncertainty relation, $\Delta E \cdot \Delta t \geq \frac{h}{2}$) and then returns to the lower state after the scattering of light is over. The energy of most of the scattered radiation has almost the same frequency as that of the incident radiation. It means that the scattering of light occurs without a change in the wavelength. This process is known as **Rayleigh scattering**. The Rayleigh lines are elastically scattered since there is no change of energy. But certain weak radiations of different wavelengths or frequencies were also visualized by them above and below that of the scattered radiation. This occurs because, during the scattering process, a molecule will also experience a change in vibrational and rotational levels. Due to this change, there will be a difference in the energy involved in the vibrational transitions. Therefore, two types of radiations with small different wavelengths are observed because of this effect. The new phenomenon of these scattered radiations of changed frequencies that occurs along with the Rayleigh scattering is referred to as **Raman scattering**. This occurs as a result of the inelastic scattering of incident photons. The process of Raman scattering has been explained schematically in Figure 7.1 [4].

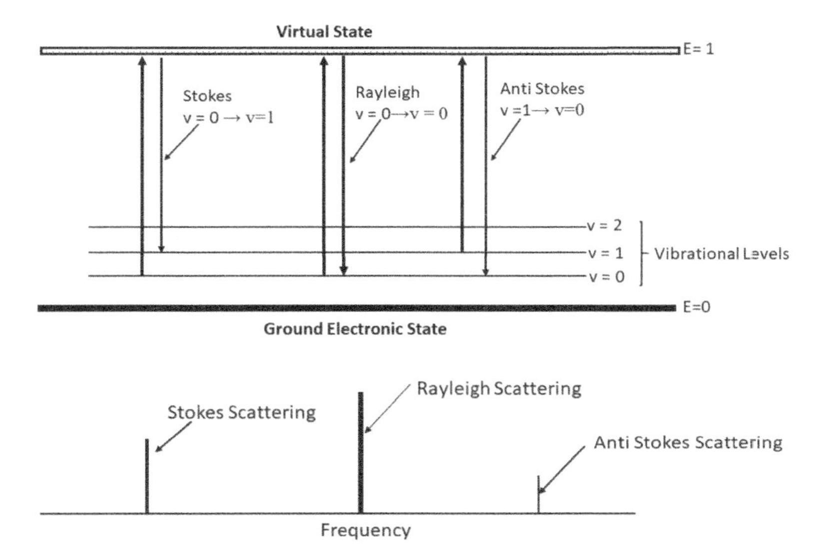

Figure 7.1 Schematic representations of Rayleigh and Raman Scattering showing equally spaced Stokes and anti-Stokes lines.

Raman scattering occurs when the molecule exchanges energy with an incident photon and the molecule then subsequently returns to one of the vibrational energy levels either above or below the initial level. The frequency shift corresponding to the difference of energy between the incident radiation and the scattered radiation is known as **the Raman shift**.

7.2 Quantum Mechanical Concept of Raman Effect

The observed Raman lines of different frequencies can easily be explained with the help of the quantum mechanical theory of radiation. The radiation of frequency υ will consist of photons (streams of particles) having energy $h\upsilon$ where h is Planck's constant. These photons will collide with the molecule if put in their way. If the collision is perfectly *elastic* (i.e., the photons are deflected unchanged) and the detector is placed at right angles to the incident beam, then it will receive the scattered (Rayleigh lines) photons having the same energy hυ and the same frequency 'υ.' But if the collision is *inelastic* there is an exchange of energy between the photon and molecule during a collision. The molecule may lose or gain certain amounts of energy as per quantum rules. If this energy change is ΔE, then it must be equal to the difference in energy between two of its allowed state levels, i.e., ΔE must represent a change in vibrational or rotational energy levels of the molecule. If the molecule after collision gains an energy ΔE, then the scattered photon will have lesser energy ($h\upsilon$ – ΔE) and the frequency of the scattered radiation equal to ($h\upsilon$ – $\Delta E/h$) will be less. On the other hand, if the molecules lose an energy ΔE, after collision then the energy of the photon will be greater ($h\upsilon$ + ΔE) and the frequency of the scattered radiation will be equal to (υ + $\Delta E/h$). When the scattered radiation has a lower frequency than that of the incident beam then it is called **Stokes Raman radiation** and this will appear as Stokes lines in the spectrum. While the higher frequency radiation appearing as anti-Stokes lines in the spectrum are known as **anti-Stokes Raman radiation**. Stoke radiations are generally more intense than anti-Stokes as they are accompanied by an increase in molecular energy while the anti-Stokes involve a decrease. It should be noted that the total scattered radiation frequency is extremely small and a very sensitive instrument is required to study the Raman effect [3, 4].

If υ' is the frequency of the scattered radiation, then one can write

$$v' = \pm\frac{\Delta E}{h} v' = \frac{\Delta E}{h} \tag{7.1}$$

This suggests that in Raman scattering, the energies of the incident and scattered radiations are different. The scattered radiation has lesser energy than the incident radiations for the Stokes lines whereas anti-Stokes lines possess higher energy than the incident radiation. In Figure 7.1b, it has been shown that the Stokes and anti-Stokes lines arise equally separated from the Rayleigh line. In Eq. (7.1), the value of the frequency v' for Rayleigh, Stokes, and anti-Stokes lines can be written as

For Rayleigh lines, $$\upsilon' = \upsilon \tag{7.2}$$

For Stokes lines, $$\upsilon' = \upsilon - \frac{\Delta E}{h} \tag{7.3}$$

For anti-Stokes lines, $$\upsilon' = \upsilon + \frac{\Delta E}{h} \quad (7.4)$$

Since the change in energy, ΔE, occurs as per quantum rules, therefore, one can write

$$\Delta E = \pm\, nh\upsilon_{\mathrm{m}} \quad (7.5)$$

where n = 1, 2, 3, etc., and υ_{m} is the frequency of the molecule due to vibrational or rotational oscillations. When the value of n = 1, Eq. (7.5) becomes

$$\Delta E = \pm\, h\upsilon_{\mathrm{m}} \quad (7.6)$$

Therefore, Eq. (7.1) can also be written as

$$\upsilon' = \upsilon \pm \upsilon_{\mathrm{m}} \quad (7.7)$$

This equation clearly indicates that the difference between frequencies of the incident and scattered radiation ($\upsilon' - \upsilon$) is equal to a characteristic frequency of υ_{m} of the molecule. Eq. (7.7) further signifies and confirms that Stokes and anti-Stokes lines (shown in Figure 7.1) are situated at equal distances on both sides of the Rayleigh lines.

7.3 Molecular Polarizability and the Raman Effect

The scattering of radiation by a molecule can be explained on the basis of the electrical nature of the matter. In Raman spectroscopy, the polarizability (α) of the molecule must change during the course of vibration. Therefore, only those molecular motions are Raman active for which the electrical polarizability of the molecule changes during the motion. Due to this, an induced electrical dipole moment is set up in the molecule whose magnitude μ_{ind} is defined by

$$\mu_{\mathrm{ind}} = \alpha E \quad (7.8)$$

where E is the magnitude of the applied electric field and α is the polarizability of the molecule.

The polarizability (α) is actually the ability of the deformation of the electron cloud of the molecule by the applied electric field. The molecular vibration will be Raman active only when it is accompanied by a change in the polarizability of the molecule under investigation.

Now, when a beam of radiation of frequency υ falls on such molecules, then the induced dipole also oscillates with a frequency υ. This oscillating dipole emits radiation of its oscillating frequency υ which provides an explanation for Rayleigh scattering. In addition to this, if a molecule also undergoes some internal motion like vibrational or rotational oscillations, it will also be superimposed on the oscillating dipole. For example, if υ_{vib} is the vibrational frequency that changes the polarizability then the frequency of the oscillating dipole will have two components either $\upsilon + \upsilon_{vib}$ or $\upsilon - \upsilon_{vib}$ along with the exciting frequency υ. The Raman radiations are then observed. However, if the vibration or rotation does not change the polarizability of the molecule then the dipole will oscillate only at the frequency of the incident radiation (υ) and no Raman lines will appear.

In all the homonuclear diatomic molecules such as H_2, N_2, O_2, etc., there is only a vibrational mode. These molecules do not exhibit IR absorption bands and are IR inactive because there is no change in the dipole moment during vibrational motions. However, there occurs a change in polarizability during vibrational motion as such, these types of molecules may be Raman active.

It has also been seen that the magnitude of polarizability depends upon the direction in which the external magnetic field has been applied. In the H_2 molecule, when the electric field is applied parallel to the H–H bond, the observed polarizability is high (Figure 7.2a) but when the electric field is applied perpendicular to the H-H bond, the value of polarizability is low (Figure 7.2b) [3].

However, polyatomic molecules involve vibrations of more than one bond. Therefore, in such molecules, the total molecular polarizability can be evaluated with the help of the polarizability of individual bonds of the molecule. Therefore, it is the sum of all the polarizability of the bonds that will decide the Raman activity of the molecule, for example, polyatomic linear molecule CO_2 that possesses a center of symmetry and four fundamental modes [$(3n - 5) = 4$] of vibrations. As explained in the figure (Section 7.6 in Chapter 7), the symmetric stretching vibrational mode (υ_1) will show a change in polarizability, thus it is Raman active. On the other hand, in the case of asymmetric vibrational mode (υ_3), the two bands stretch in different directions; thus, the polarizability is zero and this mode becomes Raman inactive. The two bending modes (υ_3 and υ_4)

are degenerate and they are also Raman inactive due to no change in polarizability.

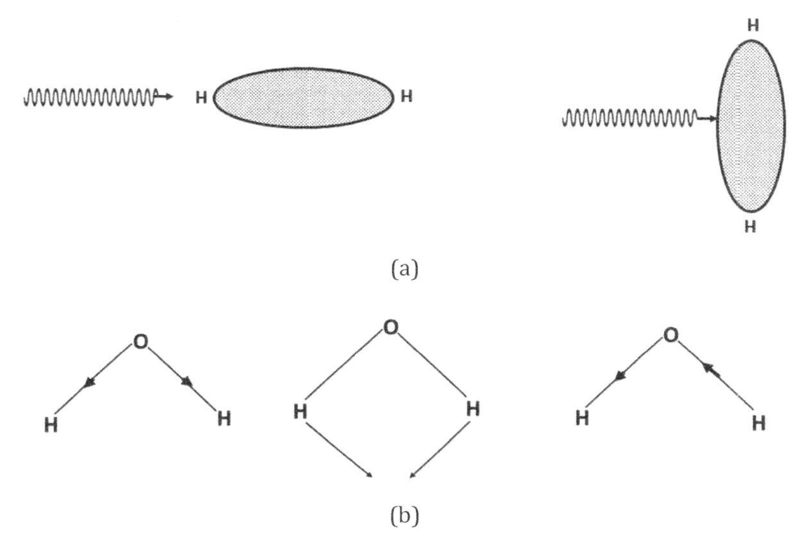

Figure 7.2 (a) Diagram showing polarizabilities in the homonuclear molecule. (b) Raman active vibrations in H_2O molecules.

7.4 Rotational Raman Spectra

The selection rule for rotational Raman spectra is written as

$$\Delta J = 0, \pm 2\Delta J \tag{7.9}$$

This rule is different from that for pure rotational changes in IR spectra, i.e., $\Delta J = \pm 1$. In Raman spectroscopy, when $\Delta J = 0$, Rayleigh scattering is observed. When $\Delta J = +2$, the Raman shift is found in the form of Stokes Raman lines, whereas $\Delta J = -2$ corresponds to anti-Stokes Raman lines.

The rotational energy levels of linear molecules for this spectroscopy can be expressed as

$$E_J = h\upsilon = \frac{h^2}{8\pi^2 I} J(J+1) \text{ (joules)} \tag{7.10}$$

or

$$\upsilon = \frac{h^2}{8\pi^2 I} J\left(J+1\right) \text{ (cm)} \tag{7.11}$$

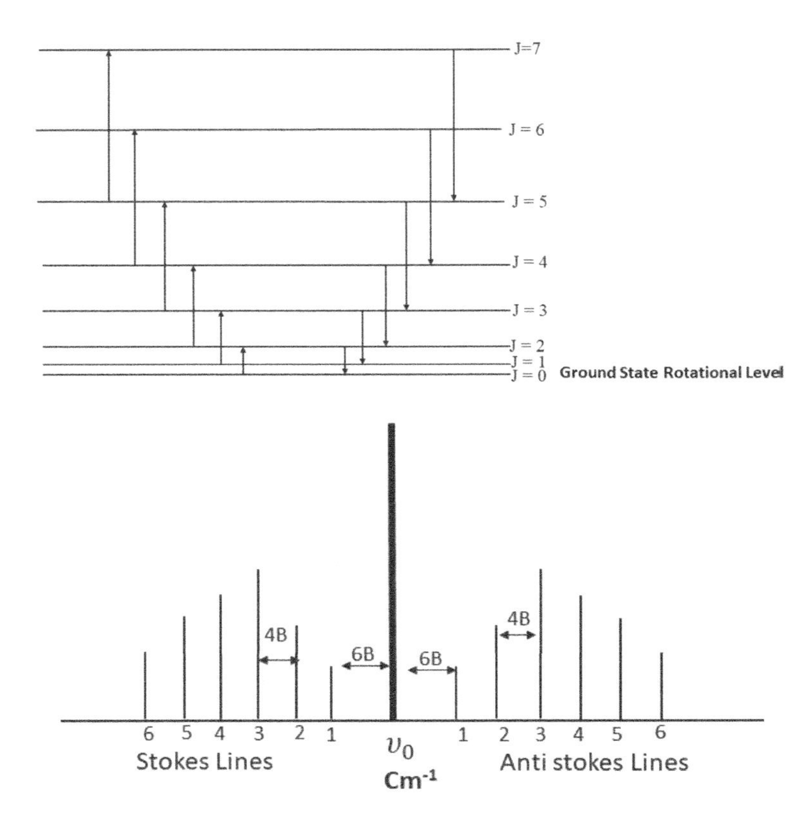

Figure 7.3 Rotational energy level of a diatomic molecule and rotational Raman spectrum arise from different transitions.

Thus, in terms of wave number one can write, $\overline{\upsilon} = BJ(J+1)\,(\text{cm}^{-1})$, where J = 0, 1, 2, etc. When the selection rule ΔJ = +2 is applied, the energy change during the transition $J \rightarrow J + 2$ for Stokes lines in terms of wave numbers can be written as

$$\overline{\upsilon}_{J\rightarrow J+2} = \overline{\upsilon}_{J+2} \rightarrow \overline{\upsilon}_{J} = B\big[(J+2)(J+3) - J(J+1)\big] = B\big[4J+6\big].$$

or
$$\overline{\upsilon} = 2B[2J+3] \quad (\text{cm}^{-1}) \tag{7.12}$$

when ΔJ = −2. A similar expression for the rotational anti-Stokes lines is observed but with a negative sign as mentioned below

$$\overline{\upsilon} = -2B[2J+3] \quad (\text{cm}^{-1}) \tag{7.13}$$

On combining both Eqs. (7.12) and (7.13) for ΔJ = ±2, finally one can write that

$$\bar{\upsilon} = \pm 2B[2J+3] \ \ (\text{cm}^{-1}) \tag{7.14}$$

Thus, the wave numbers of the Raman spectral lines can therefore easily be obtained by the expression

$$\bar{\upsilon}(\text{Raman}) = \bar{\upsilon}_0 \pm 2B[2J+3] \ \ (\text{cm}^{-1}) \tag{7.15}$$

where the (+) sign indicates **Stokes Raman lines** while the (-) sign is assigned to **anti-Stokes Raman lines**. $\bar{\upsilon}_0$ is the frequency of the incident radiation. The allowed energy transitions show the Raman spectrum and the corresponding Raman spectrum is shown in Figure 7.3.

Eq. (7.15) suggests that when J = 0, then $\bar{\upsilon} = \bar{\upsilon}_0 \pm 6B$. It means the first Raman Stokes and anti-Stokes spectral lines will appear at a distance of 6B cm^{-1} on either side of the Rayleigh line (Figure 7.3). Further, for J = 1, the frequency $\bar{\upsilon} = \bar{\upsilon}_0 \pm 10B$. This suggests the appearance of a second Raman spectral line will be at a distance of 10B cm^{-1} on either side form the Rayleigh line. Thus, there will be a separation of 4B cm^{-1} between two successive Raman spectral lines which are shown in Figure 7.3.

7.5 Vibrational Raman Spectra

Vibrational Raman spectra are obtained when the transitions in vibrational energy levels occur. The selection rule for vibrational Raman spectra is just the same as that for vibrational IR spectra.

$$\Delta v = \pm 1 \tag{7.16}$$

According to this selection rule, Stokes lines will appear in Raman spectra when the transition from one vibrational level takes place (from the lower vibrational level to the next upper vibrational level). On the contrary, when the transition takes place from a higher level to the immediate lower level. Anti-Stokes lines will be obtained. Most of the molecules since remain in the lowest vibrational level v = 0, therefore, the transitions will largely occur from the level v = 0 to v = 1. It means an intense line will appear in the spectrum. At room temperature, it is quite possible that some molecules may also be present in the first vibrational level v = 1. These molecules can undergo transitions either from the level v = 1 to the next higher level, v = 2, or from the level v = 1 to v = 0, leading to the appearance of less intense Stokes and anti-Stokes lines, respectively.

The energy change during the transition between two vibrational levels is given by the following equation (Section 7.2.4).

$$\Delta E = hv = (1 - 2x_e)hv_e \quad \text{(Joule)} \tag{7.17}$$

where $\overline{\upsilon}_e$ and x_e are wave number and internuclear distance at equilibrium, respectively.

Therefore, the frequency of this transition in terms of wave number

$$\overline{\upsilon} = \overline{\upsilon}_e(1 - 2x_e) \quad (\text{cm}^{-1}) \tag{7.18}$$

This frequency of vibrational Raman spectra is similar to the fundamental band frequency obtained in the IR spectrum of the molecule. The agreement between the values of these two spectra is quite good. Very weak transitions are also possible for transition $\Delta v = \pm 2$. Figure 7.4 gives an idea about the vibrational Raman scattering with IR transition.

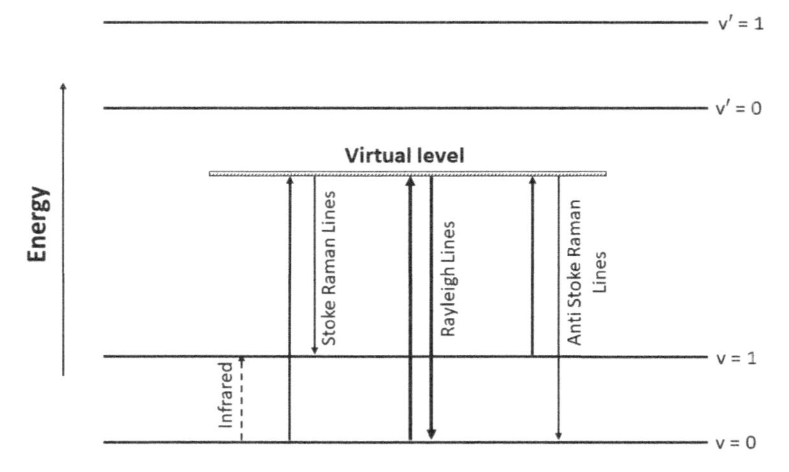

Figure 7.4 Energy level diagram for vibrational Raman scattering alongwith IR transitions.

7.6 Rotational–Vibrational Raman Spectra

When vibrational and rotational transitions simultaneously take place the selection rule for Raman spectra becomes [4].

$$\Delta J = 0, \pm 2 \text{ (for rotational energy level change)}$$

$$\Delta v = \pm 1 \text{ (for vibrational energy level change)}$$

For a non-rigid rotator diatomic molecule performing anharmonic oscillation, the energy levels are obtained by Eq. (7.19).

$$E_{v,J} = h\upsilon_e\left\{1 - x_e(v + 1/2\right\} + BhcJ(J+1) \quad \text{(Joules)} \qquad (7.19)$$

where $v = 0, 1, 2, \ldots$ and $J = 0, 1, 2, \ldots$

and the frequency of the transition in terms of wave number can be obtained with the help of following Eq. (7.20):

$$\bar{\upsilon} = \bar{\upsilon}_e\left\{1 - x_e\left(v + 1/2\right)\right\}\left(v + 1/2\right) + BJ(J+1) \quad (\text{cm}^{-1}) \qquad (7.20)$$

When the selection rule is applied for the Raman effect, i.e., $\Delta J = 0, \pm 2$, and $\Delta v = \pm 1$, and only fundamental vibrations are considered, i.e., ($v = 0$ to $v = 1$), the frequencies of the transitions are obtained for different values of J in terms of following expressions:

(i) When $\Delta J = 0$, a *Q-branch line* is obtained whose frequency is given by

$$\bar{\upsilon}(\text{Q}) = \bar{\upsilon}_e\left\{1 - 2x_e\right\} \quad (\text{cm}^{-1}) \qquad (7.21)$$

(ii) When $\Delta J = +2$, an *S-branch line* is obtained whose frequency is given by

$$\bar{\upsilon}(\text{S}) = \bar{\upsilon}_e\left\{1 - 2x_e\right\} + 2B(2J+3) \quad (\text{cm}^{-1}) \qquad (7.22)$$

(iii) When $\Delta J = -2$, an *O-branch line* is obtained whose frequency is given by

$$\bar{\upsilon}(\text{O}) = \bar{\upsilon}_e\left\{1 - 2x_e\right\} - 2B(2J+3) \quad (\text{cm}^{-1}) \qquad (7.23)$$

In rotational–vibrational Raman spectra, the frequencies of Stokes Raman lines (in wave numbers) for Q, O, and S branches are therefore given by the following expressions:

$$\bar{\upsilon}(\text{Q branch Stokes Raman line}) = \bar{\upsilon}_0 - \bar{\upsilon}(\text{Q}) \quad (\text{cm}^{-1}) \qquad (7.24)$$

$$\bar{\upsilon}(\text{S branch Stokes Raman line}) = \bar{\upsilon}_0 - \bar{\upsilon}(\text{S}) \quad (\text{cm}^{-1}) \qquad (7.25)$$

$$\bar{\upsilon}(\text{O branch Stokes Raman line}) = \bar{\upsilon}_0 - \bar{\upsilon}(\text{O}) \quad (\text{cm}^{-1}) \qquad (7.26)$$

Similarly, the frequencies of anti-Stokes lines appearing in the Raman spectra for Q, S, and O branches are given by

$$\bar{\upsilon}(\text{Q branch anti-Stokes Raman line}) = \bar{\upsilon}_0 + \bar{\upsilon}(\text{Q}) \quad (\text{cm}^{-1}) \qquad (7.27)$$

$$\bar{\upsilon}(\text{S branch anti-Stokes Raman line}) = \bar{\upsilon}_0 + \bar{\upsilon}(\text{S}) \quad (\text{cm}^{-1}) \qquad (7.28)$$

$$\bar{\upsilon}(\text{O branch anti-Stokes Raman line}) = \bar{\upsilon}_0 + \bar{\upsilon}(\text{O}) \quad (\text{cm}^{-1}) \qquad (7.29)$$

Frequencies of Stokes Raman and anti-Stokes Raman lines in the Raman spectra for Q, O, and S branches are presented in Figure 7.5.

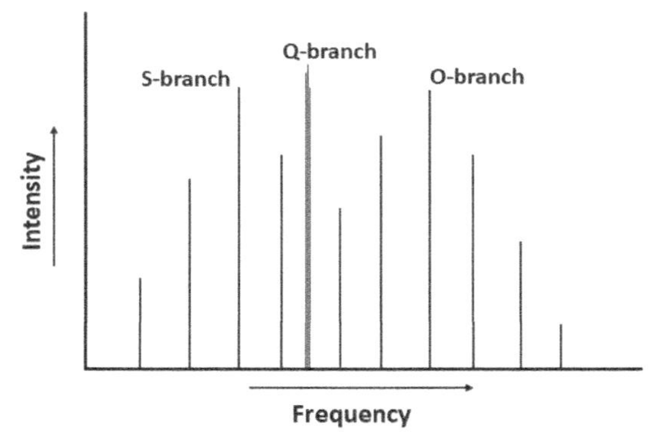

Figure 7.5 Rotational–vibrational Raman spectra for a diatomic molecule.

7.7 Instrumentation

The Raman spectrometer is made up of the following parts:

(a) A light (radiation source)
(b) A sample illumination and light collection optic system
(c) Detectors and data recording system

The experimental arrangement for Raman spectroscopy is shown in Figure 7.6 [5].

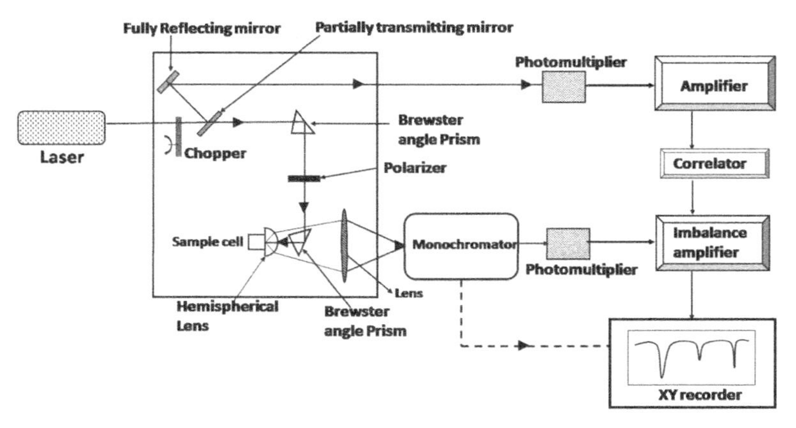

Figure 7.6 Schematic diagram of Raman spectrometer.

7.7.1 Light (Radiation Source)

In Raman spectroscopy, a polychromatic light source is not used since every line excites its own Raman spectrum which could overlap with each other. Therefore, a light source, emitting strong monochromatic radiation such as Mercury arc lamp emitting intense blue lines at 4358 Å and green lines at 5461 Å is used. In modern Raman spectrometer, lasers are used as a light source that generate highly monochromatic exciting radiations in a well-defined direction. For example, He–Ne ion laser (6328 Å), Ar-ion laser (4880 Å and 5145 Å, etc.), krypton ion (530 nm and 647 nm), diode laser (782 nm and 820 nm) are commonly used.

7.7.2 A Sample Illumination and Light Collection Optic System

The major problem that arises in Raman spectroscopy is that the Stokes and anti-Stokes signals are weak and the separation of these signals from intense Rayleigh scattering is quite difficult, therefore, optical filters are used to prevent the undesired signals and allow the weaker Raman signals for analysis in the spectrometer. The emitted radiation is then passed through the monochromators. Some commonly used filters for this purpose are

(i) Longwave pass (LWP) edge filter
(ii) Shortwave pass (SWP) edge filter
(iii) Notch filter
(iv) Laser line filter

The grating can also be used for filtering unwanted frequency signals. Typically, a holographic grating is used to produce stray lights (it is generally produced in the spectrometer due to dispersion of light on the grating and depends on the quality of grating).

7.7.3 Detectors and Data Recording System

The data recording system converts the Raman signals into electrical signals which are processed by the computer data system. The intensities of the Raman lines are determined by a photomultiplier tube (PMT) along with an amplifier and a recorder. Presently multichannel detectors like photodiode arrays or charged coupled devices (CCD) are used to detect Raman scattered light. CCD detectors show improved sensitivity and performance. The Raman spectra are recorded on a linear wavenumber scale.

7.8 Comparison Between IR and Raman Spectroscopic Techniques

Raman and IR spectroscopic techniques are used to identify chemical bonding by vibrational transitions in molecules. On the other hand, these two techniques have fundamentally different processes. Both spectroscopic techniques give complementary information and are often interchangeable. Similarities and differences between the two analytical tools are discussed below:

1. IR spectroscopy depend on the absorption of radiations whereas the results obtained from Raman spectroscopy are because of the scattering of light.
2. A change in dipole moment or charge distribution in course of vibration makes a vibration IR active. If there is any temporary distortion of electrons distribution in a molecule, then the vibration is said to be Raman active. This difference can be understood in a better way as any fundamental group which has a large change in dipoles appear as strong in the IR spectra. On the contrary, the functional group that has weak dipole changes can be better seen in Raman spectra.
3. Molecules that have polar bonds such as O-H, C=O, N-H, etc. usually show more intense bands in IR spectra. While Raman bands are generally strong because of the symmetrical charge distribution of C-C, C=C, or S=S.
4. Due to the minimum Raman scattering of water, this spectroscopy is useful for the investigation of aqueous solutions. Whereas in IR spectroscopy, water is opaque to IR radiation thus water cannot be used as a solvent in IR spectroscopy.

References

1. Smith, E. and Dent, G., *Modern Raman Spectroscopy: A Practical Approach* (John Wiley, England), 2005.

2. Banwell, C. N. and McCash, E. M., *Fundamentals of Molecular Spectroscopy*, 4th Ed. (Tata McGraw-Hill, UK), 1994.
3. Singh, N. B., Gajbhiye, N. S. and Das, S. S., *Comprehensive Physical Chemistry* (New age International, India), 2014.
4. Kafle, B. P., *Chemical Analysis and Material Characterization by Spectrometry* (Elsevier, UK), 2020.
5. Rabek, J. F., *Experimental Methods in Polymer Chemistry* (John Wiley), 1983.
6. Hendra, P. J. and Warnes, C. G., *FT Raman Spectroscopy* (Academic Press, San Diego), 1994.
7. Ferraro, J. R. and Nakamoto, K., *Introductory Raman Spectroscopy* (Academic Press, San Diego), 1994.
8. Colthrup, N. B., Daly, L. H. and Wiberley, S. E., *Introduction to Infrared and Raman Spectroscopy*, 3rd Ed. (Academic Press, San Diego), 1990.

Chapter 8

Electronic Spectroscopy

Vinay K. Verma
Department of Chemistry and Biochemistry, SBSR, Sharda University, Greater Noida, India
vinaykumar.verma@sharda.ac.in

8.1 Introduction

Ultraviolet-visible (UV-Vis) spectroscopy refers to the absorption molecular spectroscopy where the electronic transitions are associated with various electronic energy levels of the compounds. The electronic transition occurs in the wavelength range from 180 nm to 800 nm. The wavelength extending from 180 nm to 400 nm is called as ultraviolet (UV) region, however, it is the visible region from 400 nm to 800 nm. Most of the organic molecules and functional groups do not absorb and remain transparent in this electromagnetic spectral width. Therefore, this spectroscopy is of limited utility. However, if the spectral information obtained from this region when combined with other spectral data like infrared (IR), mass spectrometry (MS), and nuclear magnetic resonance (NMR),

Spectroscopy
Edited by Preeti Gupta, S. S. Das, and N. B. Singh

ISBN 978-981-4968-32-4 (Hardcover), 978-1-003-41258-8 (eBook)
www.jennystanford.com

can lead to valuable structural information about the unknown molecule. It is highly useful to measure the number of conjugated double bonds in both the aliphatic and aromatic compounds. It also finds its wide application in distinguishing conjugated and non-conjugated systems, i.e., α,β-unsaturated carbonyl compounds from β,γ-analogs, homonuclear and heteronuclear conjugated dienes, etc. [1–7].

8.2 The Nature of Electronic Excitations

When continuous electromagnetic radiation passes through the transparent solution of materials, there may be the absorption of certain wavelengths of radiation. If this happens, the spectrum obtained after passing the transmitted radiation through the prism will no longer be continuous and bears some gaps in it, which is termed as absorption spectrum. As a consequence of absorption of energy (ΔE), atoms or molecules reach the higher energy excited state (E_{excited}) from the lower energy ground state (E_{ground}). This excitation process is quantized as shown in Figure 8.1.

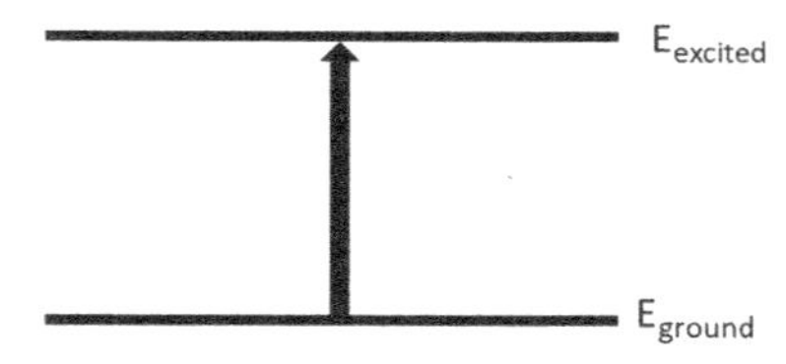

Figure 8.1 Electronic excitation in molecules.

Energy absorbed = Energy of excited state – Energy of ground state

$$\Delta E = E_{\text{excited}} - E_{\text{ground}} = h\nu \quad (8.1)$$

The electronic transition which occurs in UV-Vis spectroscopy is between the various electronic energy levels. Therefore, it is also termed electronic spectroscopy. Because of the absorption of energy, the probable electron transition is from the highest occupied molecular orbital (HOMO) to the lowest unoccupied molecular orbital (LUMO) of greater potential energy. The ΔE value for most of the molecular systems is ranging from 125 to 650 kJ/mole.

For most of the molecular systems, the occupied orbitals are denoted by σ, π, and n which possess electron pairs of σ and π bonds and non-bonded pairs of electrons, respectively. The unoccupied orbitals mostly referred to as antibonding orbitals (σ^* and π^*) are of the highest potential energy. The promotion of an electron from a π-bonding orbital to a π^* antibonding orbital is designated as $\pi \rightarrow \pi^*$. The n $\rightarrow \pi^*$ transition requires less energy compared $\pi \rightarrow \pi^*$ or $\sigma \rightarrow \sigma^*$ transition. As n-electrons do not form bonds, there are no antibonding orbitals associated with them. There are four important types of transition: (a) $\sigma \rightarrow \sigma^*$ (b) n $\rightarrow \sigma^*$ (c) $\pi \rightarrow \pi^*$ (d) n $\rightarrow \pi^*$ (Figure 8.2). The decreasing order of energies required for various electronic transitions is $\sigma \rightarrow \sigma >$ n$\rightarrow \sigma > \pi \rightarrow \pi^* >$ n $\rightarrow \pi^*$.

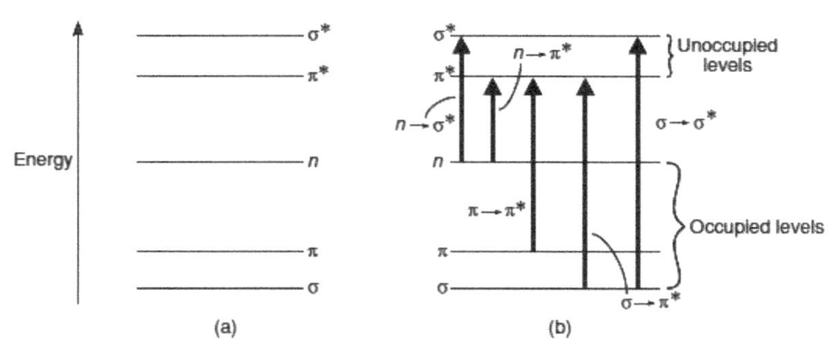

Figure 8.2 Various electronic transitions.

Table 8.1 Different transitions, their regions, and wavelengths in molecules

S.N.	Transition	Region	Wavelength	Molecules
1	$\sigma \rightarrow \sigma^*$ $\pi \rightarrow \pi^*$	Far ultraviolet	<200 nm	Alkane Alkene, carbonyl compounds, alkynes, azo compounds and so on
2	n $\rightarrow \sigma^*$	Ultraviolet	≈ 200 nm	Oxygen, nitrogen, sulfur and halogen compounds
3	n $\rightarrow \pi^*$	Near UV and visible	300–600 nm	Carbonyl compounds

8.2.1 $\sigma \rightarrow \sigma^*$ Electronic Transition

This transition is shown by molecules such as alkane and saturated hydrocarbons which contain only a single bond and lack atoms bearing non-bonded pair of electrons. The transition of an electron from sigma bonding to sigma antibonding molecular orbital is shown as $\sigma \rightarrow \sigma^*$ (Figure 8.3). Sigma electrons are strongly bonded between two bonded nuclei. Therefore, $\sigma \rightarrow \sigma^*$ electronic transition is associated with the absorption of very short wavelengths (~150 nm), shorter than the wavelengths that are experimentally accessible using typical spectrophotometers. Such transitions cannot be studied in ambient environmental conditions because oxygen present in the air begins to absorb below 200 nm wavelength. Therefore, it is studied in the vacuum UV region. Since this transition is associated with shorter wavelengths, therefore, saturated hydrocarbons which do not absorb in the near UV region can be used as one of the good solvents for recording the spectrum of compounds. Figure 8.3 shown below illustrates this type of transition.

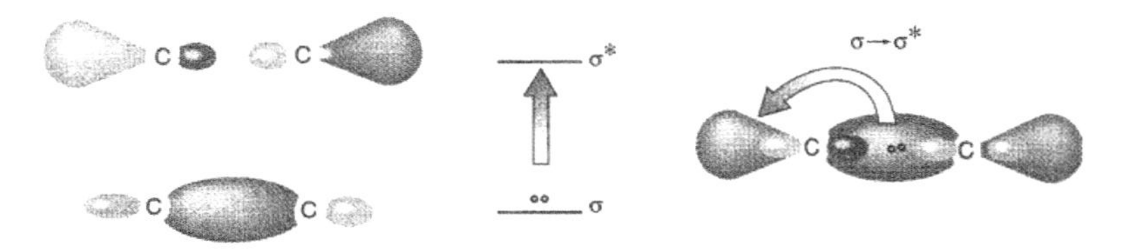

Figure 8.3 Illustration of $\sigma \rightarrow \sigma^*$ electronic transition.

8.2.2 n $\rightarrow \sigma^*$ Electronic Transition

This transition is generally shown by alcohol, ether, amine, halogen, and sulfur compounds. The saturated hydrocarbons containing C–X [X = atoms bearing non-bonded pair of electrons (O, N, S, Cl, etc.)] type of bond show n $\rightarrow \sigma^*$. This transition is also associated with a shorter wavelength, however, absorbs the radiation which lies in an experimentally accessible range. The energy of n $\rightarrow \sigma^*$ transition in alcohols and amines (λ = 175–200 nm) is relatively higher than the thiols and sulfides (λ = 200–220 nm). Most of the absorptions are <200 nm, i.e., below the solvent cut-off points for common solvents.

Therefore, they do not show any absorption in the solution spectra. This electronic transition is depicted below in Figure 8.4.

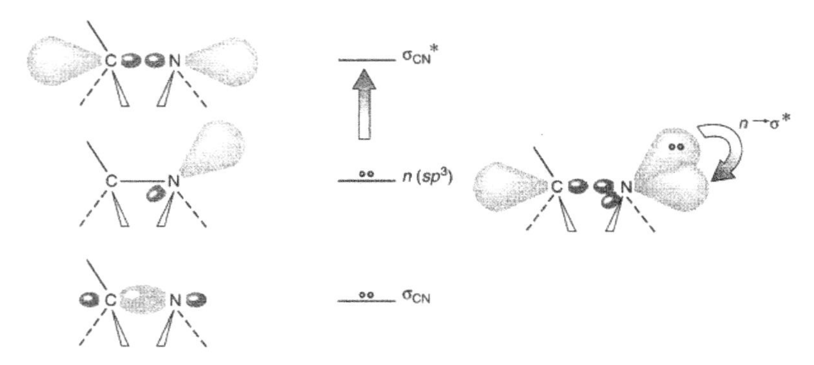

Figure 8.4 Schematic diagram showing **n** $\rightarrow \sigma^*$ electronic transition.

8.2.3 $\pi \rightarrow \pi^*$ Transition

This electronic transition is shown by unsaturated compounds like simple alkene, alkynes, aromatics, carbonyl compounds, etc., which consist of an electron in the π bonding molecular orbital. This transition involves the transition of an electron from the highest occupied π bonding molecular orbital to the lowest unoccupied π^* antibonding molecular orbital. $\pi \rightarrow \pi^*$ transition is also rather high energy but their position, i.e., λ_{max} is highly sensitive to the substitution. The olefinic double bond is non-polar in nature and hence its band intensity is not affected by the solvent.

In conjugated molecular systems like butadiene, mesityl oxide, etc., the band due to $\pi \rightarrow \pi^*$ electronic transition is known as the k-band. These electronic transitions are characterized by high molar absorptivity, ε_{max}. Benzenoid molecules bearing substituents that are further extending the conjugation of the benzene ring such as styrene, benzaldehyde, or acetophenone also show the k-band where the band intensity of $\pi \rightarrow \pi^*$ transition is dependent on the extent of conjugation. The $\pi \rightarrow \pi^*$ electronic transition has been presented in Figure 8.5.

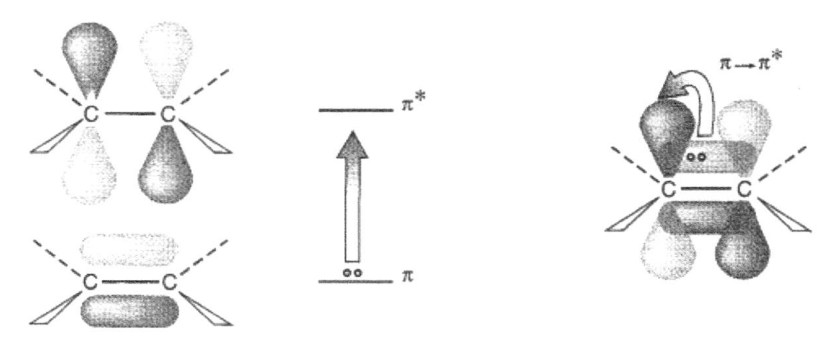

Figure 8.5 Schematic diagram showing $\pi \rightarrow \pi^*$ electronic transition.

8.2.4 n → π^* Transition

This transition is obtained in an unsaturated molecule that consists of unshared pair of electrons on a heteroatom like oxygen, nitrogen, or sulfur. In n → π^* transition, an electron from non-bonded pair of N, O, or S is excited to π^* antibonding molecular orbital. This transition is one of the lowest energy transitions among all possible electronic transitions which gives the absorption band at a longer wavelength in the spectrum and is known as the R-band. In carbonyl compounds, this transition is one of the most studied transitions. In molecules containing >C=O group, band at ~280 to 290 nm is due to n → π^* transition. The molar absorptivity (ε_{max} = 15) of this transition is very low because most of the n → π^* transition is forbidden by symmetry consideration. Carbonyl compounds also show one more band due to the $\pi \rightarrow \pi^*$ transition. This transition appears at ~188 nm with appreciably high molar absorptivity (ε_{max} = 15) in comparison to the n → π^* transition. Figure 8.6 shows the n → π^* and to $\pi \rightarrow \pi^*$ transition.

8.3 Selection Rule for the Electronic Transition

Molecule shows various electronic transitions by following the selection rules as given below.

1. Simultaneous excitation of more than one electron is not allowed (forbidden).

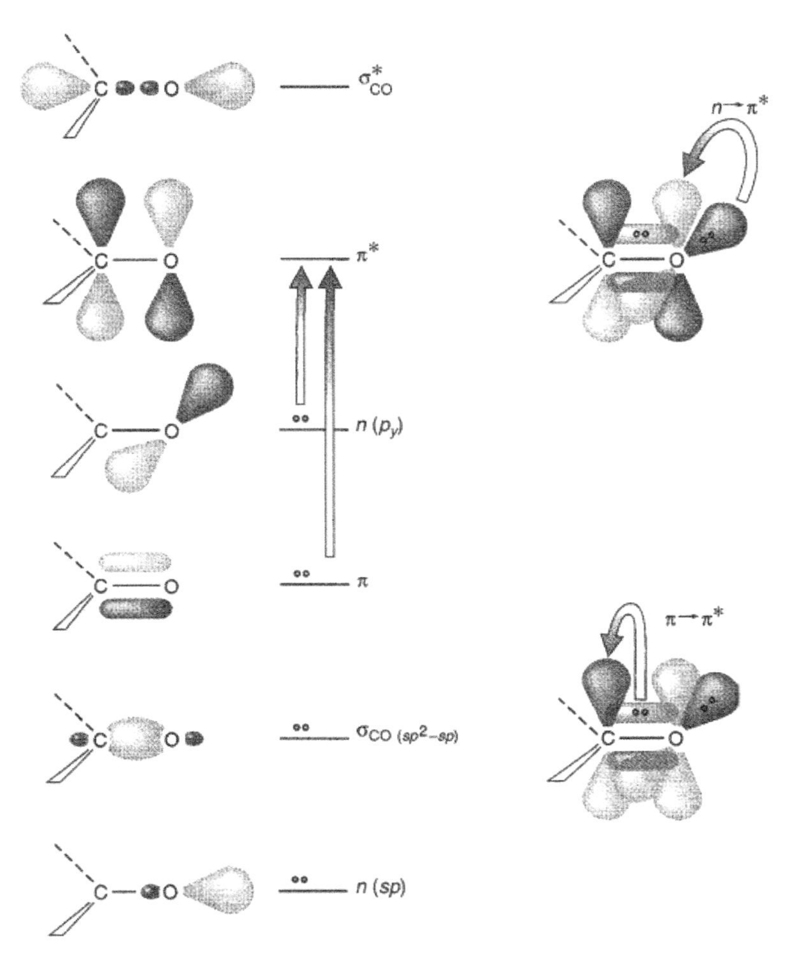

Figure 8.6 Schematic diagram showing $\pi \rightarrow \pi^*$ and n $\rightarrow \pi^*$ electronic transition.

2. **Spin selection rule:** Transition between the same spin multiplicity is allowed, i.e., an electronic transition from singlet ground state to single excited state or triplet ground state to triplet excited state. The electronic transition from singlet ground state to triplet excited state or vice versa is forbidden. The selection rule is $\Delta S = 0$, which says that only states of the same spin multiplicity combine.

3. **Laporte selection rule:** Molecules having a center of symmetry, the transition between two gerade or two ungerade states (g $\rightarrow$ g or u $\rightarrow$ u) are Laporte forbidden. The transition between gerade and ungerade (g $\rightarrow$ u or u $\rightarrow$ g) allows the electronic transitions. Therefore, for the permitted electronic transition, parity must change.

The molar absorptivity of the allowed transition is high (strong band), however, it is low (weak band) for the forbidden electronic transition (Table 8.2).

Table 8.2 Selection rules, nature of the band, and molar absorptivity

S. No.	Symmetry	Spin (multiplicity)	Nature of band	Molar absorptivity
1	Allowed	Allowed	Very strong band	Very high
2	Forbidden	Allowed	Strong band	High
3.	Allowed	Forbidden	Weak band	Low
4.	Forbidden	Forbidden	Very weak band	Very low

This selection rule is not applicable in all situations and may fail also for certain cases, such as (i) d–p mixing (in the case of tetrahedral complexes) and (ii) vibronic coupling (intensity stealing).

8.4 Beer–Lambert Law

When a beam of monochromatic radiation passes through a homogeneous absorbing medium, the rate of decrease of intensity of radiation with the thickness of the absorbing medium is proportional to the intensity of the incident radiation. This is known as Lambert's law.

Mathematically,
$$-\frac{dI}{dx} = kI \tag{8.2}$$

where I = Intensity of incident radiation
dI = Infinitesimally small decrease in the intensity of radiation on passing through infinitesimally small thickness and dx of the medium

$-\frac{dI}{dx}$ = Rate of decrease of intensity of radiation with the thickness of absorbing medium

k = Constant termed as absorption coefficient. Its value depends upon the nature of the absorbing medium.

Let I_0 be the intensity of radiation before entering the absorbing medium (x = 0). Then I, the intensity of radiation after passing through any thickness, say x, of the medium, can be calculated as

$$\int_{I_0}^{I}\frac{dI}{I} = -\int_{x=0}^{x=x} kdx \tag{8.3}$$

$$\ln\frac{I}{I_0} = -kx \text{ or } \frac{I}{I_0} = e^{-kx}$$

$$I = I_0 e^{-kx} \tag{8.4}$$

The intensity of the radiation absorbed, I_{abs} is given by

$$I_{abs} = I_0 - I = I_0(1 - e^{-kx}) \tag{8.5}$$

The above Lambert's law equation can also be written by changing the natural logarithm to the base 10.

$$I = I_0 10^{-ax} \tag{8.6}$$

where a = molar coefficient of the absorbing medium

$$a = \frac{k}{2.303}$$

Beer's law states that when a beam of monochromatic radiation is passed through a solution of an absorbing substance, the rate of decrease of intensity of radiation with the thickness of the absorbing solution is proportional to the intensity of incident radiation as well as the concentration of the solution.

Mathematically, $$-\frac{dI}{dx} = kIc \tag{8.7}$$

where c = concentration of the solution in moles/liter, k' = molar absorption coefficient and its value depends upon the nature of the absorbing substance.

Suppose I_0 is the intensity of the radiation before entering the absorbing solution. When x = 0, then the intensity of radiation, I after passing through the thickness x, of the medium can be calculated d:

$$\int_{I_0}^{I} \frac{dI}{I} = -\int_{x=0}^{x=x} kcdx \quad (8.8)$$

$$\ln\frac{I}{I_0} = -kcx \text{ or } \frac{I}{I_0} = e^{-k'cx} \quad (8.9)$$

$$I = I_0 e^{-k'cx} \quad (8.10)$$

$$I = I_0 e^{-a'cx}$$

Here where $a' = \frac{k'}{2.303}$ is the molar absorption coefficient of the absorbing solution.

Therefore, Beer's law can also be stated as, when monochromatic light is passed through a solution of an absorbing substance, its absorption remains constant when the concentration and the thickness of the absorption layer (x) are changed in the inverse ratio.

The integrated Beer–Lambert law can be defined as the intensity of emitted light decreases exponentially as the thickness and concentration of the absorbing medium increase arithmetically. Mathematically, it is

$$A = \log(I_0 / I) = \epsilon cl \text{ for a given wavelength} \quad (8.11)$$

where A = absorbance, I_0 = intensity of light incident upon sample cell, I = intensity of light leaving sample cell, c = molar concentration of solute, l = length of sample cell (cm), and ε = molar absorptivity.

The term log (I_0/I) is also known as absorbance (optical density in older literature) and is represented by A. The molar absorptivity (formerly known as molar extinction coefficient) is a property of the molecule undergoing an electronic transition and is not a function of the variable parameters involved in preparing a solution. Its unit is cm^2 $mole^{-1}$. However, the unit is by convention never expressed. The size of the absorbing system and the probability that the electronic transition will take place control the absorptivity, which ranges from 0 to 10^6. Values above 10^4 are termed high-intensity absorption while values below 10^3 are low-intensity absorptions. The forbidden transition has absorptivity in the range from 0 to 1000.

Beer–Lambert law is rigorously obeyed when a single species gives rise to the observed absorption. The law may not be obeyed, however, when different forms of the absorbing molecule are in equilibrium, when solute and solvent form complexes through some sort of association, when thermal equilibrium exists between the ground electronic state and low-lying excited state, or when fluorescent compounds or compounds changed by irradiation are present.

8.5 Chromophore and Auxochrome

UV-Vis spectroscopy deals with the excitation of electrons from the lower energy ground state to the higher energy excited state. However, there is a very important role of nuclei that hold the electrons together in a bond. The nuclei determine the wavelength of radiation that will be absorbed. It is only the nuclei that determine the strength of the bond and hence affect the energy spacing between the excited and ground states. Therefore, it can be stated that the wavelength of electromagnetic radiation absorbed is the characteristic of a group of atoms rather than of electrons themselves. The colored substances show the color because of the presence of one or more unsaturated linkages. Thus, the group of atoms producing such an effect is known as chromophores. Some examples of chromophores are C=C, C=O, N=N, NO_2, etc.

When structural change occurs in the chromophore, the wavelength and molar absorptivity of absorption both are expected to change accordingly. However, many times it is very difficult to judge theoretically how the nature of absorption will change as the structure of the chromophores is modified.

A saturated group with non-bonded electrons that when attached to a chromophore, alter both the position as well as the intensity of the absorption band of the chromophore is known as auxochrome. Some examples of the auxochromes are –OH, $-NH_2$, $-NR_2$, –NHR, etc.

The attached substituents may not show the absorption themselves, but their presence changes the intensity and the wavelength. For example, benzene is a colorless chromophore. However, introducing the $-NO_2$ group in the benzene that is nitrobenzene as a chromophore is pale yellow. The *p*-aniline, which bears both $-NO_2$ and $-NH_2$ groups as a chromophore and an auxochrome, respectively, is dark yellow in color.

Bathochromic shift is known as redshift which is a shift to lower energy and longer wavelength. Shift to higher energy or shorter wavelength is termed **hypsochromic shift** or blue shift. **the hyperchromic effect** is an effect where the intensity of the absorption band increases. **The hypochromic effect** is characterized by a decrease in the intensity of the absorption band. Table 8.3 represents the transition, λ_{max}, and molar absorptivity of chromophores.

Table 8.3 Different transitions, λ_{max}, and molar absorptivity of various chromophores

Typical absorptions of simple isolated chromophores							
Class	**Transition**	**λ_{max} (nm)**	**log ε**	**Class**	**Transition**	**λ_{max} (nm)**	**log ε**
R–OH	$n \rightarrow \sigma^*$	180	2.5	$R-NO_2$	$n \rightarrow \pi^*$	271	<1.0
R–O–R	$n \rightarrow \sigma^*$	180	3.5	R–CHO	$\pi \rightarrow \pi^*$	190	2.0
$R-NH_2$	$n \rightarrow \sigma^*$	190	3.5		$n \rightarrow \pi^*$	290	1.0
R–SH	$n \rightarrow \sigma^*$	210	3.0	R_2-CO	$\pi \rightarrow \pi^*$	180	3.0
$R_2C=CR_2$	$\pi \rightarrow \pi^*$	175	3.0		$n \rightarrow \pi^*$	280	1.5
R–C≡C–R	$\pi \rightarrow \pi^*$	170	3.0	RCOOH	$n \rightarrow \pi^*$	205	1.5
R–C≡N	$n \rightarrow \pi^*$	160	<1.0	RCOOR	$n \rightarrow \pi^*$	205	1.5
R–N=N–R	$n \rightarrow \pi^*$	340	<1.0	$RCONH_2$	$n \rightarrow \pi^*$	210	1.5

It is recorded as a plot of wavelength at the *x*-axis versus absorbance at the *y*-axis. It is customary to then replot the data with either ε or logε on the ordinate and wavelength on the abscissa. However, very few electronic spectra are reproducible and hence described by indicating λ_{max} and ε of the principle absorption band. Spectral characteristics of benzoic acid are mentioned in Table 8.4 and the spectra are shown in Figure 8.7.

Table 8.4 λ_{max} and molar absorptivity of benzoic acid

λ_{max} (nm)	**log ε**
230	4.2
272	3.1
282	2.9

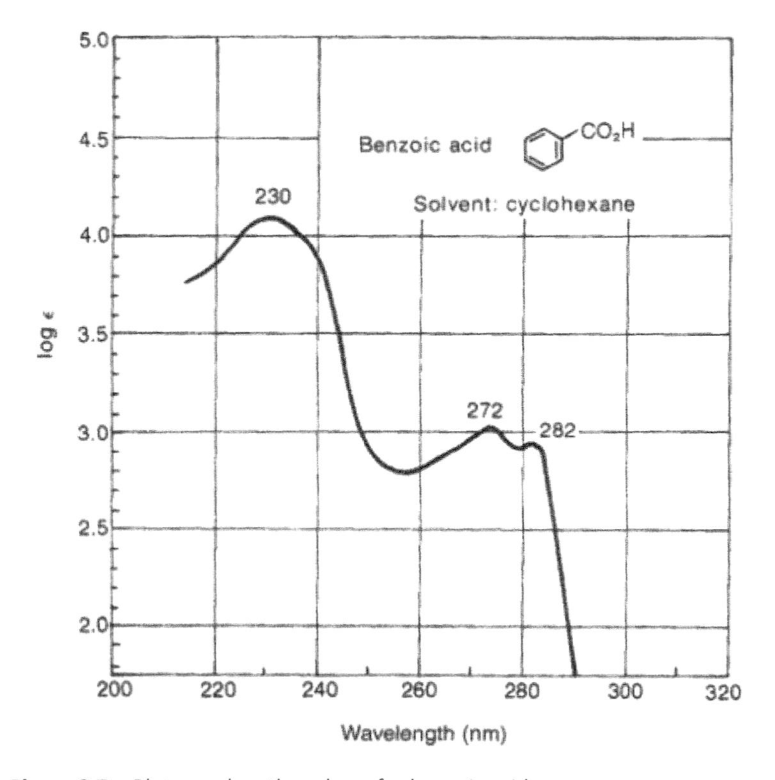

Figure 8.7 Plot wavelength *vs.* log ε for benzoic acid.

8.6 Effect of Solvent on Electronic Transition (Solvent Effect)

The choice of solvent to be used for UV spectroscopy is quite important. Therefore, the following criteria must be kept in mind prior to the recording of the spectra.

1. It should not absorb in the same region as the substance whose spectral study is to be made. Generally, the solvents which consist of conjugation in their structure are avoided. Therefore, the most commonly used solvents are water, 95% ethanol, and hexane as they remain transparent in the ultraviolet spectrum where interesting peaks from sample molecules are likely to occur.

Table 8.5 λ_{max} values of various solvents

Solvernt Cutoffs			
Acctonitrile	190 nm	*n*-Hexane	201 nm
Chloroform	240	Methanol	205
Cyclohexane	195	Isocctane	195
1,4-Dioxane	215	Water	190
95% Ethanol	205	Trimethyl phosphate	210

2. Solvents should not alter the natural fine structure of an absorption band. The effect of polar and non-polar solvents on an absorption band is shown below in Figure 8.8.

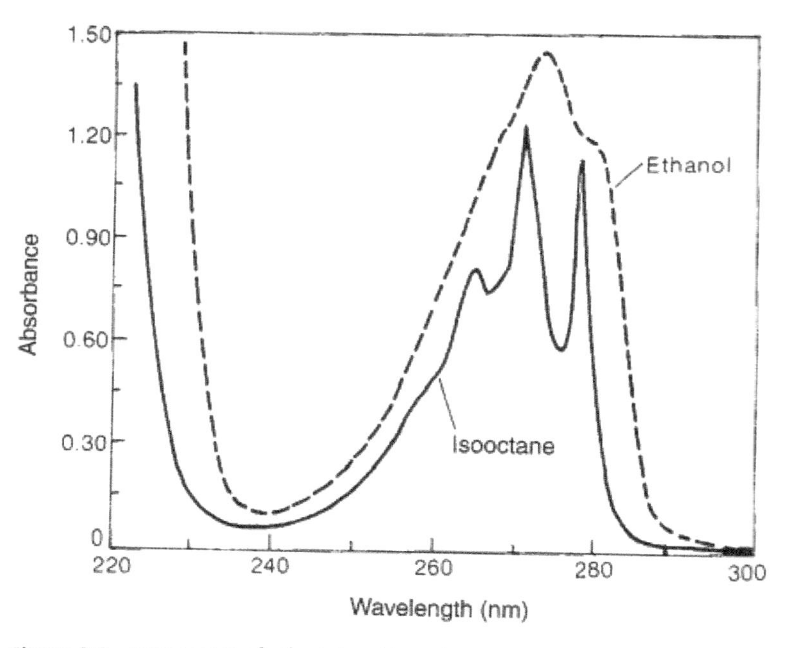

Figure 8.8 UV spectra of ethanol and isooctane.

The absorption spectrum in non-polar solvents closely resembles the spectrum of compounds recorded in the gaseous state. In this situation, the fine structure of the bands is observed. This is due to the fact that a non-polar solvent does not make a hydrogen bond with the molecule. However, polar solvents generally form the hydrogen bond with the solute via the formation of the solute–solvent complex. Therefore, in polar solvents, the fine structure may disappear.

3. Solvents used for recording the spectrum of the compound may alter the wavelength of absorbed radiation. This may be due to the stabilization of either ground or excited state via interaction with the polar solvents. Polar solvents make the hydrogen bond very readily with the ground state as compared to the excited state. Due to the fact, that the wavelength of the ultraviolet radiation absorbed for the electronic transition decreases, i.e., blue shift or hypsochromic shift is observed. Carbonyl compounds show two types of electronic transition which are weak $n \rightarrow \pi^*$ and $\pi \rightarrow \pi^*$ transition. In the case of $n \rightarrow \pi^*$ transition, the non-bonded state is much more easily stabilized by polar solvent as compared to the π^* antibonding state. This results in the increase of energy spacing between the two states and hypsochromic shifts are observed. For example, acetone in a non-polar solvent like hexane shows the λ_{max} at 279 nm which is at 264.5 nm if the solvent is polar. The blue shift of $n \rightarrow \pi^*$ transition with and without the solvent is depicted in Figure 8.9.

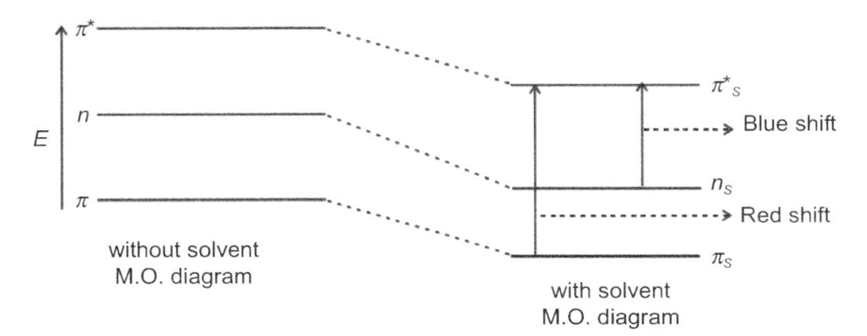

Figure 8.9 Transition for blue shift and red shift in the presence and absence of the solvent.

Table 8.6 Effect of solvents on electronic transitions

Solvent shift on the $n \rightarrow \pi^*$ transition of acetone					
Solvent	H_2O	CH_2OH	C_2H_5OH	$CHCl_3$	C_6H_{14}
λ_{max} (nm)	264.5	270	272	277	279

In the case of $\pi \rightarrow \pi^*$ transition, if the excited state is more polar than the ground state, dipole-dipole interaction is more pronounced with the excited state which lowers its energy state more than that of the ground state. Due to the above fact, spacing between π bonding and π^* antibonding orbitals is reduced and bathochromic shift or redshift is observed (Figure 8.10). The red shift may be of the order of 10 to 20 nm in case of a change of solvent from hexane to ethanol.

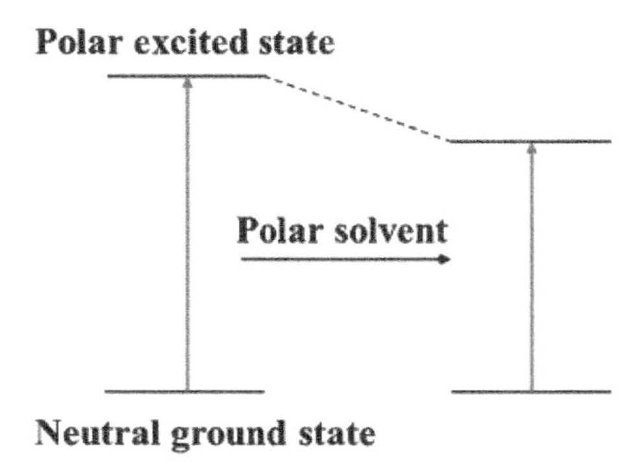

Figure 8.10 $\pi \rightarrow \pi^*$ transition of the compound in a polar solvent.

8.7 Effect of Conjugation

Increases in conjugation decrease the spacing between HOMO and LUMO. Therefore, the energy required for the electronic transition in the conjugated system is less, and hence the bathochromic shift results from an increase in the length of a conjugated system. This can easily be explained using molecular orbital theory (MOT). Consider the example of the ethylene molecule. In an ethylene molecule, two pure *p* atomic orbitals (AOs), ϕ_1 and ϕ_2 lie perpendicular to the σ bond. By the linear combination of atomic orbitals (LCAOs), these *p* orbitals form two π molecular orbitals (MOs), ψ_1 and ψ_2*. The π bonding orbital, ψ_1 is formed by the addition of the wave functions of two pure *p* AOs. However, the π antibonding orbital, ψ_2* is formed by the subtraction of these two wave functions. The bonding molecular orbital formed has lower energy than that of the pure *p* AOs. Similarly, the antibonding molecular orbital has higher energy than the pure *p* AOs (Figure 8.11).

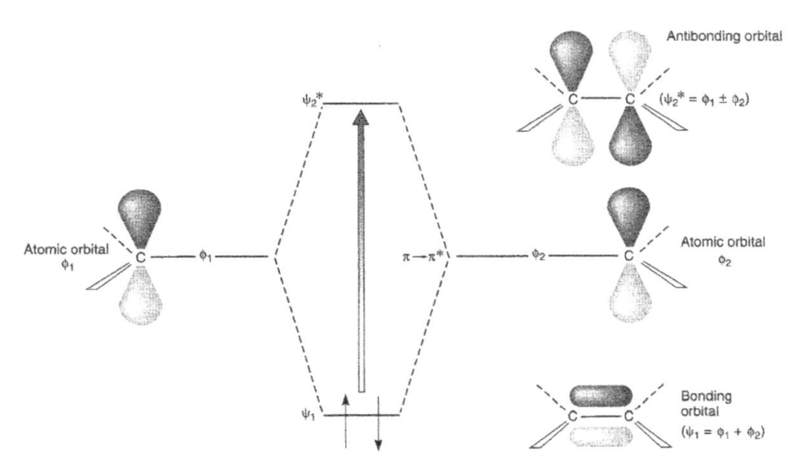

Figure 8.11 Representation of conjugation in ethylene molecule.

It is noticed that in the ethylene molecule, two *p* AOs combine to form two MOs. As each *p* AO consists of one electron, hence the new π system contains two electrons. The bonding molecular orbital, ψ_1 is of lower energy hence both the electrons are filled in it. Therefore, the $\pi \rightarrow \pi^*$ electronic transition in this system is from ψ_1 to ψ_2^*.

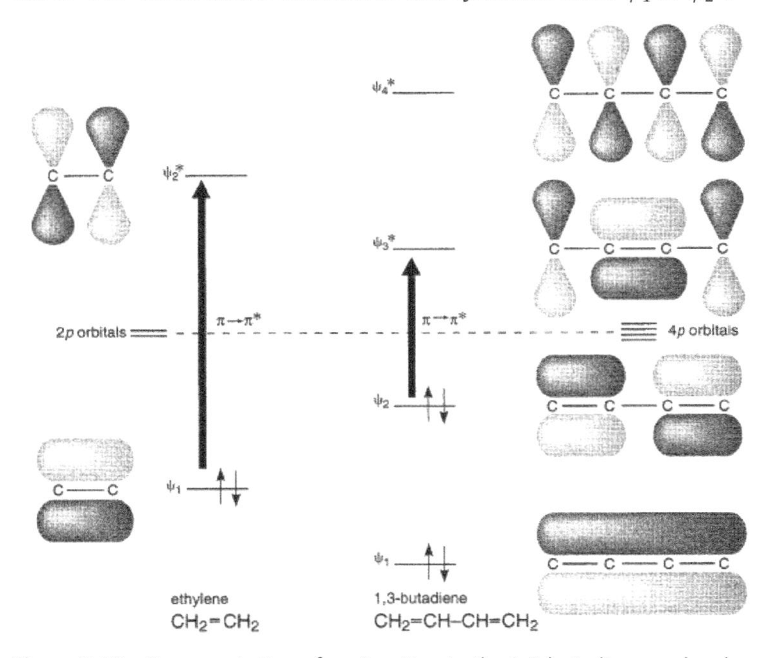

Figure 8.12 Representation of conjugation in the 1,3-butadiene molecule.

Conjugation increases on moving from two-orbital ethylene case to four *p* orbital 1,3-butadiene that form its π system of two conjugated double bonds. Four AOs build four MOs as shown in Figure 8.12 together with MOs of ethylene molecule for comparison purposes. The $\pi \rightarrow \pi^*$ electronic transition in 1,3-butadiene is from ψ_2 to $\psi_3{}^*$ which is lower in energy than the transition in ethylene from ψ_1 to $\psi_2{}^*$.

This implies that on the progressive increase in conjugation, the wavelength of absorption progressively increases, i.e., a redshift is observed. Consequently, the energy of the electronic transition from HOMO to LUMO decreases (Figure 8.13).

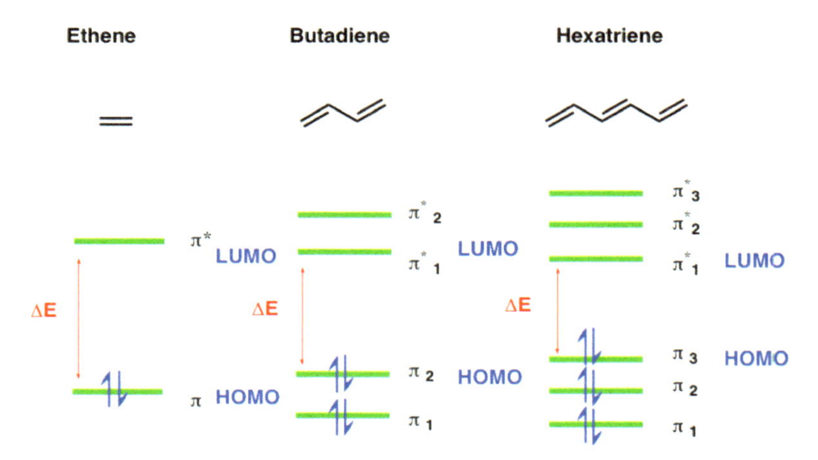

Figure 8.13 Electronic transition between HOMO and LUMO in a different molecule.

Generally, many strong auxochrome bear loan pair of electrons which exert their bathochromic shift by extension of the length of conjugation. Atom B is heteroatom bearing a pair of non-bonded electrons. The unshared pair of electrons is in resonance with the double bond thereby increasing the length of the conjugated system by one extra orbital.

$$\left[>C{=}C{-}\ddot{B} \longleftrightarrow >\bar{\ddot{C}}{-}C{=}B^{+} \right]$$

Due to the interaction of the non-bonded pair of electrons, it becomes part of the π system and increases the length by one extra orbital. The unspecified atom B in Figure 8.14, can be any of the auxochromic groups which bear non-bonded pairs of electrons such as –OH, –OR, –X, or $-NH_2$. In the system where ethylene π electrons are in conjugation with the non-bonded electron pair of auxochrome, the energy of $\pi \rightarrow \pi^*$ electronic transition will always be lower as compared to the system which is without the interaction of the non-bonded electron pair.

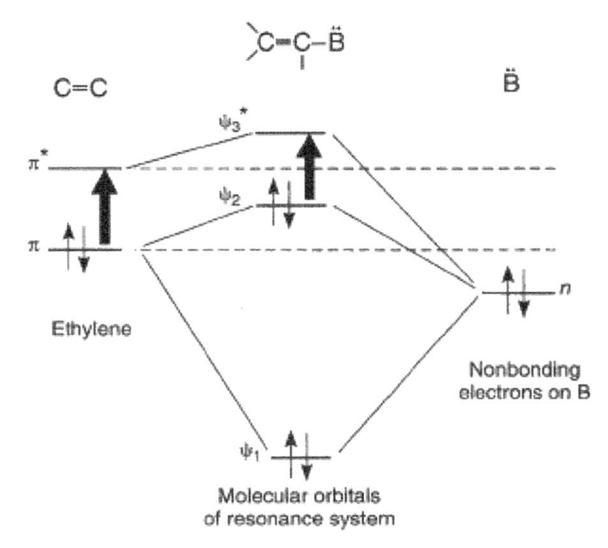

Figure 8.14 Molecular orbital diagram of ethylene and atom B showing $\pi \rightarrow \pi^*$ electronic transition.

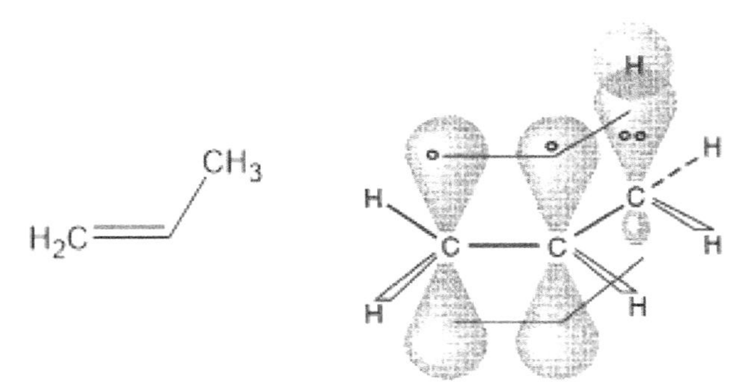

Figure 8.15 Ethylene molecule showing hyperconjugation.

If the methyl group is present as auxochrome in the ethylene molecule, it also produces the bathochromic shift. The methyl group does not possess non-bonded electron pair, however, it extends the conjugation in the ethylene molecule by no-bond resonance which is often called hyperconjugation as shown in Figure 8.15.

8.8 Woodward–Fisher Rule

8.8.1 For Dienes

1,3-Butadiene has four MOs. These are ψ_1, ψ_2 bonding MOs and ψ_3^*, ψ_4^* as antibonding MOs. As there are two antibonding MOs available for electronic transition from ψ_2 bonding molecular orbital, therefore 1,3-butadiene shows two possible $\pi \rightarrow \pi^*$ electronic transitions. The electronic transition from ψ_2 to ψ_3^* is allowed transition and its absorption wavelength lie near 230 nm. However, the electronic transition from ψ_2 to ψ_4^* is not often observed. This is mainly because of two reasons. Firstly, it lies in the solvent cut-off point of common solvents, i.e., near 175 nm. Therefore, it is not observed in the spectrum. Secondly, this transition is forbidden for *trans* conformation of butadiene. Further, it is also not detectable because *trans* conformation is energetically more favorable for 1,3-butadiene as compared to *cis* conformation.

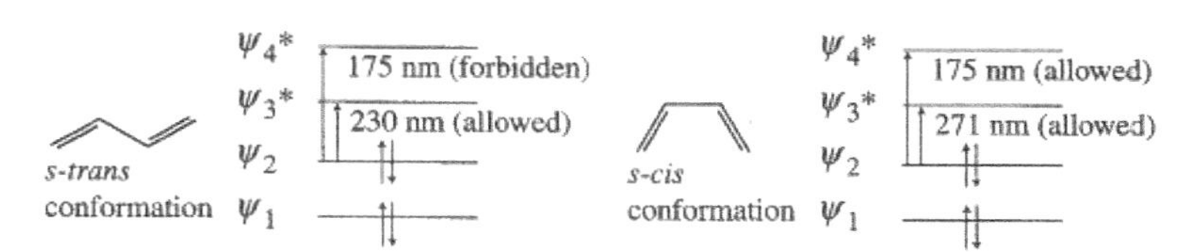

Figure 8.16 Various transitions in the *cis* and *trans* conformation of 1,3-butadiene.

The molar absorptivity (ε = 2 × 10^4 to 2.6 × 10^4) of $\pi \rightarrow \pi^*$ electronic transition is very high, therefore appears as an intense band. The absorption wavelength for $\pi \rightarrow \pi^*$ transition lies in the range of 217–245 nm which is quite insensitive to the nature of the solvent.

Generally, alkyl substitution in simple butadiene and conjugated diene produces bathochromic shifts and hyperchromic effects. However, for a certain pattern of alkyl substitution like methyl substitution at first and third carbon atoms of 1,3-butadiene, bathochromic shift but the hypochromic effect is observed despite the hyperchromic effect. 1,3-Dialkyl butadiene shows much crowding due to the close interaction among two methyl groups of trans confirmation and hence converts in cis conformation by rotation along the C–C single bond. Therefore, it shows a bathochromic shift and hypochromic effect.

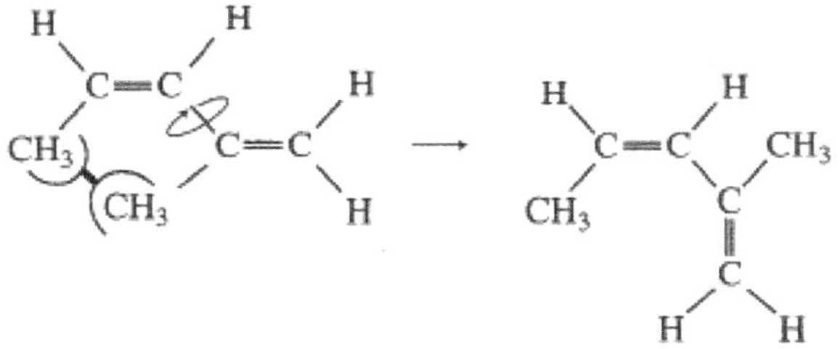

s-trans → s-cis

Table 8.7 Woodward–Fisher rule for diene system

Molecular system	Homoannular (cisoid)	Heterannular (transoid)
Conjugated diene system		
Parent	λ = 253 nm	λ = 214 nm for acyclic dienes and 245 nm for acyclic triene
Increments for each substituent		
If double bond extending conjugation	30	30
Alkyl substituent or ring residue	5	5
Exocyclic double bond	5	5
$-OCOCH_3$	0	0
–OR	6	6
–Cl, –Br	5	5
$-NR_2$	60	60

After studying a large number of dienes, Woodward and Fisher have devised an empirical correlation of structural variations that helps to calculate and predict the absorption wavelength in the conjugated diene system. The summary of the rule is given in Table 8.7.

For example, let us consider some examples.

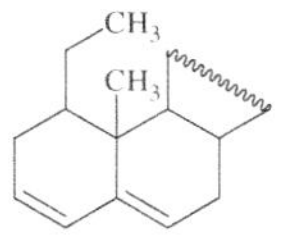 CH_3 CH_3	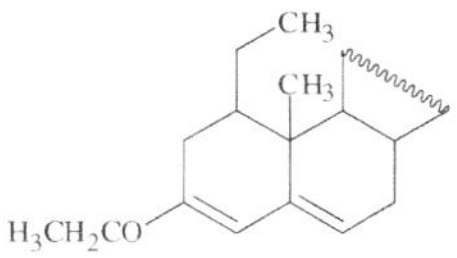 CH_3 CH_3 H_3CH_2CO
Transoid: 214 nm Ring residues: 3 × 5 = 15 Exocyclic double bond: 5 ------------ 234 nm Observed: 235 nm	Transoid: 214 nm Ring residues: 3 × 5 = 15 Exocyclic double bond: 5 $-OCH_2CH_3$: 6 ------------ 240 nm Observed: 241 nm
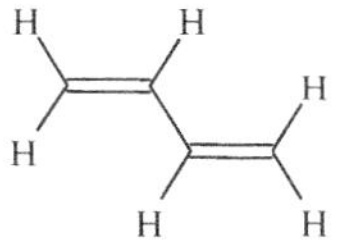 H H H H H H	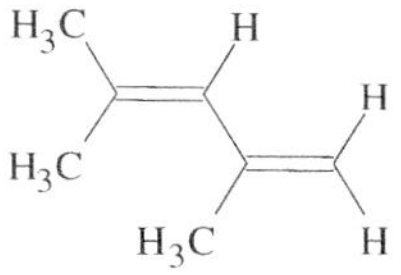 H_3C H H H_3C H_3C H
Transoid: 214 nm Observed: 217 nm	Transoid: 214 nm Alkyl groups: 3 × 5 = 15 ---------- 229 nm Observed: 228 nm
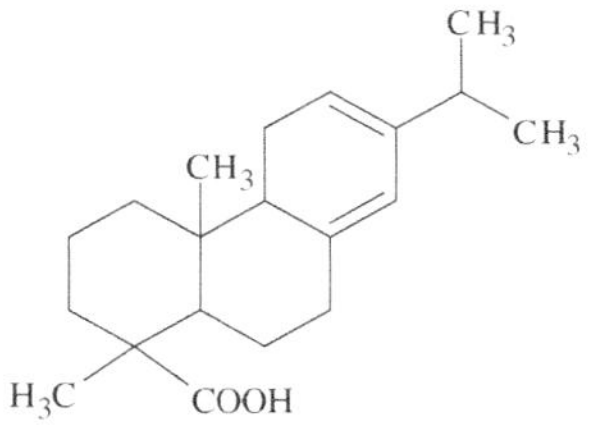 CH_3 CH_3 CH_3 H_3C COOH	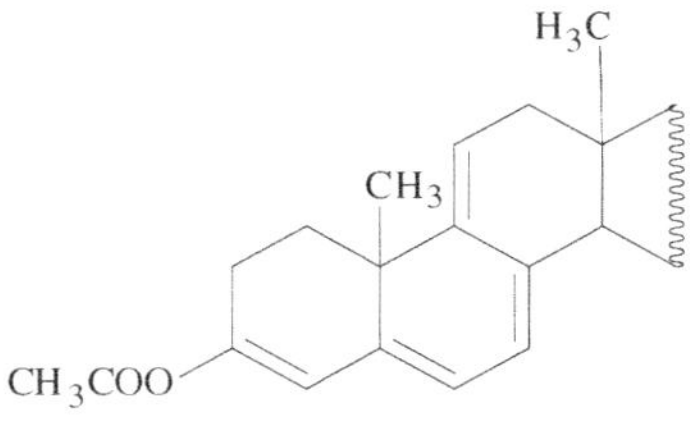 H_3C CH_3 CH_3COO

Cisoid:	253 nm	Cisoid:	253 nm
Alkyl substituent:	5	Double bond extending conjugation:	
Ring residues:	3 × 5 = 15		2 × 30 = 60
Exocyclic double bond:	5	Ring residues:	5 × 5 = 25
	------------	Exocyclic double bond:	3 × 5 = 55
	278 nm	$-OCOCH_3$:	0
Observed:	275 nm		------------
			353 nm
		Observed:	355 nm

8.8.2 For Carbonyl Compounds: Enones

The carbonyl group consists of two eight electrons with two electrons in each σ and π bonding MOs. However, n nonbonding MO has four electrons. The molecular orbital diagram is shown below.

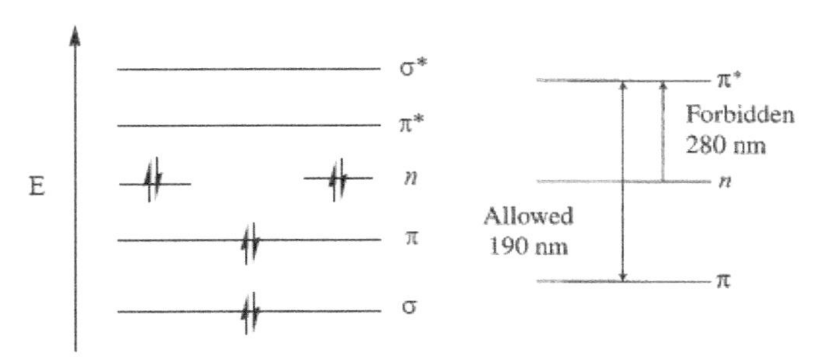

As discussed earlier, it shows two principal transitions, i.e., $\pi \rightarrow \pi^*$ transition which is allowed, and $n \rightarrow \pi^*$ transition as forbidden. The $n \rightarrow \pi^*$ transition shows the wavelength absorbed above the solvent cut-off points while the $\pi \rightarrow \pi^*$ transition is below 200 nm. On the extension of conjugation by the substitution of the carbonyl group with auxochrome containing non-bonded electron pair, lessor bathochromic shift in $\pi \rightarrow \pi^*$ transition and a pronounced hypsochromic effect in $n \rightarrow \pi^*$ transition is observed (Figure 8.17). Hypsochromic effects of the acetyl group on $n \rightarrow \pi^*$ transition of a carbonyl group are tabulated below (Table 8.8).

Table 8.8 λ_{max} and molar absorptivity of compounds in different solvents

Substitution	λ_{max}	ε_{max}	Solvent
CH_3-CO-H	293 nm	12	Hexane
CH_3-CO-CH_3	279	15	Hexane
CH_3-CO-Cl	235	53	Hexane
CH_3-CO-NH_2	214	–	Water
CH_3-CO-OCH_2CH_3	204	60	Water
CH_3-CO-OH	204	41	Ethanol

The hypsochromic shift in the $n \rightarrow \pi^*$ transition is due to the inductive effect of the substituent like O, N, or halogen atoms. These groups withdraw the non-bonded electron pair of the carbonyl oxygen and make the carbonyl oxygen more electronegative in absence of the non-bonded electron. Thereby the non-bonded pair of electrons are held more firmly by the carbonyl oxygen than they would be without the inductive effect.

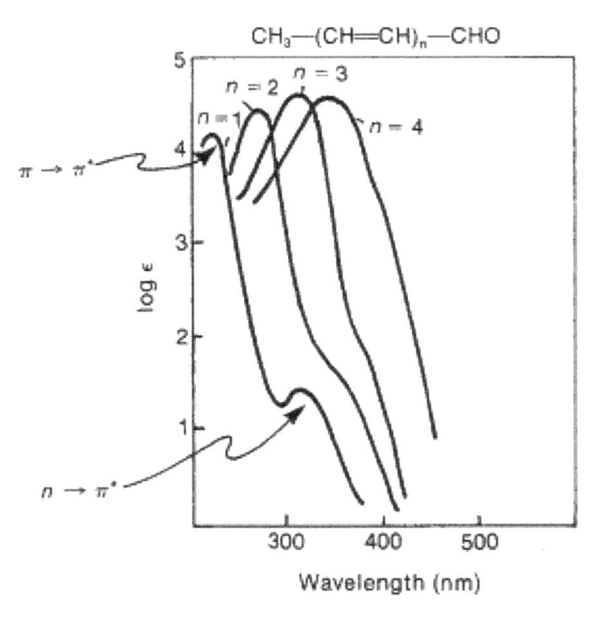

Figure 8.17 UV-visible spectra of polyene aldehyde [CH3-(CH=CH)$_n$-CHO] with the value of n =1 to 4.

In case, where the carbonyl group is in conjugation with double bonds, the energy required for $\pi \rightarrow \pi^*$ and $n \rightarrow \pi^*$ transition is lower. However, the energy of $\pi \rightarrow \pi^*$ transition decreases rapidly with the hyperchromic effect which is very low for $n \rightarrow \pi^*$ transition. Sometimes $n \rightarrow \pi^*$ transition is buried inside the $\pi \rightarrow \pi^*$ transition where the intensity increases very high.

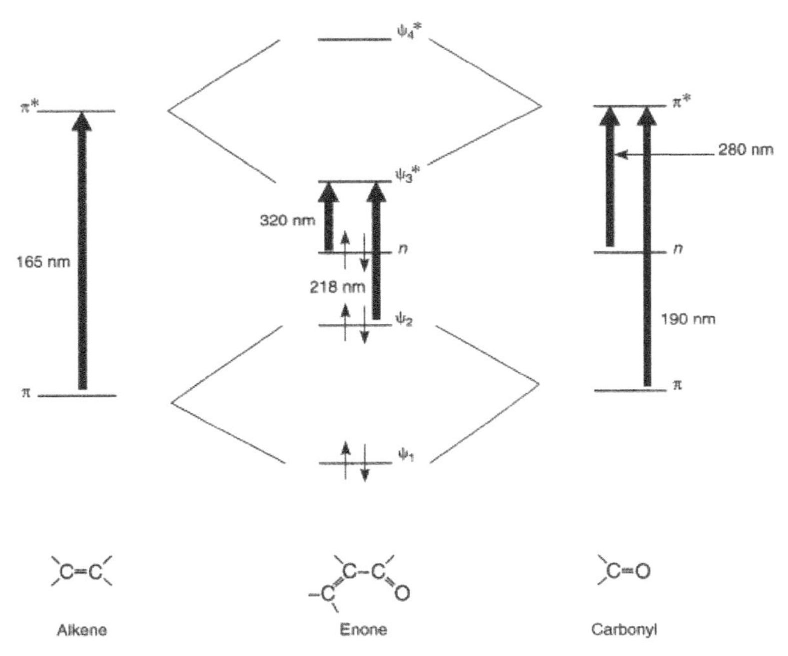

Figure 8.18 Diagram representing orbitals of enone system with the alkene and carbonyl.

8.8.3 Woodward Rule for Enone System

The $\pi \rightarrow \pi^*$ transition in the enone system shows the intense absorption band with ε values ranging from 8000–20,000. On the structural modification of chromophore, $\pi \rightarrow \pi^*$ transition is affected in a predictable manner. However, $n \rightarrow \pi^*$ transition does not show such predictable behavior. Woodward has studied several enone systems and devised an empirical rule for the prediction of the wavelength of $\pi \rightarrow \pi^*$ transition of an unknown molecule which is given in Table 8.9.

Table 8.9 Woodward rule for enone system

Empirical rule for enones	
β–C(–β)=C(–α)–C=O	δ–C(–δ)=C(–γ)–C(β)=C(–α)–C=O
Base value:	
Six membered ring or acyclic parent enone	215 nm
Five membered ring parent enone	202 nm
Acyclic dienone	245 nm
Increment for:	
Double bond extending conjugation	30
Alkyl group or ring residue	α 10 β 12 γ and higher 18
Polar groupings:	
–OH	α 35 β 30 δ 18
$-OCOCH_3$	α, β, δ 6
$-OCH_3$	α 35 β 30 γ 17 δ 31
–Cl	α 15 β 12
–Br	α 25 β 30
$-NR_2$	β 95
Exocyclic double bond	5
Homocyclic diene component	39
Solvent correction	Variable

Some examples of this rule are given below.

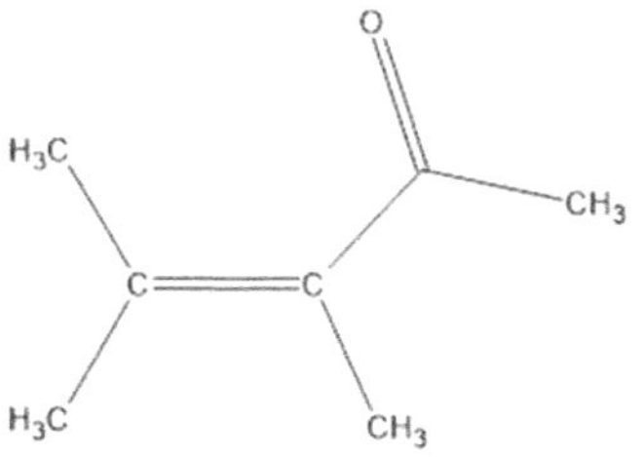

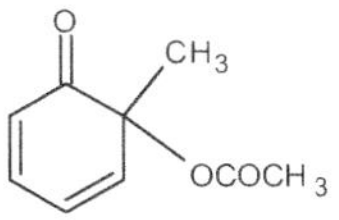

Acyclic enone: 215 nm α-CH_3: 10 β-CH_3: 2 × 12 = 24 ------------ 249 nm Observed: 249 nm	Six-membered enone: 215 nm Double bond extending conjugation: 30 Homocyclic diene: 39 δ-Ring residue: 18 --------------- 302 nm Observed: 300 nm
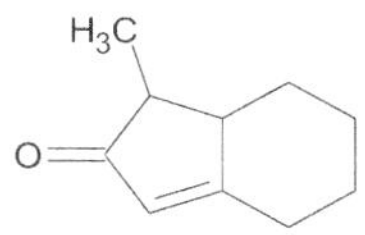	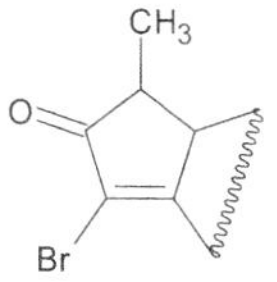
Five membered enone: 202 nm β-Ring residue: 2 × 12 = 24 exocyclic double bond: 5 ------------- 231 nm Observed: 231 nm	Five membered enone: 202 nm α-Br: 25 β-Ring residue: 2 × 12 = 24 exocyclic double bond: 5 ------------- 256 nm Observed: 251 nm
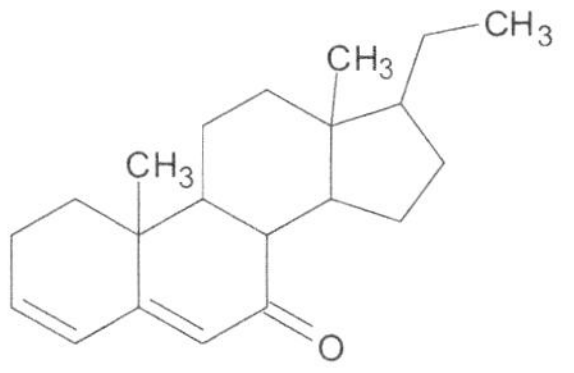	
Six-membered enone: 215 nm Double bond extending conjugation: 30 β-Ring residue: 24 δ-Ring residue: 18 exocyclic double bond: 18 ---------- 280 nm Observed: 280 nm	

8.8.4 Woodward's Rule for *α*, *β*-Unsaturated Aldehydes, Acids, and Esters

The empirical rule for the prediction of absorption wavelength in *α*, *β*-unsaturated aldehydes, acids, and esters has been developed by

Nielsen. It follows the same rules as enones. However, the absorption wavelength α,β-unsaturated aldehydes is ~5–8 nm shorter than the corresponding ketones. The empirical formula for α,β-unsaturated aldehydes, acids, and esters is given in Table 8.10.

Table 8.10 Woodward rule for α,β-unsaturated aldehydes, acids, and esters

β-C(β)=C(-α)-C=O	
Parent	208 nm
With α or β alkyl groups	220
With α, β or β, β alkyl groups	230
With α, β, β alkyl groups	242
β-C(β)=C(-α)-COOR β-C(β)=C(-α)-COOH	
With α or β alkyl groups	208 nm
With α, β or β, β alkyl groups	217
With α, β, β alkyl groups	225
For exocyclic α, β double bond	Add 5 nm
For endocyclic α, β double bond in five or seven membered ring	Add 5 nm

Consider the case of 2-cyclohexenoic and 2-cycloheptenoic acids as an example.

COOH		COOH	
α, β- Dialkyl:	217 nm	α, β- Dialkyl:	217 nm
double bond in a six-membered ring:	0	double bond in a seven-membered ring:	5
	---------------		---------------
	217 nm		222 nm
Observed:	217 nm	Observed:	222 nm

8.9 Absorption in Aromatic Compounds

Benzene is the simplest chromophore in the aromatic system. However, the electronic transition benzene molecule is quite complex. It shows three absorption bands due to the $\pi \rightarrow \pi^*$

transition. Sometimes its spectrum consists of many fine structures. The molecular orbital diagram suggests that there should be four possible electronic transitions with each transition having the same energy. This concludes that the ultraviolet spectrum of benzene consists of only one absorption band. However, due to symmetry consideration and electron-electron repulsion, the energy for which the electronic transition occurs is slightly different. Therefore, three electronic transitions occur in the benzene ring. The MOs and energy states of benzene are given in Figure 8.19.

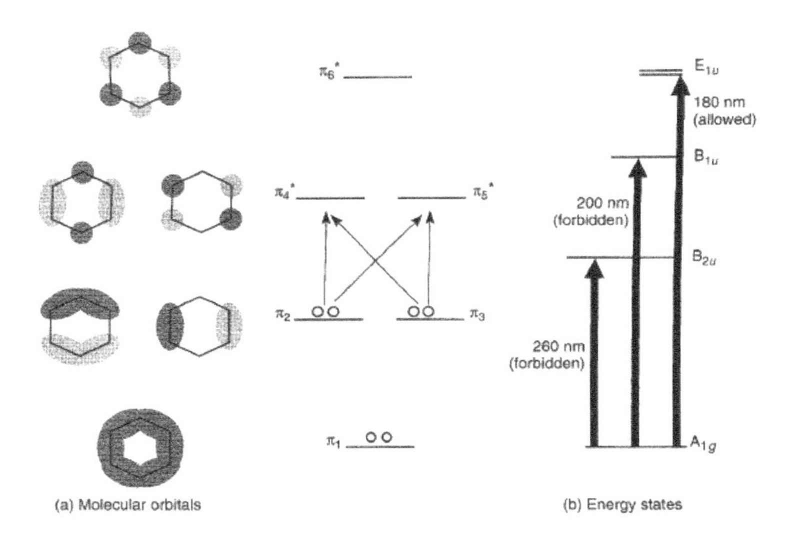

Figure 8.19 MOs and energy states for benzene.

The absorption bands in the benzene are described by primary bands that appear at 184 and 202 nm (Figure 8.20). However, the secondary band with fine structure appears at 255 nm. Benzene absorbs very strongly at 184 nm ($\varepsilon = 4.7 \times 10^4$) but this transition is not observed in the spectrum as it is a vacuum ultraviolet region and beyond the wavelength range of many of the commercial spectrophotometers. The second primary band that appears at 202 is much less intense and corresponds to forbidden transition. The absorption at 255 nm is very weak ($\varepsilon = 200$). The fine structure in the band of 255 nm is mainly because of the vibrational sublevels and their influence on the electron transition. This transition is symmetry forbidden electronic transition. If the spectrum of benzene is recorded in polar solvents, the fine structures are lost and appear as a broad band without any structural information. This also happens in the case of monosubstituted benzene.

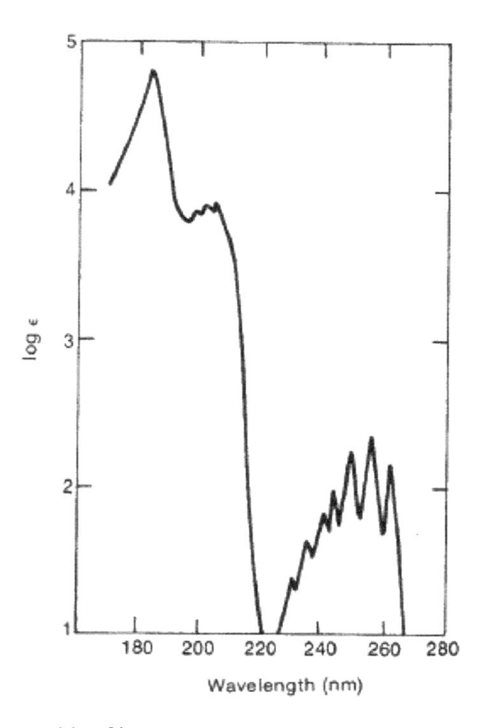

Figure 8.20 UV-visible of benzene.

Although, the substitution in the benzene ring causes the redshift and hyperchromic effect but the formulation of any empirical rule to predict the absorption wavelength.

If the benzene is monosubstituted with groups containing nonbonding electrons, shifts in both the bands take place. This is primarily because of the extension of π conjugation through resonance.

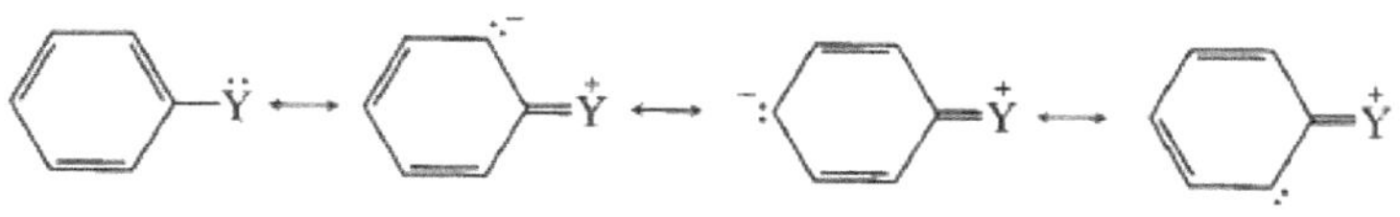

8.10 Color in Transition Metal Complex

When the compound is exposed to white light, mainly three phenomena take place. Firstly, white light gets completely absorbed and the compound looks black. Secondly, no absorption of light takes place where the compound is colorless (white color is treated as colorless). Finally, there may be a certain fraction of light get absorbed and hence it is colored. Color shown by the compounds is completely dependent on the energy required for the *d*–*d* transition (Figure 8.21). Color in the transition metal complexes can be rationalized by crystal field theory and MOT. Both the theories are based on one fact that the energy of *d* orbitals of transition metal in solution is not identical. Therefore, the transition of electrons takes place from lower energy to higher energy *d* orbital. In absence of the ligands, *d* orbitals degenerate. However, it increases in presence of the ligand, and hence, *d* orbitals split into different energy levels designated by *the symbol* Δ as given below. The magnitude of the spitting energy of the d orbital depends mainly on the oxidation state of metal ions. The splitting energy, Δ, of the *d* orbital is also dependent on the ligand field strength. Therefore, the strong field ligands cause a high value of Δ. The ligand field strength of some of the ligands is organized in increasing order and given below.

$I^- < Br^- < SCN^- < Cl^- < OH^- < -NH_2 < NH_3 < H_2N—(CH_2)_2—NH_2 <$ *o*-phenanthriline $< NO_2^- < CN^-$

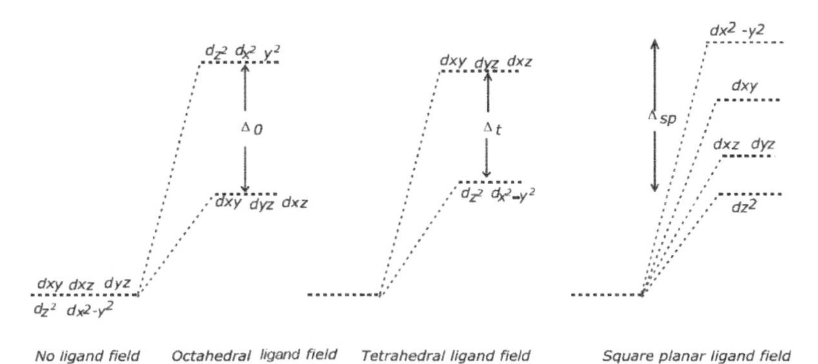

Figure 8.21 *d*–*d* transition in a different complex.

With the increase of Δ, the energy required for *d*-*d* transition is more therefore, the wavelength of absorption maximum decreases.

As shown in Figure 8.22, Δ_o is higher in the case of the complexes having a strong field ligand, i.e., CN^- as compared to the weak field ligand (NH_3). Hence wavelength of absorption maximum is higher in the case of complex bearing ammonia ligand. The transition metal complexes accordingly show the color complementary to a particular wavelength of the absorbed light.

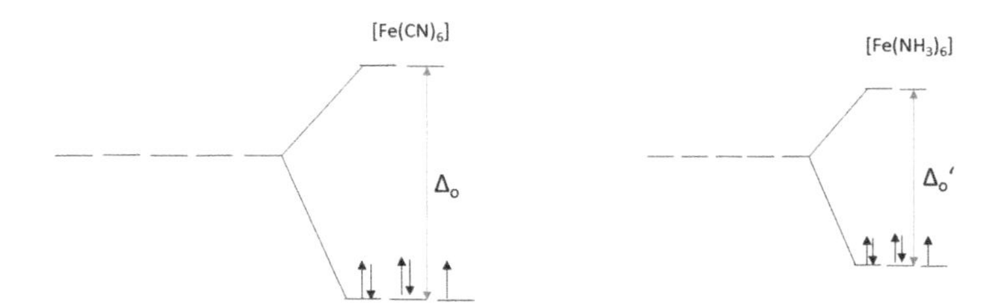

Figure 8.22 *d–d* transitions and Δ_o values of the strong field and weak field complex.

8.11 Charge Transfer Transition

The coordination complex shows intense color which is because of the charge transfer. Charge transfer refers to the transfer of electrons from the metal to the ligand or vice versa. These transitions are Laporte as well as spin-allowed transitions. They are redox transition i.e. the system transfer the electron is oxidized and the other which receives it is reduced. Mostly it occurs in the ultraviolet region. However, it may also occur in the visible region of the electromagnetic spectrum. Transfer in the visible region is characterized by intense color. The most important electronic transitions are given below.

8.11.1 Ligand to Metal Charge Transfer

Transfer of electron takes place from field ligand orbital to the vacant metal ion in the transition metal complexes. This type of transition is generally shown by π donor ligands such as halides, sulfides, selenides, azides, alkoxides, etc. To make electron transition feasible, the metal must be in a high oxidation state possessing high-energy vacant orbital. Metal shows this property may belong to the main group, transition, or inner transition. However, the ligand should have low energy and possess filled orbitals. In this process, internal oxidation of the ligand and internal reduction of the metal ion takes place. Figure 8.23 shows the transition of an electron from the filled orbital of the ligand to the high-energy vacant orbital of the metal ion. As the t_{2g} energy level of the metal ion is completely occupied hence transition takes place in e_g energy level.

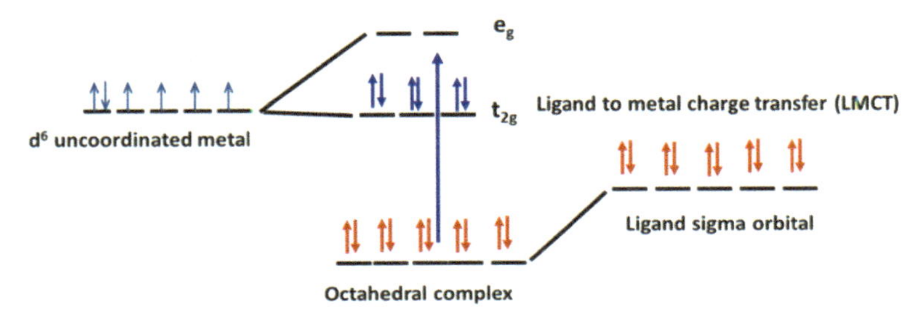

Figure 8.23 Schematic diagram of L→M charge transfer transition.

The energy of ligand to metal charge transition increases down the group, however, it decreases across the period shown in the figure. Across the period, effective nuclear charge increases, which decreases the size of the orbital. This causes the energy to reduce in the order of V > Cr > Mn.

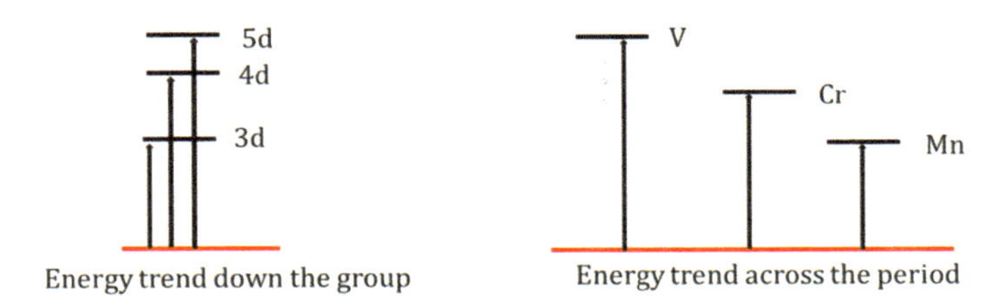

8.11.2 Metal to Ligand Charge Transfer

In this process, electron transfer takes place from the filled metal orbital to the ligands bearing vacant π orbital. Therefore, metal gets oxidized with simultaneous reduction of the ligand. Metal orbital should be in a low oxidation state and have lower potential energy. The ligand should have high-energy vacant orbital. Such types of ligands are CO, CN, NO, bipyridyl, pyridyl, imine, etc.

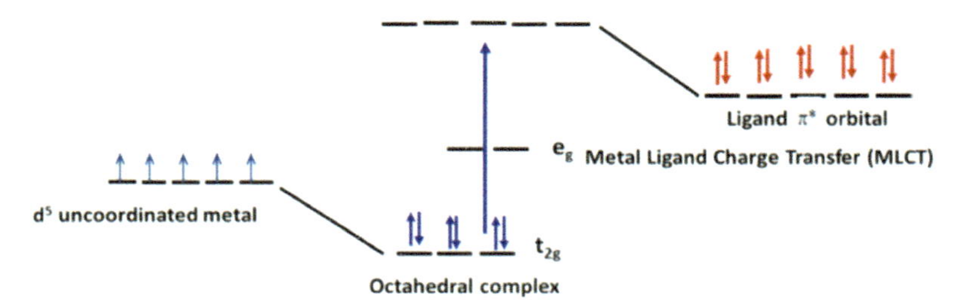

Figure 8.24 Schematic diagram of M →L charge transfer transition.

References

1. Friedel, R. A. and Orchin, M., *Ultraviolet Spectra of Aromatic compounds* (John Wiley, New York), 1951.
2. Hershenson, H. M., *Ultraviolet Absorption Spectra: Index for 1954–1957* (Academic Press, New York), 1959.
3. Parikh, V. M., *Absorption Spectroscopy of Organic Molecules* (Additicn-Wesley Publishing, Reading, MA), 1974.
4. Scott, A. I., *Interpretation of the Ultraviolet Spectra of Natural Products* (Pergamon Press, New York), 1964.
5. Silverstein, R. M. and Webster, F. X., *Spectrometric Identification of Organic Compounds*, 6th Ed. (John Wiley, New York), 1998.
6. Stern, E. S. and Timmons, T. C. J., *Electronic Absorption Spectroscopy in Organic Chemistry* (St. Martin's Press, New York), 1971.
7. Pavia, D. L., Lampman, G. M., and Kriz, G. S., *Introduction to Spectroscopy*, 3rd Ed. (Thomson Learning Academic Resource Center), 2001.

Chapter 9

Fluorescence and Phosphorescence Spectroscopy

M. Swaminathan
Department of Chemistry, Kalasalingam Academy of Research and Education, Krishnankoil 626126, India
m.swaminathan@klu.ac.in

9.1 Introduction

Luminescence is generally applied for any emission from the excited state of an atom or molecule. If it occurs from a molecule excited by a photon, it is called molecular photoluminescence. When a molecule is excited chemically, we get chemiluminescence, and similarly, thermal excitation gives thermoluminescence. Photoluminescence can be classified as fluorescence and phosphorescence based on the nature of the excited state from which emission occurs. Fluorescent molecules absorb light in the UV-Visible region of 200 to 700 nm and so UV-Visible spectroscopy preludes fluorescence spectroscopy. Though UV-Visible absorption and fluorescence spectroscopy are electronic spectral techniques used for analysis, fluorescence

Spectroscopy
Edited by Preeti Gupta, S. S. Das, and N. B. Singh

ISBN 978-981-4968-32-4 (Hardcover), 978-1-003-41258-8 (eBook)
www.jennystanford.com

spectroscopy has wide applications. Fluorescence spectroscopy plays a vital role in the analysis of trace contaminants in the environment, industries, and animal cells because the fluorescence of compounds gives high sensitivity and specificity. Since it is the study of the wavelength dependence emission from the excited singlet state (S_1) it reflects the properties of the molecule in the S_1 state. Hence fluorescence spectroscopy is a powerful research tool for the investigation of excited state molecular behavior. Fluorescent compounds are known as fluorophores and the first known fluorophore was quinine [1] reported in 1845, and this compound is responsible for the development of the spectrofluorimeter in the 1950s. The fluorimetric technique was developed only in the mid-fifties. But today it has wide applications due to its high sensitivity and sophisticated instrumentation.

9.2 Principle of Fluorescence and Phosphorescence

Principles of fluorescence and phosphorescence spectroscopy were discussed by several authors in detail [2–5]. When a molecule is electronically excited to any vibrational state of singlet states S_1, S_2, or S_3 it may lose energy by a number of processes. Jablonski diagram in Figure 9.1 shows the various types of radiative and non-radiative processes [5, 6]. In general, the excited molecule comes immediately to the $S_1(v_o)$ state by vibrational relaxation, called an internal conversion. The radiative emission from the $S_1(v_o)$ state to any of the vibrational states of the ground state (S_O) is called fluorescence. Since the fluorescence originates always from v_o of S_1 state it is independent of excitation wavelength. The energy gap for the fluorescence emission is less than the energy gap for the absorption as some of the energy is lost due to vibrational relaxation. So, fluorescence is always stokes shifted (red-shifted) to the absorption. Fluorescence spectra are not much affected by a small change in temperature. At very low temperatures (77 K) the sample is frozen and so the emission is more structured and blue shifted to the normal temperature spectrum. The lifetime times of the excited singlet state of fluorophores are in the range of nano (10^{-9}) to femto (10^{-15}) seconds.

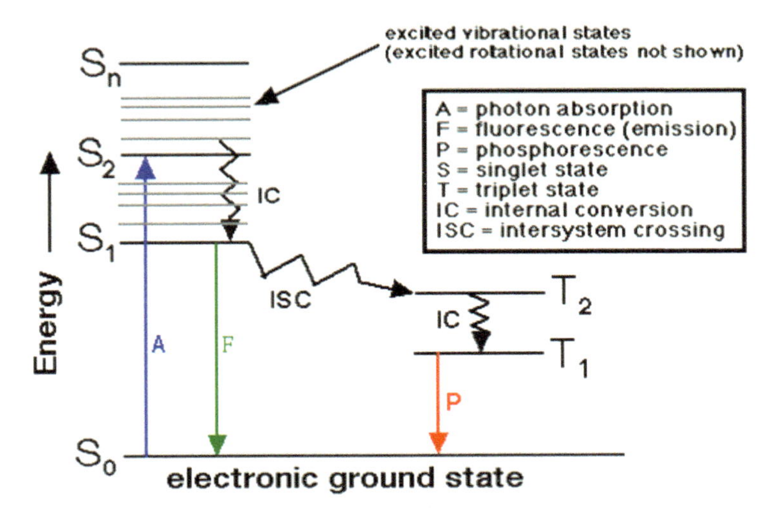

Figure 9.1 Jablonski diagram.

By intersystem crossing, the molecule from the $S_1(\nu_o)$ state may go to the triplet state (T_1) and it may lose energy from the T_1 (ν_o) state to any of the vibrational states of S_o by the radiative or non-radiative process. The emission from the $T_1(\nu_o)$ state to the ground state is called phosphorescence. Since the T_1 state is very close to the S_o state, the molecule loses the energy only through the non-radiative process at room temperature. So, phosphorescence under normal conditions cannot be observed at room temperature. At a very low temperature (77 K) it is observed. Phosphorescence is red-shifted to fluorescence and it is a delayed emission as the transition from triplet to singlet is a forbidden transition. The lifetimes of a triplet state are in the range of 10^{-2} to 10^{-4} seconds.

9.2.1 Rules Governing Fluorescence

9.2.1.1 Kasha's rule

A molecule is highly fluorescent if the electronic transition is π–π^*. If a molecule undergoes n–π^* transition, it is weakly fluorescent and highly phosphorescent [7]. In condensed media, generally same fluorescence is observed irrespective of the excitation wavelength, which is clearly understood from the Jablonski diagram (Figure 9.1).

9.2.1.2 Frank–Condon (FC) principle

Electronic transitions are very fast compared with nuclear motions. According to this FC principle, an electronic transition is most likely to occur without changes in the positions of the nuclei in the molecular entity and its environment [5]. The resulting state is called a Franck–Condon state, and the transition involved a vertical transition. Vertical excitation and emission according to the FC principle are shown in Figure 9.2.

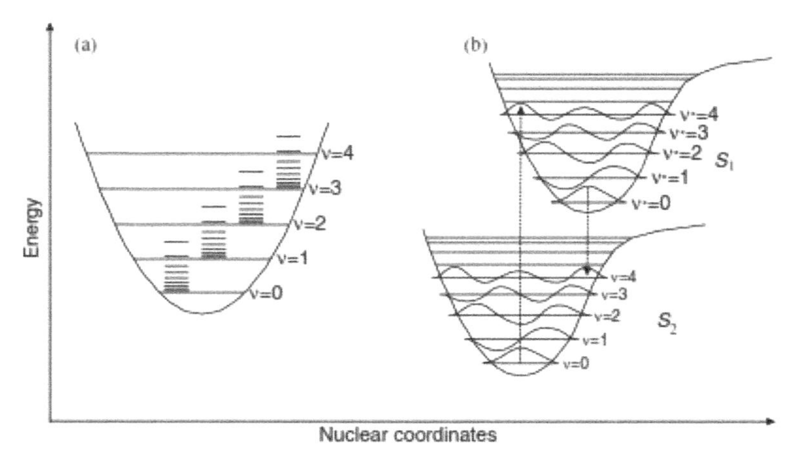

Figure 9.2 Frank–Condon principle.

9.3 Instrumentation

The Block diagram of a simple steady-state spectrofluorimeter, fabricated in our laboratory [8], is shown in Figure 9.3. Details of the fabricated instrument are described elsewhere [9].

Xe–Hg lamp (150 W) is used for excitation. Two grating monochromators one for excitation (M1) and the other for emission (M2) are used. Emission occurs at all sides and observation is made at a right angle to avoid the transmitted light entering the emission monochromator. The wavelength selected for excitation using M1 is focused on the sample, placed in a quartz cuvette and the emission is passed through M2, sensed by a photomultiplier tube

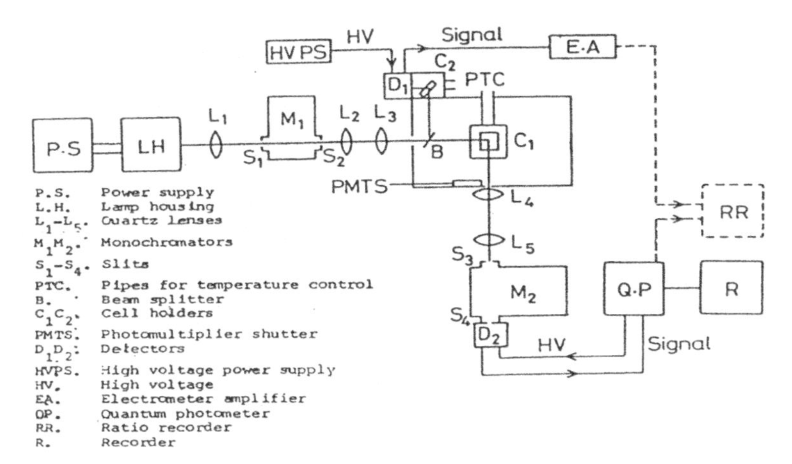

Figure 9.3 Block diagram of spectrofluorimeter.

(D_2 detector). The signal is analyzed by a quantum photometer (QP) and recorded to get a fluorescence (emission) spectrum. Cell holder C1 has a provision for water circulation to keep the temperature constant. If the emission is fixed at a wavelength by M2, and M1 is scanned, an excitation spectrum is obtained. The instrumental parameters such as excitation light intensity, the response of the monochromators, and photomultiplier sensitivity are wavelength dependent. Lamps, monochromators, and photomultipliers are wavelength dependent. Hence the spectra obtained from a steady-state fluorimeter are not true spectra. It becomes necessary to determine the correction factors for an instrument and modify the observed spectrum to get the corrected spectrum. As shown in Figure 9.3, a corrected excitation spectrum can be obtained by splitting the light into a reference cell with a solution of rhodamine B as a quantum counter and recording the ratio of the output signal from the sample detector D_2 to that of the reference detector D_1, which monitors the emission of a quantum counter. Nowadays, instrumental correction factors are fed into the computer attached to the instrument to get the corrected emission spectrum. Low-temperature fluorescence at liquid nitrogen temperature can be obtained using a low-temperature accessory shown in Figure 9.4.

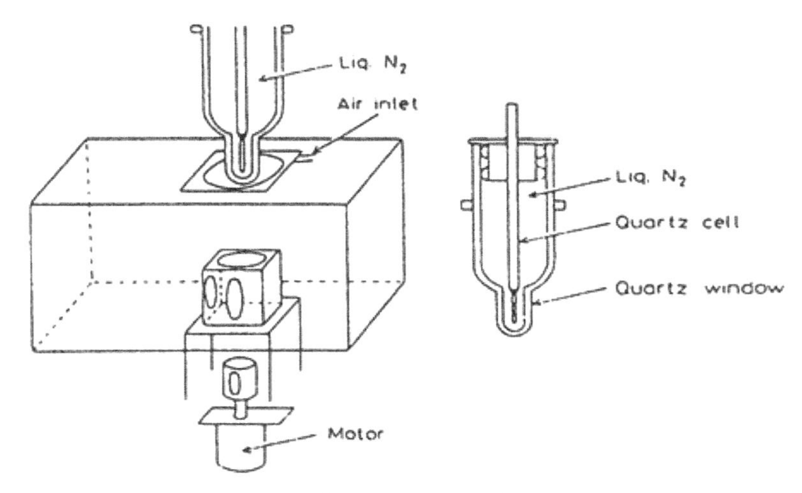

Figure 9.4 Low temperature fluorescence and phosphorescence accessory.

A quartz tube is used as a sample cell and it is placed in the Dewar flask containing liquid nitrogen. This setup is placed inside the cell container. Phosphorescence can be observed only at very low temperatures and it is a delayed emission. A chopper motor with a window is fitted at the bottom of the cell compartment and it allows the light first and after a fraction of a second depending on the speed of rotation allows the emission at 90°.

9.4 General Properties of Fluorescence and Phosphorescence

9.4.1 Fluorescence

- Fluorescence is a radiative emission due to electron transfer from $S_1(\nu_0) \rightarrow S_0$ (any vibrational state).
- Since the energy gap for emission is less than that of absorption (Figure 9.1) fluorescence is Stokes shifted to absorption.
- According to Kasha's rule, fluorescence is independent of excitation wavelength (λ_{ex}).
- Fluorescence lifetime is 10^{-9} to 10^{-15} sec.

9.4.2 Phosphorescence

- T_1 (ν_0) → S_0 (any vibrational state)
- T_1 state is close to S_0 (less energy gap)
- Lifetime → 10^{-2} to 10^{-4} sec
- Triplet → Singlet (forbidden transition)

  Phosphorescence can be distinguished from fluorescence by its delayed emission, longer lifetime, and larger redshift.

  When compared to the UV-Vis absorption technique, fluorescence has more characteristics and parameters. Hence applications of absorption spectra are limited. Some of its applications are
- For quantitative analysis (analytical technique) using the equation

  $A = \varepsilon Cl$ or calibration graph
- Determination of the composition of complexes
- Determination of ground state pK_a value
- Study of charge transfer complexes

  But fluorescence can be used widely for more applications since it has many parameters and measurements.

9.5 Fluorescence Parameters

Some of the fluorescence measurements are:

9.5.1 Fluorescence Emission Spectra (λ_{ex}: Fixed)

Fluorescence intensity, shape, and peak position can be used to infer the properties of molecules in the excited state.

The intensity of fluorescence I_f varies with the instrument

$$I_f = IA\phi \tag{9.1}$$

where I, the intensity of excitation light; A, absorbance at λ_{exi}; and ϕ, quantum yield.

A calibration graph is a must for the quantitative analysis of a fluorescent compound. Shape and peak position are characteristics of a compound.

9.5.1.1 *Stokes's shift*

Except for atoms in the vapor phase, the emission is always shifted to a lower wavelength (red-shifted) relative to absorption. The difference in frequency of absorption and emission maxima is called a stokes shift. Stokes shift gives an idea of the extent of relative stabilization of the ground state and the emissive state in the solvent medium. Stoke's shift also reveals the change in some physical characteristics such as the dipole moment of molecules on excitation from ground to excited state [5]. Applications of the fluorescence spectrum will be discussed later.

9.6 Fluorescence Excitation Spectra [Wavelength of Emission (λ_{em}): Fixed]

The corrected excitation spectrum should exactly match the emission spectrum if there is no change in the shape and structure of the molecule during excitation. In this case, we get mirror image symmetry of absorption and emission spectrum. Mirror image symmetry exhibited by anthracene molecule is shown in Figure 9.5.

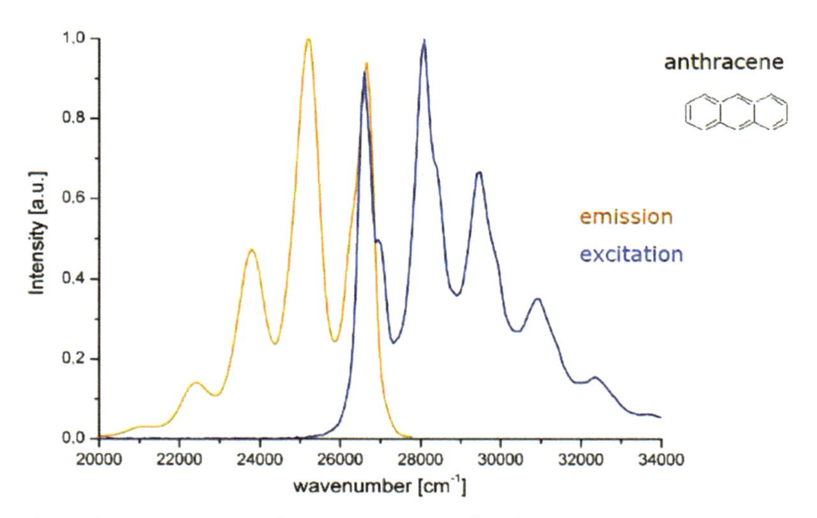

Figure 9.5 Excitation and emission spectra of anthracene.

Since anthracene is a condensed fused ring system, it does not undergo any change during excitation. Some molecules, which change their shape or undergo some interactions during excitation, will not have mirror image symmetry. Deviations from the mirror image symmetry indicate a change in geometric arrangement on excitation from the ground to the excited state. Comparison of excitation and absorption spectra helps in understanding such geometry changes.

9.7 Fluorescence Quantum Yield

Fluorescence quantum yield is an important parameter and characteristic of fluorescing molecules (fluorophores).

$$\text{Absolute quantum yield} = \frac{\text{Flourescence of all } \lambda \text{ and at all directions}}{\text{Light absorbed}}$$

As it is very difficult to measure the absolute quantum yield of fluorescence, relative quantum yield is used for practical applications. From the absolute quantum yield of some representative fluorophores (1), the relative quantum yield is determined for other fluorophores (2) by applying Eq. (9.1) for each fluorophore. Quinine sulfate is taken as a standard representative fluorophore for which quantum yield was determined and reported.

$$I_f \text{ of } F_1 = I_1 A_1 \phi_1$$

$\phi_1 = \phi_f = 0.55$ (quinine sulfate in 0.1 N HCl); ϕ_f = fluorescence quantum yield

$$F_2 = I_2 A_2 \phi_2$$

When λ_{ex} is the same for both fluorophores, $I_1 = I_2$

$$\frac{F_2}{F_1} = \frac{A_2\phi_2}{A_1\phi_1}\ ;\ \phi_2 = \frac{F_2 A_1 \phi_1}{F_1 A_2} \tag{9.2}$$

F_1 and F_2 are determined from the area of fluorescence curves. Substituting absorbencies at the excitation wavelength (A_1 and A_2), ϕ_1, F_1, and F_2, the relative quantum yield of unknown fluorophore (ϕ_2) can be obtained.

9.8 Fluorescence Lifetime

The lifetime of excited singlet state τ_o is measured easily by a single photon counting method using a time-resolved spectrofluorimeter. Fluorescence lifetime (τ_f) is less than the singlet state lifetime because the molecule may lose energy from the singlet state by both radiative and non-radiative processes as shown in the Jablonski diagram.

$$F_t = F_0 e^{-t/\tau_0} \tag{9.3}$$

The fluorescence decay curve (F_t vs. t) is obtained from the instrument. τ_o is the time required for fluorescence to decay to $1/e$ of its initial value. The fluorescence decay curve is shown in Figure 9.6 and the value of τ_o is indicated in Figure 9.6. If we plot the curve of $\log F_t$ vs. t, then the slope is equal to $1/\tau_o$.

The fluorescence lifetime, $\tau_f = \tau_o \phi_f$, if $\phi_f = 1$, $\tau_f = \tau_o$

Fluorescence lifetime is also an important parameter of a fluorophore and its change under different environments may be used for different applications. Quantum yield and lifetime are used as powerful tools for investigating the dynamical processes of the excited state.

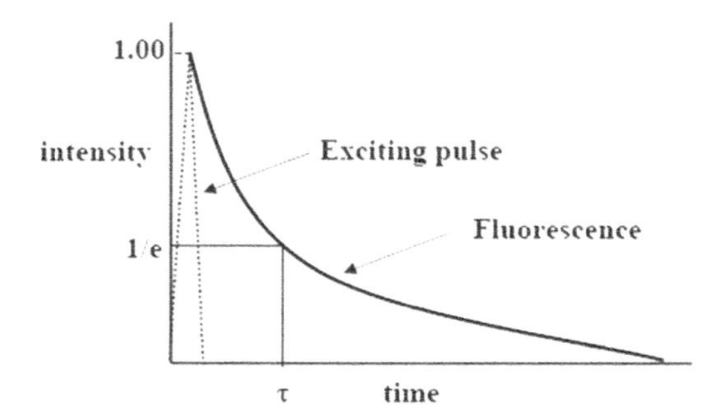

Figure 9.6 Fluorescence decay curve.

9.9 Synchronous Fluorescence Spectroscopy

Apart from the conventional emission and excitation spectra in which either λ_{ex} or λ_{em} is varied, there is a possibility to vary both

λ_{ex} and λ_{em} simultaneously while recording the spectrum. This technique has three variants, based on the scan rates of excitation and emission monochromators: (i) Synchronous excitation fluorescence spectroscopy (constant $\Delta\lambda$ between λ_{ex} and λ_{em}), (ii) constant energy synchronous fluorescence spectroscopy (constant $\Delta\boldsymbol{\upsilon}$ between λ_{ex} and λ_{em}), and (iii) Variable separation synchronous fluorescence spectroscopy (different rates for λ_{ex} and λ_{em} variation). Conventional and synchronous scan processes are displayed in Figure 9.7.

Synchronous fluorescence spectroscopy

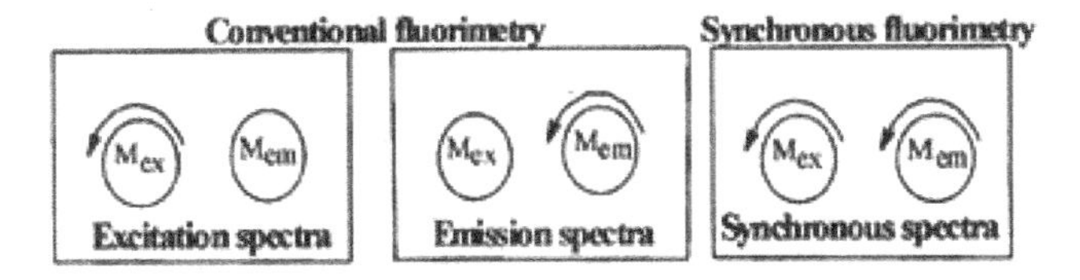

Figure 9.7 Comparison between conventional excitation/emission scan process from conventional fluorimetry and synchronous fluorimetry.

Among these three, synchronous excitation fluorescence spectroscopy with a fixed $\Delta\lambda$ is the simple and most frequently used mode. Synchronous spectra are simpler, with narrower peaks and hence more characteristic than conventional spectra. The intensity of synchronous fluorescence, I_s, depends on the character of normal excitation and emission spectra as well as on the wavelength difference, $\Delta\lambda$, between λ_{ex} and λ_{em}. It is given by the equation:

$$I_s = KcdE_x(\lambda_{ex})E_m(\lambda_{ex} + \Delta\lambda) \quad \text{in terms of excitation function} \quad (9.4)$$

or, alternatively as

$$I_s = KcdE_x(\lambda_{em})E_m(\lambda_{em} - \Delta\lambda) \quad \text{in terms of emission function}$$

where E_x is the excitation function at a given wavelength ($\lambda_{em} - \Delta\lambda$), E_m is the normal emission intensity at the corresponding emission wavelength ($\lambda_{ex} + \Delta\lambda$), c is the analyte concentration, d is the thickness of the sample and K is a luminescence constant comprising the "instrumental geometry factor" and other related parameters. The synchronous signal can be taken either as an emission or excitation spectrum, with a synchronously scanned excitation or emission wavelength respectively. In conventional spectrofluorimetry, it is possible to increase the intensity of all bands at the same time whereas the synchronous technique allows the stronger peaks to be increased selectively by use of a suitable $\Delta\lambda$. The maximum fluorescence intensity for a particular component in the mixture occurs when $\Delta\lambda$ is equal to the difference between the wavelengths of absorption and emission maxima.

9.10 Fluorescence and Structure

All molecules which absorb light are not fluorescent. There is no hard and fast rule for the molecules to be fluorescent, but there are some generalizations for molecules to be fluorescent

- A molecule with π electron system of increased conjugation. Vitamin A is fluorescent. Fluorescence increases from benzene to naphthalene and also shifted to a longer wavelength.
- A system with less number of substituents.

Benzene is fluorescent, but 1,3,5-trimethyl benzene is non-fluorescent. If the number of substituents is more, there is a chance of losing energy by vibrational relaxations. The intensity of fluorescence decreases as loose bolts in an engine decreases its efficiency.

- The lowest S_1 state is π–π^*
  Naphthalene is fluorescent, but quinoline in non-polar solvents is non-fluorescent as the lowest S_1 state is n–π^*.
- *o*- and *p*-directing groups increase fluorescence efficiency.
  Benzaldehyde is non-fluorescent whereas aniline and phenol are fluorescent.
- Molecular rigidity increases the fluorescence emission
  Phenolphthalein is non-fluorescent; rigidly structured fluorescein is highly fluorescent. Similarly, malachite green is non-fluorescent, but rhodamine B is highly fluorescent.
- Heavy atom substitution reduces the fluorescence efficiency. Heavy atoms induce the transition from singlet to triplet state and hence heavy atom substitution favor phosphorescence.

$S_1 \rightarrow T_1$ transition is induced

	ϕ_p/ϕ_f
1-Fluoronaphthalene	0.068
1-Chloronaphthalene	5.2
1-Bromonaphthalene	6.4
1-Iodonaphthalene	>1000

9.11 Fluorescence and Temperature

Fluorescence is not much affected by a small change in temperature, but at a very low temperature, the sample gets frozen and a structured spectrum, blue shifted to normal spectrum is obtained. This is because, the emission occurs from the Franck–Condon states before undergoing vibrational relaxation. Fluorescence at a liquid nitrogen temperature of 77 K can be taken using a phosphorescence accessory without a chopper motor. Low temperature is very useful to investigate the emission of fluorophore in its ground state characteristics before undergoing any change in the excited singlet state.

9.12 Intrinsic and Extrinsic Fluorophores

Intrinsic fluorophores are naturally occurring fluorescent molecules such as aromatic amino acids, neurotransmitters, porphyrins, and green fluorescent protein. Extrinsic fluorophores are synthetic dyes or modified biochemicals that are added to a specimen to produce fluorescence with specific spectral properties. They are extensively used for fluorescent labeling. Nowadays fluorescent labeling is of paramount importance to biological studies and numerous chemical dyes are used extensively to label biological specimens. In this process, small organic fluorophores are incorporated into proteins by genetic fusion to produce a fluorescent label. The development of new extrinsic fluorophores becomes an essential element for the design of new fluorescent probes. Rhodamine B is used to label proteins.

9.13 Applications

Fluorescence is a highly sensitive technique and about fourteen parameters have been identified, which can be combined to increase fluorescence selectivity and sensitivity for various applications. Nowadays the use of fluorescence measurements covers the entire area from understanding the basic photophysics of isolated molecules to chemical reaction dynamics to complex biological and environmental processes. We focus our discussion on some of the commonly used parameters, discussed earlier.

9.14 Fluorophores: Intrinsic

9.14.1 Concentration Determination

Since fluorescence is a highly sensitive technique, the concentration of a fluorophore can be determined up to 10^{-6} M or even at less concentrations depending on its quantum yield. As the intensity of fluorescence is an arbitrary unit and differs for each instrument, a calibration graph of I_f vs. [C] is a must for the determination of fluorophore concentration. A mixture of two components having an overlap in the absorption spectra can be analyzed. Two samples having similar absorption spectra may be distinguished if they fluoresce at different wavelengths. Similarly, two analyte molecules having similar fluorescence spectra may be differentiated by the proper choice of excitation wavelength (selective excitation). Fluorescence spectra of three amino acids tryptophan, tyrosine, and phenylalanine, are given in Figure 9.8. Concentration of tyrosine (Tyr), tryptophan (Trp), and phenylalanine (Phe) can be determined by choosing their emission wavelength with no or less overlap and by choosing different excitation wavelength.

Fluorescence can be used to check the purity of the compound. As the fluorescent spectrum is independent of excitation wavelength, the emission spectrum is the same for different excitation wavelengths, if the compound is pure. The presence of fluorescing impurities can be detected by this method. But this method has some limitations as it cannot detect non-absorbing or non-fluorescing impurities.

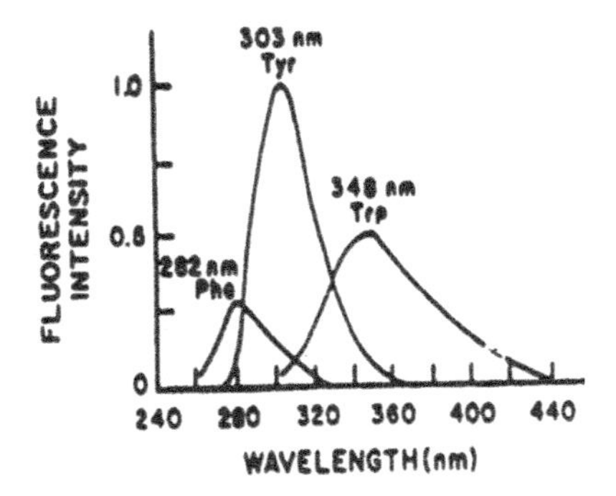

Figure 9.8 Fluorescence spectra of phenylalanine, tyrosine, and tryptophan.

9.14.2 Study of Excited State Properties

As fluorescence originates from the excited singlet state, it reflects the physical and chemical properties of a fluorophore in its excited singlet state. The changes in physical and chemical properties such as acidity/basicity, geometry, dipole moment, etc. on excitation can be studied.

9.14.3 Excited State Acidity Constants

Weber [10] was the first to report the abnormal fluorescence of due to the protolytic dissociation of 1-naphthylamine-4-sulfonate in 1931. Later in 1950, Forster [11] showed this effect to arise in protolytic dissociation in solutions containing hydroxyl and aminopyrene sulfonates in their first excited singlet state. Similar results were observed with hydroxyl and other naphthalene derivatives [12, 13]. It was found that the ground state pK_a of 9.8 of 1-naphthol changed to pK_a^* of −1 in the excited singlet state [14]. The acidity of 1-naphthol increased so much on excitation. Then a number of organic compounds with acidic and basic groups were analyzed fluorimetrically and found that compounds containing −OH, −SH, $-NH_2$, >NH ($-\phi$) groups become stronger acids and weaker bases in the S_1 state than S_0 state and −COOH, −COR, −CSR, =N− ($-\phi$) become stronger bases in S_1 state than in S_0 state. There are two methods to determine the excited singlet state pK_a values.

9.14.3.1 *Forster cycle method: Theoretical*

Forster cycle method [11, 12] uses absorption or fluorescence maxima in wave number for acid (BH, ν_{BH}) and its conjugate base (B^-, ν_{B^-}) for the calculation of ΔpK_a ($pK_a - pK_a^*$) using the following equation.

$$pK_a - pK_a^* = 2.1\times10^{-3}\left(\nu_{BH} - \nu_{B^-}\right) \text{ at } 25^\circ C \tag{9.5}$$

Even if the compound is non-fluorescent, this method can make use of absorption maxima for the calculation of ΔpK_a.

9.14.3.2 *Fluorimetric titration method (pH dependence of fluorescence)*

Weller [14] developed a method to find out the excited state pK_a value using fluorescence intensity as a function of pH of the system. The fluorescence of a molecule depends on the pH, which decides whether the molecule exists as an acid or conjugate base in the excited state. Spectrophotometric titration of pH dependence of the absorption of an acid and its conjugate base will reveal the ground state pK_a value, whereas fluorimetric titration of pH dependence of fluorescence will give excited state pK_a (pK_a^*) value. A typical fluorescence titration curve obtained for 2-amino-7-bromofluorene [15] is given in Figure 9.9.

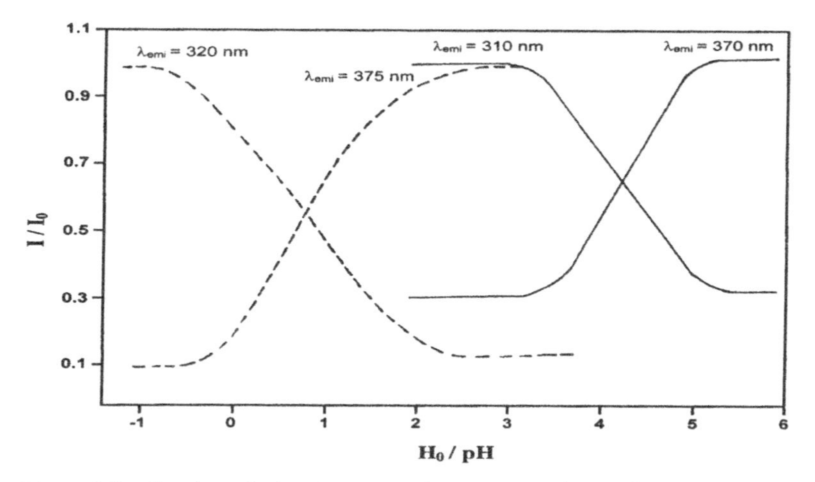

Figure 9.9 Fluorimetric titration curves for 2-amino-7-bromofluorene without and with β-cyclodextrin (CDx) (——without β-CDx, -----with 0.002 M β-CDx).

In case acid–base equilibrium is not attained within the lifetime of the excited singlet state fluorimetric titration curve will give only the ground state pKa value, because before prototropic reaction, the excited molecule loses the energy by emission. Hence, the fluorimetric titration curves should be properly investigated. Dogra et al. studied the excited state acidity constants of a number of organic acids and bases elaborately [16–20]. We studied the excited state prototropism in detail and discussed various types of pH-dependent fluorimetric titration curves [21–25].

This excited state prototropism may lead to phototautomerism, in which a proton is shifted from one atom to another atom within the molecule (intramolecular) or another atom with a different molecule (intermolecular) and they are called intramolecular phototautomerism and intermolecular phototautomerism. The abnormal fluorescence of salicylic acid was first reported by Weller [26] in 1956. In 1968, Schulman explained the reason for the highly red-shifted fluorescence of salicylic acid relative to *o*-anisic acid, in which –OH is replaced by $-OCH_3$ + group [27]. When salicylic acid is excited, emission occurred from the salicylate anion. On excitation, salicylic acid immediately forms a zwitterion in the excited state as OH becomes more acidic and C=O in COOH becomes more basic in the excited state. This is illustrated in Figure 9.10.

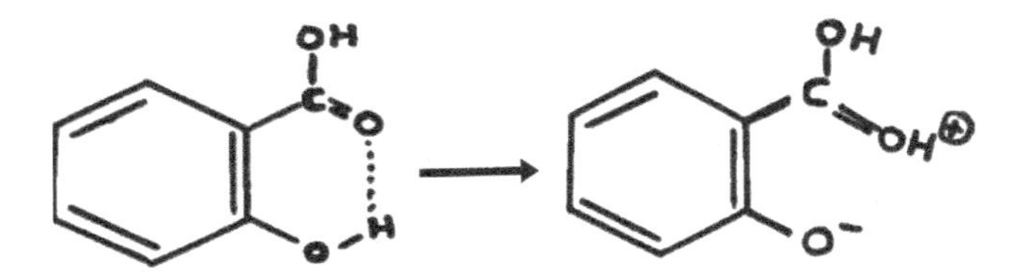

Figure 9.10 Proton transfer from –OH to C=O of COOH on excitation.

Swaminathan and Dogra reported phototautomerism in 5-aminoindazole [20]. In this molecule excited state acid–base equilibrium is entirely different from the ground state. As seen in Figure 9.11, the amino group is basic and pyrrolic nitrogen is acidic in the ground state, but in the excited state amino group becomes acidic and pyridinic nitrogen is more basic. Hence intramolecular phototautomerism is observed with the monocation.

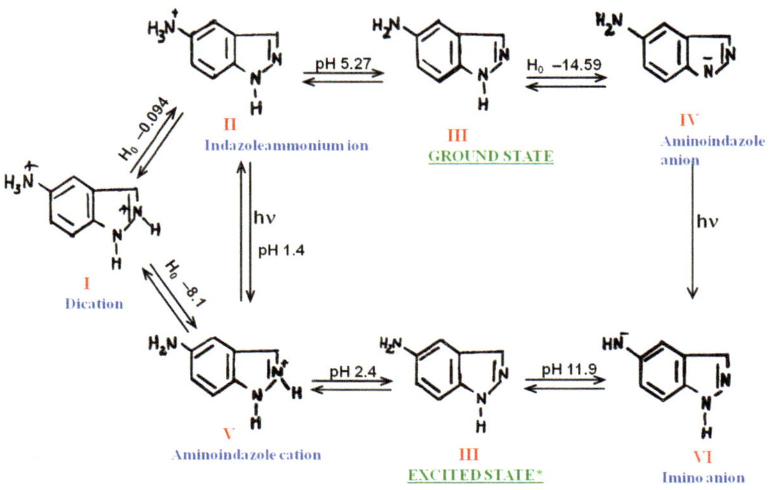

Figure 9.11 Scheme of ground and excited state equilibria of 5-amino indazole at different H_0/pH/H^- (I) dication, (II) indazole ammonium ion, (III) neutral form, (IV) 5-aminoindazole anion, (V) 5-aminoindazole cation, and (VI) imino anion [28].

The ground state indazole ammonium ion is changed to aminoindazole cation in the excited state. This is also discussed in the book on proton applications in chemistry [28].

The excited state geometry of a molecule, if different from ground state geometry can be analyzed by the mirror image symmetry of absorption and fluorescence. As seen in Figure 9.5, the mirror image symmetry of anthracene absorption and emission revealed the same geometry in the ground and excited state. But in the case of biphenyl molecule, it is non-planar in the ground state and attains planarity in the excited singlet state and this change is shown by the absence of mirror image symmetry of the absorption and fluorescence (Figure 9.12).

Absorption is broad and fluorescence is structured due to the attainment of planarity. We reported a change in geometry of the 4,4′-diaminobiphenyl molecule in the excited state [29]. In this compound, biphenyl moiety attains planarity in the excited singlet state.

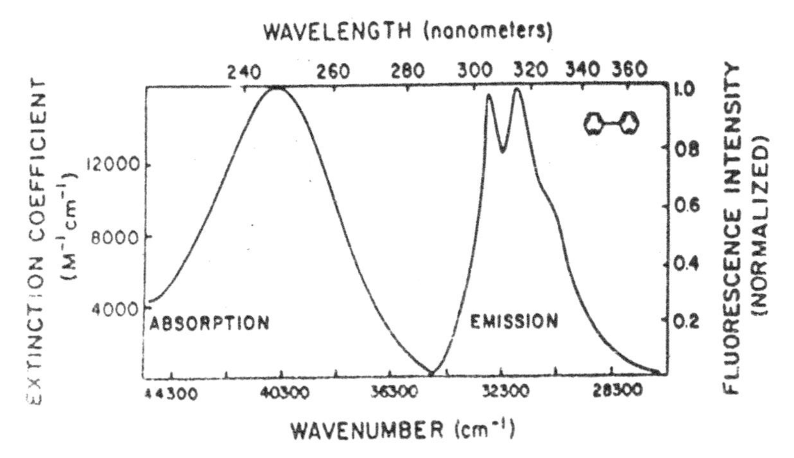

Figure 9.12 Absorption and emission spectra of biphenyl.

9.15 Solvotochromic Shifts

The interaction between the solvent and fluorophores affects the energy difference between ground and excited electronic levels, which in turn shifts the absorption and emission maximum of the fluorophore. They are known as solvatochromic shifts and their value depends on the polarity of the solvent and change in the polarity of the fluorophore on excitation. Lippert equation given below correlates the stokes shift with the dipole moment of fluorophore and solvent parameters (dielectric constant and refractive index).

Lippert equation

$$\Delta\upsilon_{ss} = \left(\frac{\Delta\mu^2}{hca^3}\right)\text{F1} \tag{9.6}$$

$$F1 = \left\{2(\epsilon - 1)(2\epsilon + 1\right\}\left\{2(\tilde{\eta}^2 - 1)/(2\tilde{\eta}^2 + 1\right\}$$

$$\Delta\mu = \mu_e - \mu_g$$

a = onsagar cavity radius

By using this equation excited state dipole moment of fluorophore can be determined. Fluorescent molecules with large solvatochromic shifts have been developed for use as fluorescent probes in estimating micropolarity in a micro heterogeneous medium, helping elucidate protein structure. Calculation of excited state dipole moment and the use of fluorescent probes for estimating micropolarity are elaborately discussed in Chapter 7 of the book *Principles of Fluorescence Spectroscopy* by J. R. Lakowicz [5].

Some of the fluorescent molecules undergo intramolecular charge transfer on excitation and so emission is observed from a twisted intramolecular charge transfer state (ICT). Since this state is more polar, it is more stabilized by polar solvents. Solvotochromic shifts can reveal the formation of the ICT state in fluorophores. ICT state was reported in 4-dimethylaminobenzonitrile containing electron withdrawing and electron releasing groups at p-position. Locally excited emission (LE) and emission from ICT of a molecule in less polar and polar solvents are shown in Figure 9.13.

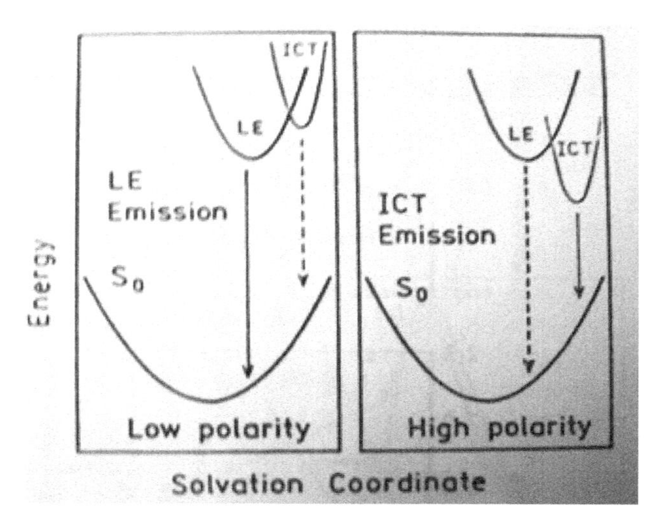

Figure 9.13 Emission from the locally excited state (LE) and intramolecular charge transfer state (ICT) in low- and high-polar solvents.

9.16 Fluorescence Quenching

Fluorescence quenching by suitable quenchers has been widely used as one of the common methods for studying the energy of the excited state and the interactions between fluorophore and quenchers.

Fluorescence enhancement and quenching are also used for sensing metal ions. Basically, there are two types of quenching, static and dynamic. In static quenching, the interaction of the quencher in the ground state makes the fluorophore into a non-fluorescent molecule or complex. In the dynamic quenching process, the quencher interacts with the excited fluorophore resulting in fluorescence quenching. Quantum yield or intensity of fluorescence at wavelength maximum is correlated to the concentration of quencher by Stern–Volmer equation.

$$I_0/I = 1 + K_{SV}[Q] \quad (9.7)$$

I_0, the intensity of fluorescence without quencher; I, the intensity of fluorescence with a quencher, K_{SV} Stern–Volmer quenching constant. A plot of I_0/I vs. [Q] gives K_{SV} a slope. Linearity of SV plot reveals (i) only one quenching mechanism is operative, i.e., either static or dynamic; (ii) quenching is bimolecular.

As $K_{SV} = k_q\ \tau_{o,}$ the quenching rate constant, k_q can be determined from the lifetime (τ_o) of the excited singlet state.

Non-linearity in Stern–Volmer plot shows that more than one mechanism is operative. If the mechanism involves both static and dynamic quenching, SV curve will not be linear. A mixed trend for quenching of 4-aminodiphenyl fluorescence was observed and a static-dynamic model was proposed and discussed [30]. To differentiate between static and dynamic models, a quenching curve can be drawn with the lifetimes of the fluorophores at different concentrations of quencher (τ_o/τ) vs. [Q]. This curve is not affected by static quenching, if present. For pure static quenching, there will not be any change in a lifetime with quencher concentration and the curve will be a horizontal line. For both static and dynamic quenching, a number of mechanisms like collisional, electron transfer, formation of non-fluorescent complex, excimer, exciplex, resonance energy transfer, photoinduced electron transfer (PET), metal complex formation, etc., have been reported and discussed. Fluorescence quenching can be used as an analytical means for determining the concentration of compounds, metals, and gases which quench fluorescence and it is also used to reveal the localization of fluorophores in proteins, membranes, and other biological systems. Based on the mechanism of quenching, quenching study has been used for a variety of applications. Fluorophores

2, 3-diaminophenazine, 1, 2-diamino-anthraquinone, and 2, 4-dinitrophenylhydrazine have been used as fluorescent probes for selective detection of Cu^{2+} ions in an aqueous medium over other metal ions [31]. Quenching mechanisms and the applications of quenching are discussed elaborately in Chapters 8 and 9 of the book by J. R. Lokowicz [5].

Fluorimetry is used in sensing air and water pollutants. Selective probes have been developed for the detection of metal ions, and chloride ions in environmental and biological samples. The presence of trace amounts of metal pollutants in water samples can also be analyzed by fluorimetry [5]. Methods based on the phenomenon of chemiluminescence are being effectively used for the determination of air pollutants NO-NO_2 and SO_2 up to the level of 1 ppb.

Fluorescence of inclusion complexes: Cyclodextrins and their derivatives have the property of forming inclusion complexes with many drugs. The inclusion complex formed between beta-cyclodextrin and dapsone is shown in Figure 9.14.

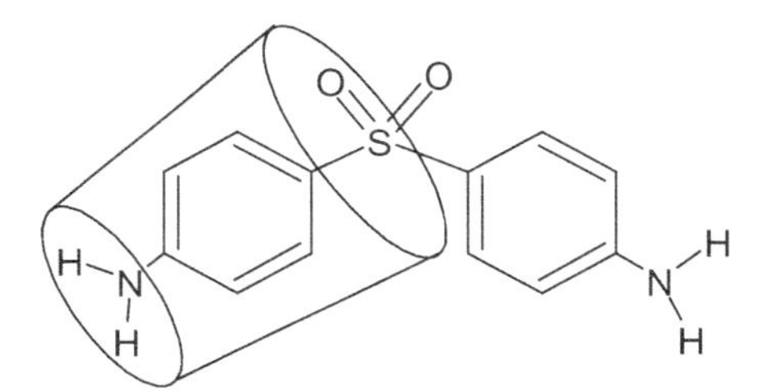

Figure 9.14 Inclusion complex of dapsone.

This drug complexation has the following advantages:

- Liquid compounds can be transformed into a crystalline form which is suitable for tablet manufacturing.
- The physical and chemical stabilities of drugs are improved.
- The bioavailability of poorly soluble drugs can be enhanced.

The stoichiometry of the complex and solubility of the drug can be well studied by the fluorometric technique [32–35]. Since the complexation enhances the fluorescence, this method is found to be an easy and highly sensitive one for the assay of fluorometric drugs in pharmaceutical preparations.

9.17 Synchronous Fluorescence Spectroscopy Applications

Synchronous fluorescence spectroscopy (SFS) has been successfully used for the simultaneous determination of polycyclic aromatic compounds in the environment [36–39]. Petroleum products contain highly fluorescent polycyclic hydrocarbons (PAH). Fluorescent properties of these PAH have been used to find out the adulteration of petrol or diesel with cheaper kerosene. Adulteration of petrol and diesel with kerosene using SFS was investigated by several researchers [40, 41]. A synchronous spectrum of petrol with different concentrations of kerosene is shown in Figure 9.15.

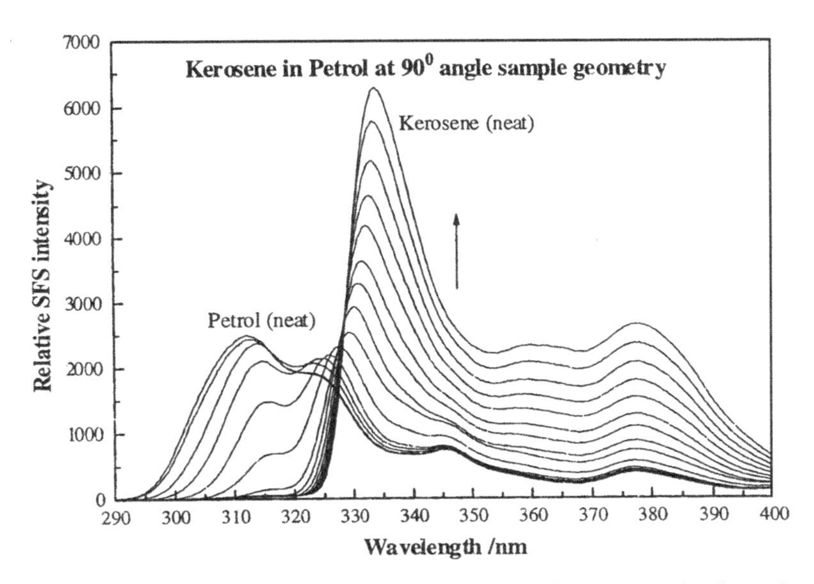

Figure 9.15 Synchronous fluorescence scan spectra of a neat sample of petrol, petrol contaminated by kerosene at 0.1, 1, 2.4, 5, 10, 20, 30, 40, 50, 60, 70, 80, and 90% v/v and a neat sample of kerosene at 90° angle sample geometry. $\Delta\lambda$ = 30 nm. The excitation and emission slit widths are kept at 5.0 nm [40].

Based on the calibration curve, the concentration of kerosene adulterated in petrol can be determined. The synchronous spectrofluorometric method has been reported for the simultaneous analysis of a binary mixture of metoprolol (MTP) and felodipine

(FDP) using the synchronous fluorescence intensity of the two drugs at $\Delta\lambda$ of 70 nm in an aqueous solution [42]. SFS has the ability to analyze complex multi-component mixtures without resorting to tedious separation procedures.

9.18 Phosphorescence Spectroscopy

As seen in the Jablonski diagram (Figure 9.1) phosphorescence occurs when a molecule in the lowest triplet T_1 loses its energy by the radiative transition. But at room temperature under normal conditions molecule loses its energy by thermal relaxation because the energy difference between T_1 and S_0 is very small. Phosphorescence can be observed from the molecules under frozen conditions, i.e at a very low temperature of (77 K) liquid nitrogen. Phosphorescence is observed at a longer wavelength, red-shifted to fluorescence. Excitation, fluorescence, and phosphorescence spectra of phenanthrene are shown in Figure 9.16.

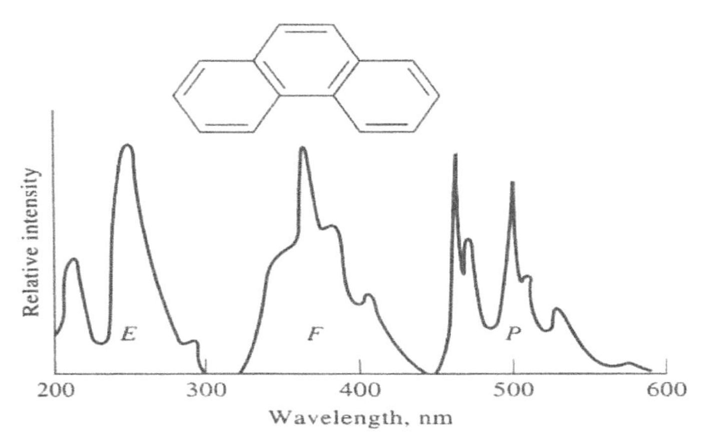

Figure 9.16 Excitation (left), fluorescence (middle), and phosphorescence (right) of phenanthrene.

Phosphorescence sensitivity is comparable to fluorescence but its analytical use is limited due to the practical aspect of measuring the signal at 77 K. It is possible to observe phosphorescence at room temperature if the molecule is made rigid by some method. Swaminathan and Dogra analyzed the phosphorescence spectra of pyrazoles at 77 K and determined excited triplet state pK_a^* (T_1) values using the Forster cycle method [43]. Solid-surface room temperature phosphorescence (SSRTP) has been generally observed by immobilizing the lumiphor on a solid support such as filter paper or polymeric resins [44]. SSRTP has been shown to be a sensitive and selective approach for the analysis of trace components in biological, pharmaceutical, and environmental samples. Room temperature phosphorescence (RTP) can be observed in solution immobilizing the phosphors in ordered media like micelles, cyclodextrins (CDs) and other caging molecules. RTP has focused on sensitized phosphorescence (SP) the triplet energy of a donor molecule is transferred to an acceptor molecule, which phosphoresces in fluid solution and the emission of the acceptor is monitored. SP provides only quantitative information. The long-lived triplet state is protected by micelle or CD from the quenching processes and allows both quantitative and qualitative measurements. Phosphorescence and fluorescence methods tend to be complementary because strongly fluorescing compounds exhibit weak phosphorescence and vice versa. Condensed-ring aromatic hydrocarbons containing heavier atoms such as halogens or sulfur phosphoresce strongly. The presence of heavy atoms like iodine, silver, lead, etc. in the sample or the solid matrix is found to improve the sensitivity of the technique and helps in the analysis of complex mixtures by RTP.

Phosphorescence measurements are mainly applied in the fields of drugs and pharmaceutical and pesticide analysis. Drugs like phenobarbital, cocaine, procaine, chlorpromazine, and salicylic acid as well as sulphonamide drugs exhibit phosphorescence [45, 46]. Several metal ions like transition elements (Cu, Zn, Nb, Gd) as well as s-block elements (Be) have been analyzed by phosphorimetry [44, 47].

9.19 Conclusions

In this chapter, we discussed the basic principle and characteristics of fluorescence and phosphorescence. Sophisticated instrumentation, and advances in optics, light sources, and electronics resulted in high sensitivity and specificity, increasing the dimensionality of fluorescence measurements. More than fourteen different parameters have been identified, which have been used for clinical, biological, environmental, and sensor applications. We focused only on commonly used parameters such as excitation spectrum, emission spectrum, lifetime, and quantum yield and discussed the applications of fluorescence. The book *Principles of Fluorescence Spectroscopy* by J. R. Lakowicz [5] discusses various types of fluorescence and their applications in different fields elaborately. This book is very useful for researchers working on fluorimetry. Nowadays, RTP is used widely for analytical applications.

References

1. Herschel, Sir J. F. W., *Phil. Trans. R. Soc.* (London), 135 (1845), 143–145.
2. Schulman, S. G., *Fluorescence and Phosphorescence Spectroscopy: Physicochemical Principles and Practice* (Pergamon Press, New York), 1977.
3. Wolfbeis, O. S., *Fluorescence Spectroscopy: New Methods and Applications* (Springer-Verlag, Berlin), 1983.
4. Wehry, E. L., *Molecular Fluorescence Spectroscopy* (Plenum Press, New York), 1981.
5. Lakowicz, J. R., *Principles of Fluorescence Spectroscopy* (Springer, New York), 2006.
6. Jaffe, H. H. and Miller, A. L., *J. Chem. Educ.*, 43 (1966), 469.
7. Kasha, M., *Discuss. Faraday Soc.*, 9 (1950), 14.
8. Swaminathan, M., Ph.D. Thesis, Indian Institute of Technology, Kanpur, 1983.
9. Swaminathan, M. and Dogra, S. K., *IJC-A*, 22 (1983), 853.
10. Weber, K., *Z. Phys. Chem.* (Leipzig), B15 (1931), 18.
11. Forster, Th., *Z. Elektrochem.*, 54 (1950), 42.
12. Forster, Th., *Z. Elektrochem.*, 54 (1950), 531.
13. Forster, Th., in *Photochemistry in the Liquid and Solid States*, F. Daniels, ed. (John Wiley, New York), 1960.
14. Weller, A., *Z., Elektrochem.*, 56 (1952), 662.
15. Enoch, I. M. V. and Swaminathan, M., *J. Fluor.*, 14 (2004), 751.
16. Dey, J. and Dogra, S. K., *Can. J. Chem.*, 69 (2011), 1539.
17. Mishra, A. K. and Dogra, S. K., *J. Photochem.*, 29 (1985), 435.
18. Mishra, A. K. and Dogra, S. K., *Ind. J. Chem.*, 24 (1985), 364.

19. Swaminathan, M. and Dogra, S. K., *IJC-A*, 22 (1983), 278.
20. Swaminathan, M. and Dogra, S. K., *J. Am. Chem. Soc.*, 105 (1983), 6223.
21. Swaminathan, M., *Curr. Sci.*, 62 (1992), 365.
22. Kothainayaki, S. and Swaminathan, M., *J. Photochem. Photobiol. A*, 84 (1994), 13.
23. Rajendran, N. and Swaminathan, M., *Spectrochim. Acta A: Mol. Biomol. Spectrosc.*, 52 (1996), 1785.
24. Kothainayaki, S. and Swaminathan, M., *J. Photochem. Photobiol. A*, 102 (1997), 217.
25. Kothainayaki, S. and Swaminathan, M., *Spectrochim. Acta A: Mol. Biomol. Spectrosc.*, 64 (2006), 631.
26. Weller, A. Z., *Electrochem.*, 60 (1956), 1143.
27. Schulman, S. G. and Gershon, H. J., *Phys. Chem.*, 72 (1968), 3297.
28. Stewart, R. *The Proton: Applications to Organic Chemistry*, Vol. 46 (Elsevier), 1985, 238.
29. Rajendiran, N. and Swaminathan, M., *Bull. Chem. Soc. Japan*, 68 (1995), 2797.
30. Radha, N. and Swaminathan, M. *Spectrochim. Acta A: Mol. Biomol. Spectrosc.*, 60 (2004), 1839.
31. Udhayakumari, D., Velmathi, S., Sung, Y. M. and Wu, S. P., *Sens. Actuators B Chem.*, 198 (2014), 285.
32. Enoch, I. M. V. and Swaminathan, M. *J. Fluoresc.*, 16 (2006), 697.
33. Enoch, I. M. V. and Swaminathan, M. *J. Chem. Res.* (2006), 523–526.
34. Rajamohan, R., Kothai, N. S. and Swaminathan, M., *Spectrochim. Acta A: Mol. Biomol. Spectrosc.*, 69 (2008), 371.
35. Rajamohan, R., Kothai, N. S. and Swaminathan, M., *J. Incl. Phenom. Macrocycl. Chem.*, 73 (2012), 99.
36. De AldaVillaizan, M. J. L., Lonzano, J. S. and Yusty, M. A. L., *Talanta*, 42 (1995), 967.
37. Villaizan, D. A., Garcia Falcon, M. J. L, Gonzalez Amogo, M. S., Lanzo, S. J. S. and Yusty, M. A. L., *Talanta*, 43 (1996), 1405.
38. Vilchez, J., Olmo, L. M. D., Avidad, R. and Capitan-Vallvey, L. F., *Analyst*, 119 (1994), 1211.
39. Andrade E., Vazquez Blanco, A., Lopez Mahia, E., Muniategui Lorenzo, P., Prada, S. and Rodriguez, D., *Analyst*, 123 (1998), 2113.
40. Patra, D. and Mishra, A. K., *Analyst*, 125 (2000), 1383.

41. Taksande, A. and Hariharan, C., *Spectrosc. Lett.*, 39 (2006), 345.
42. Walash, M., Belal, I., El-Enany, F. F., N. M. and El-Maghrabey, M. H., *Chem. Cent. J.*, 5 (2011), 70.
43. Swaminathan, M., Dogra, S. K. *Spectrochim. Acta A: Mol. Biomol. Spectrosc.*, 39 (1983), 973.
44. Dıaz Garcıa, M. E. and Badıa, R. "Phosphorescence Measurements, Applications of" In *Encyclopedia of Analytical Chemistry*, Online (John Wiley), 2006.
45. Alava Moreno, F., Valencia Gonzalez, M. J. and Dıaz Garcıa, M. E., *Analyst*, 123 (1998), 151.
46. Alava Moreno, F., Dıaz Garcıa, M. E., Sanz-Medel, A., *Anal. Chim. Acta*, 281 (1993), 637.
47. Sanz-Medel, A., Martınez Garcıa, P. L. and Dıaz Garcıa, M. E., *Anal. Chem.*, 59 (1987), 774.

Chapter 10

X-ray Fluorescence and Its Applications for Materials Characterization

Nand Lal Mishra
Fuel Chemistry Division (retired), Bhabha Atomic Research Centre, Mumbai 400085, India
nlmisra@yahoo.com

10.1 Introduction

X-rays were discovered in the year 1895 by famous German scientist W. C. Roentgen. X-rays are electromagnetic radiations of high energy. Their energies fall in the range of ultraviolet and gamma rays. Since the discovery of X-rays, these radiations have been used in different areas of science and society for the benefit of mankind, e.g., X-ray imaging for the detection of diseases, X-ray radiography for the detection of cracks in pipes and other industrial bodies, X-ray diffraction (XRD) for studying the lattice structure of materials, X-ray fluorescence (XRF) for the chemical analysis of materials, X-ray absorption spectroscopies for structural analysis and speciation of even amorphous materials and several more. All the above applications depend on the way how X-rays are produced and interact with the object being probed.

Spectroscopy
Edited by Preeti Gupta, S. S. Das, and N. B. Singh

ISBN 978-981-4968-32-4 (Hardcover), 978-1-003-41258-8 (eBook)
www.jennystanford.com

10.2 Production of X-rays

There are two main ways through which X-rays are produced. When energetic charged particles fall on some target material and are stopped, they lose their energy either completely or partially due to de-acceleration arising because of the Coulomb field of the nucleus of the target material. This loss of energy transforms into electromagnetic radiation in the range of energy of X-rays. The range of energies of such X-rays can vary from zero to the energy of charged particles or electrons falling on the target. Since X-rays produced by such phenomenon have energies in the continuous range starting from zero to the maximum energy value of the charged particles used, these X-rays are called '*Bremsstrahlung (in German meaning continuous)*' or '*continuous X-rays.*' Normally electrons are used for such production of X-rays.

If the energies of the charged particles or electrons falling on the target material are very high, these can create a vacancy in the inner shell by knocking out one of the inner cell electron. In this process, the target material becomes unstable and stabilizes by filling this vacancy with outer shell electrons and thus emitting X-rays of specific energies with a value of the difference of the energies of the two shells involved in this transition. This energy shall be fixed for a particular atom and particular shell involved in the transition and is thus characteristic of that atom. Due to this reason, such X-rays are also characteristic of the atom and are called "characteristic X-rays." There are several characteristic X-rays of a particular atom. The energies of these X-rays do not depend on the energy of the impinging charged particle but the energy of the orbitals involved in the transition and producing the X-rays.

The main sources of production of X-rays for lab-based instruments are X-ray tubes. An X-ray tube consists of a cathode and anode sealed in a glass vacuum tube. The anode is a material of high melting point and connected to the +ve end of the electrical circuit. The cathode is in the form of a small filament. The filament is heated to produce electrons by passing a current in it. A high voltage applied across the cathode and anode accelerates the electrons, thus produced to fall on the cathode and produce X-rays as explained above. The production of X-rays is a very inefficient process and most of the energy of electrons falling on the target is used in heating it, thus, a lot of heat is produced. Due to this reason, the target material is heated by circulating water around it and is made of high melting point metal or alloy. A diagram of an X-ray tube alongwith a picture of X-ray tube are shown in Figure 10.1.

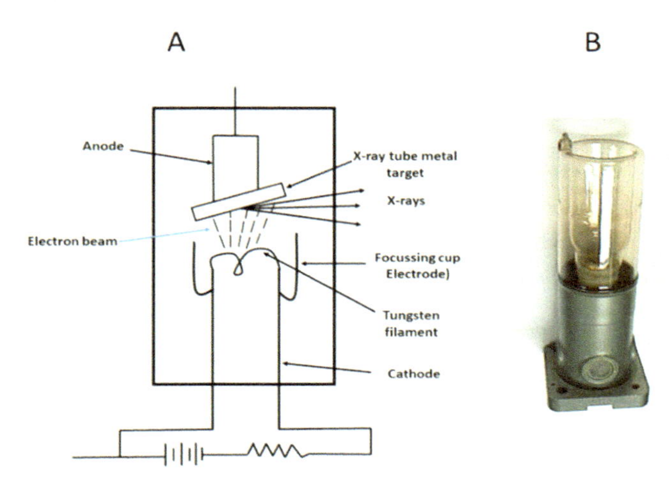

Figure 10.1 A cross-section of copper target X-ray tube [1].

Radioisotopes are other sources of X-rays. Here, electrons are produced due to electron capture or inter orbital electron conversion. The isotope sources are also used for XRF analysis. The biggest advantage of these sources is that these sources are compact and can be used in portable instruments. The main disadvantages are that these sources remain on always and the intensity of X-rays produced decreases with time.

Synchrotron radiation (SR) sources are an advanced source of X-ray production. These sources have brought several advancements in X-ray techniques for material research. SR sources work on the principle that charged particles, when accelerated, emit electromagnetic radiation due to necessary rearrangement of the electric field around the charge, causing a perturbation and this process emits electromagnetic radiation. Normally, the SR sources contain a storage ring having electrons stored and kept on a circular path, producing SR for use in various experiments. These radiations are selected or monochromatized and used for a different types of experiments in beamline stations of the SR sources. The SR offers tunability, is polarized and is a very intense radiation source. Using these sources, one can perform the X-ray spectroscopic measurements in a very accurate and precise manner. Several difficult problems have been solved using experiments of XRF, XRD, X-ray photoelectron spectroscopy (XPS), high-pressure XRD (HPXRD), and several more on SR sources. However, such sources require a lot of money for the creation of experimental facilities and are limited all over the world. Some of the SR sources in the world are Elettra-Sincrotrone Trieste, Italy; European Synchrotron Radiation Facility (ESRF), Grenoble, France; Swiss Light Source (SLS), Villigen Paul Scherrer Institut (PSI), Switzerland; Photon Factory (PF), Tsukuba, Japan. In India, the Indus 2 SR source is in Raja Ramanna Centre of Advance Technology, Indore, and has several beamlines available for the researchers.

10.3 Interaction of X-rays with Matter

The applications of X-rays in various scientific areas depend on their behavior and the changes made in the objects being probed when the X-rays fall on the objects being probed. While falling on an object, the X-rays may get scattered or absorbed. Various material analyses and medical applications of X-rays originate from this phenomenon. When absorbed, if the X-ray beam falling on the sample has a high enough energy than the binding energy of the inner orbital electron, it may knock out the tightly bound electron from the inner shell. This knocked-out electron is called a photoelectron. The photoelectron, after coming out of the atom creates an electron vacancy in the atom and thereby makes the atom unstable. The atom tries to stabilize itself by bringing electrons from outer shells and thereby emitting characteristic X-rays. If the vacancy has been created in the K shell and it is being filled by an electron coming from the L shell, the photon emitted shall have the energy equal to the difference of the energies of the K and L shells and shall be denoted by Kα. However, if this vacancy is being filled by an electron from the M shell the photon emitted shall have the energy equal to the difference in the energies of the K and M shell and shall be denoted by Kβ. Similarly, Mα, Mβ photons, and similar other photons are present. In each shell, there are subshells, and the nomenclature changes to $\alpha 1$, $\alpha 2$, $\beta 1$, $\beta 2$, $\beta 3$, and so on. These photons are known as characteristic X-ray photons and their energies for a particular atom are fixed and can be used to identify that atom if the atom in a sample can be excited by an X-ray energy beam produced from any source. The falling X-ray may get scattered or absorbed, giving rise to scattering and absorption spectroscopic techniques. The interaction of the X-rays with matter is shown in Figure 10.2 given below. The intensity of the emitted characteristic X-rays shall depend on the number of atoms excited and thereby the concentration of the particular analyte in the sample. This forms the basis of XRF analysis. The scattering of X-rays from the sample causes XRD, and this technique is used for the determination of crystal structure, phase identification, and several other applications. There are some other techniques, e.g., XPS extended X-ray absorption fine structure (EXAFS), X-ray absorption near-edge spectroscopy (XANES), etc. based on the different types of interactions, e.g., photoelectron production and absorption. The details about the above brief description and that of XRF coming in the next few paragraphs can be seen in detail in various books [1–3]

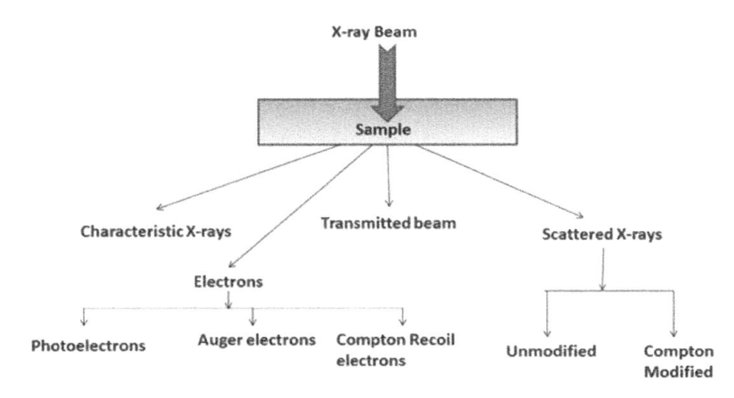

Figure 10.2 Interaction of X-rays with matter: effects produced.

10.4 Characteristic X-rays

The continuous X-rays are caused by the deceleration of electrons by the target element, the characteristic X-rays originate from a different mechanism. A brief introduction about characteristic X-rays is given in the preceding sections. If an X-ray beam of sufficiently high energy falls on a sample it may excite the characteristic X-rays. If the energy of a neutral atom is considered zero, the production of characteristic X-rays can be demonstrated as given in following Figure 10.3.

The emission of characteristic X-rays is governed by certain selection rules and all transitions are not feasible. As an example, there are three L subshells LI, LII, and LIII. Although the transition from K shell to LII and LII is possible K to LI transition is forbidden. The selection rules can be summarized as

$$\Delta l = \pm 1, \Delta j = 0 \text{ and/or } \pm 1$$

The X-ray photons generated from K to LII and K to LIII transition are called K α_2 and K α_1, respectively. In low atomic number elements, the energy, (or wavelength) difference in K α_2 and K α_1 is very small.

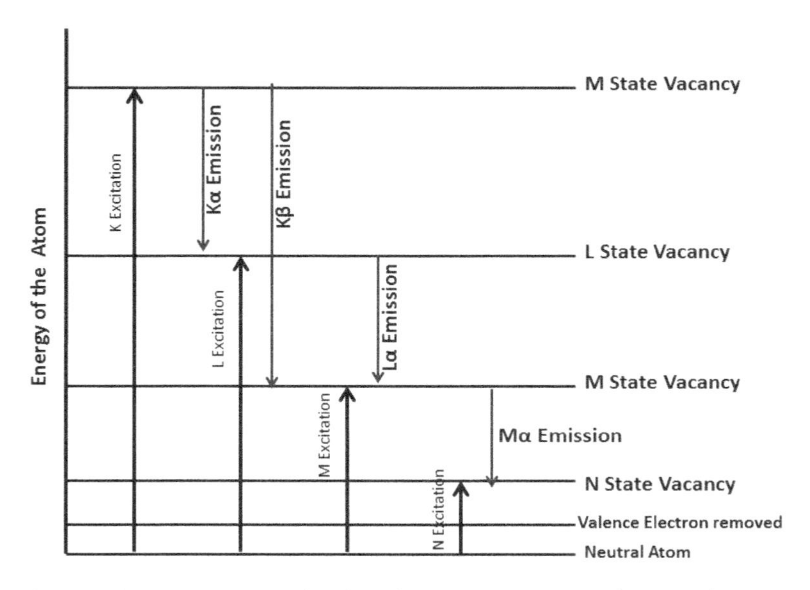

Figure 10.3 Atomic energy levels and emission processes of X-rays. The inset at the top right shows the fine structure of the L state.

10.5 X-ray Fluorescence Principle

XRF or X-ray secondary emission spectroscopy is based on the famous Moseley's law which correlates the energy of the characteristic X-rays with the atomic number of the element as given below:

$$\sqrt{C} = C(Z-6) \tag{10.1}$$

where v is the frequency of characteristic X-ray line, C and 6 are constants and are different for different characteristic X-rays. Another phenomenon may also occur where the excited atom may not emit a characteristic X-ray photon, e.g., K X-ray, but this K X-ray may eject less tightly bound electrons from the outer shell say LIII shell. The ejected electron is known as the Auger electron. After ejecting an Auger electron, the atom is in a doubly ionized state and may return to a normal state by the transition of a single electron or double-electron jumps involving the emission of the diagram (characteristic) or satellite lines.

The probability that an electron vacancy created in the atom by an X-ray or electron beam in an orbital is filled through the emission of characteristic X-ray quanta is called fluorescence yield. The emission of characteristic X-rays and Auger electrons are competitive processes. The fluorescence yield ω_K is defined as:

$$\omega_K = \frac{I_K}{n_K} \tag{10.2}$$

where I_K = total number of K X-ray photons emitted, and n_K = number of primary K shell vacancies created. The fluorescence yield of other X-ray lines are slightly complicated [1–3].

10.5.1 X-ray Fluorescence Analysis

As stated earlier, the characteristic X-rays of a particular atom have fixed energies (or wavelength) and their intensity is proportional to the number of atoms, if a particular sample is excited with a sufficient high energy X-ray beam and the energies of characteristic X-rays emitted are measured, the elements present in the sample can be identified by these energies. If the intensity of characteristic X-rays is measured, these shall give an idea of the concentration of that element in the sample. This forms the basis of XRF analysis.

For XRF analysis, normally a few standards with known concentrations of the analytes in form of pellets or beads or solutions are excited by the X-ray beam. The energy and intensity of the characteristic X-rays emitted from the sample are measured with the help of a suitable detector. The characteristic X-rays are assigned to the respective elements and their intensities are plotted against the respective concentrations to make a calibration plot. A typical calibration plot looks like that shown in Figure 10.4.

Now for XRF analysis, the samples are also made in a similar manner as the standards and their X-ray spectra are measured. The spectra are processed and the intensity of the peak of the analyte lines is calculated. These intensities are converted to the respective concentrations using the correlation equation of the calibration plot.

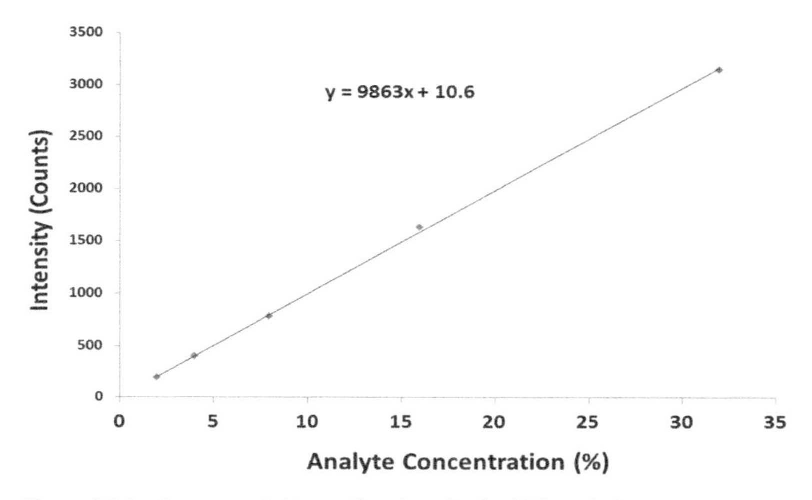

Figure 10.4 A representative calibration plot for XRF analysis.

10.6 Instrumentation and Geometrical Consideration

In order to do XRF analysis, one requires standards for making the calibration plot and samples for analysis. The sample and standard specimens are made by pressing the powders in the form of pellets or fusion beads. Alternatively, the samples can be dissolved in acids and taken in sample cups. Liquid samples and standards can be analyzed as such by placing them in sample cups. The details of sample preparation are given in various books and research papers [2–4].

Once the sample specimens are made, their XRF spectra have to be measured. This involves excitation of the sample by an X-ray beam of a suitable source, separation of the various characteristic X-rays on the basis of their energies, detection of these X-rays and measuring their intensities by a suitable detector, making calibration plots, and analyzing the samples qualitatively or quantitatively. Thus, the major steps of XRF analysis can be summarized in the following steps:

1. Sample preparation
2. Excitation
3. Detection, dispersion, or separation of energies
4. Spectra recording and processing
5. Calibration and quantification

On the basis of the above steps, there are mainly two branches of XRF spectrometry: wavelength dispersive XRF (WDXRF) spectrometry and energy dispersive XRF spectrometry. There is further sub-classification on the basis of geometry and beam size as shown in Figure 10.5.

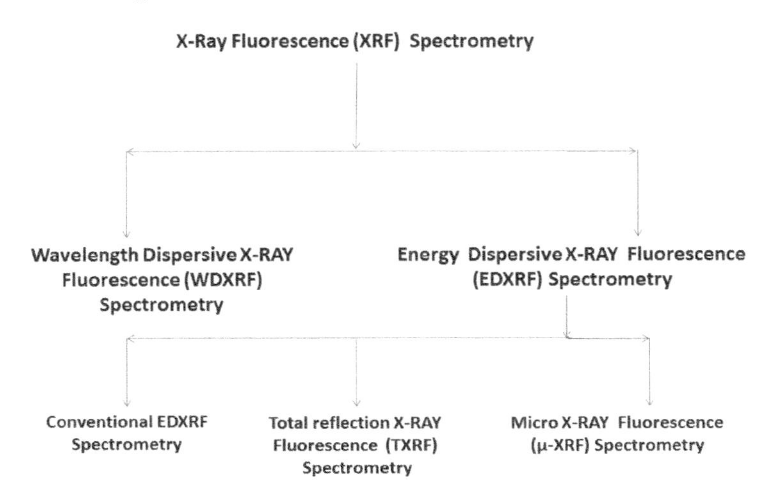

Figure 10.5 Broad classification of XRF spectrometry.

Thus, EDXRF and WDXRF are two main branches of XRF spectrometry. In WDXRF X-rays from an X-ray tube or any other source fall on the sample and excite it to emit characteristic X-rays of the elements present in the sample. The emitted X-rays are dispersed with the help of a resolving crystal, e.g., LiF, Ge, Ovonyx, etc. The crystal is rotated on its axis to change the angle between the X-ray beam and the crystal to satisfy the brag angle of different X-rays sequentially so that these X-rays are diffracted after satisfying Bragg's law sequentially as per the given equation of Bragg's law:

$$n\lambda = 2d\sin\theta \tag{10.3}$$

The diffracted X-rays are detected by a detector positioned to receive these diffracted X-rays by rotating it at a speed double that of a crystal. The corresponding spectra are obtained in form of a plot of 2Θ (angle of rotation of detector which is double the Θ: angle of rotation of crystal). A brief instrumentation of WDXRF and a representative WDXRF spectrum are shown in Figures 10.6 and 10.7.

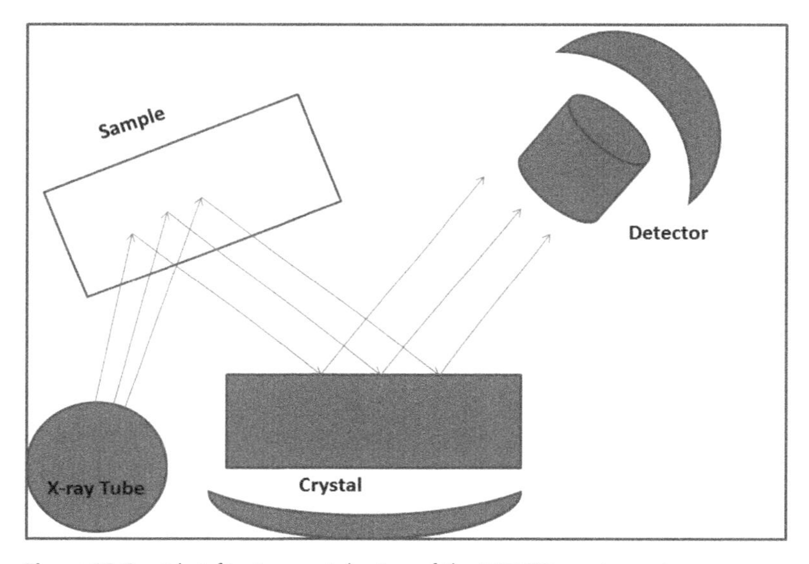

Figure 10.6 A brief instrumental setup of the WDXRF spectrometer.

In EDXRF, X-rays fall on the sample at an angle of about 45°, and the emitted X-rays after sample excitation enter a detector at the almost same angle. Thus, the angle between the incident beam and the detector is 90°. This geometry is called 45°–45° geometry and is chosen to reduce the background to a minimum. The excited X-rays are resolved with the help of the detector as well as a multi-channel analyzer (MCA) which has several electronic channels. Each channel is sensitive to a particular range of X-rays. The spectra achieved in EDXRF are in form of a plot of channel number against the intensity. The channel numbers are normally calibrated in terms of energy and the spectra are shown as a plot between the energy of emitted X-rays versus their intensity. A simple EDXRF instrumentation setup and representative EDXRF spectra are shown in Figures 10.8 and 10.9, respectively.

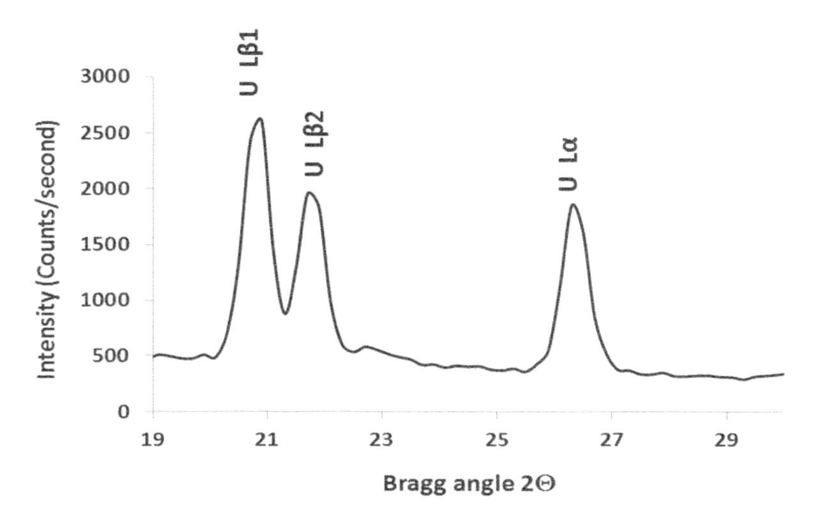

Figure 10.7 A typical WDXRF spectrum of a UO2 sample.

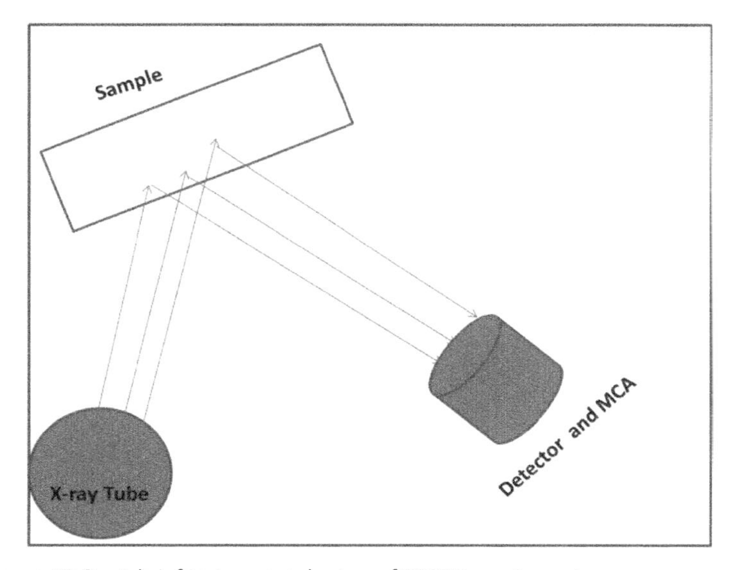

Figure 10.8 A brief instrumental setup of EDXRF spectrometer.

The EDXRF and WDXRF differ in instrumentation aspects and have some advantages as well as disadvantages with respect to each other. The WDXRF resolution is very nice, whereas the EDXRF detection limit is better. With the advancement of time, EDXRF with simple instrumentation has become more popular.

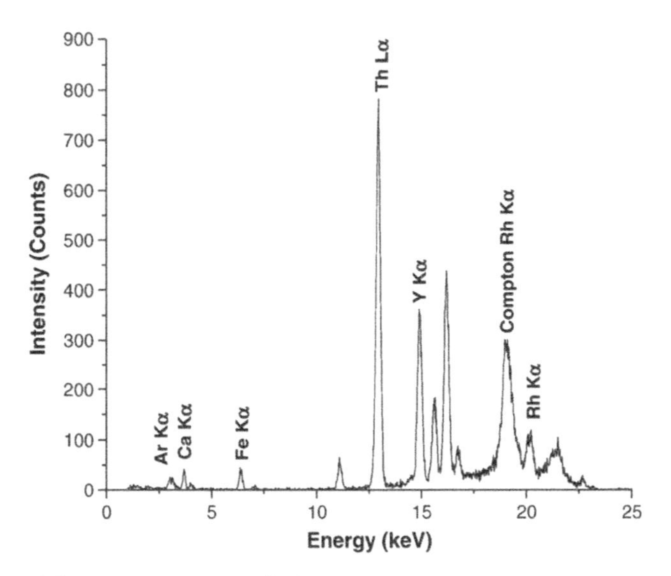

Figure 10.9 A representative EDXRF spectrum.

Total reflection XRF (TXRF) is a variant of EDXRF where the incoming beam falls at a very small angle less than the critical angle (nearing zero degrees) on a flat polished surface containing a thin film of the sample and is totally reflected. Since the X-rays do not penetrate deep into the sample support and the sample is in the form of a thin film of a few nanometers, the spectral background is very less. Also, the incident and totally reflected beams excite the sample. The angles of the incident as well as the totally reflected beam angle with sample support, are about zero degrees. This makes easier to place a detector very near the sample support. These three features of TXRF make its detection limit comparable to other well-established trace determination techniques. The other features, e.g. the requirement of only a few microliters of sample, multi-element analytical possibility, and determination of metals and non-metals alike make this technique better than other well-established trace determination techniques.

Micro-XRF is also a special kind of XRF where a focused micron size beam falls on the sample and excites it. The sample can be moved in steps of a micron in the x-, y-, and z-axis, and thus elemental mapping on the surface as well as in three dimensions can be achieved. Also, a very small object, e.g., an insect body part, grains, or a small area of paint, can be analyzed using micro-XRF. The micro-XRF technique is very useful in the study of small objects and non-uniform objects on micron levels for getting information about the elemental analysis with surface and 3D mapping.

10.7 Application of XRF in Materials Characterization

The main application of XRF techniques is the chemical analysis of the materials on a quantitative or qualitative basis, which is a very important parameter to assess the quality of materials in industrial production, scientific research, magnetic materials, coal, nuclear fuel, etc. Although the WDXRF and EDXRF are not very much suitable to analyse the materials for elements at ultra-trace/trace levels as other conventional trace element analytical techniques, e.g. Inductively Couplead Plasma Mass Spectrometery (ICP-MS), Inductively Compled Plasma Atomic Emission Spectroscopy (ICP-AES), Neutran Activation Analysis (NAA) etc., these techniques are well suited for major element analysis in a non-destructive as well as online sample analysis in production processes, biological and other research areas. Due to severe matrix effects, these techniques require matrix matched standards for the analysis. However, their application to unknown samples gives an idea about the elements present in a non-destructive manner, and later the same can be analyzed by other techniques for trace elements.

EDXRF and WDXRF have been used for quality control of processed material in production plants, environmental samples, nuclear fuel, coal, solid analysis, agricultural products, etc.

The suitability of WDXRF and EDXRF to analyze trace elements poses a limitation for their application for ultra-trace and trace analysis which is very much required in quality control of industrial, biological, and environmental samples and for the scientific research in air pollution, nuclear material quality control, biological materials, agricultural products water samples, etc. The requirement of a small amount of sample and other features make TXRF, very much suitable for the study of precious, radioactive, forensic, and toxic materials. The main disadvantage of TXRF is the requirement of sample dissolution before analysis. However, such limitation has been taken care of by recent researchers after developing almost non-destructive sample preparation methods for nuclear and other materials [5–6].

Not only the amount of a particular analyte present in the sample but also its oxidation state or speciation is important to know the property of the material, e.g., the presence of As is very toxic in water samples. However, it is the As^{+3} that makes the water sample toxic and this information can be found by measuring the XRF spectra at different energies with the help of X-ray Absorption Near Edge (XANES) spectroscopy. TXRF application in XANES mode uses a synchrotron light source, which gives very intense tunable, polarized, X-ray beam. Such sources are several all over the world and are used for different X-ray analysis applications. However, assess to such sources is limited. TXRF or EDXRF can be combined with XANES and TXRF spectra for a short duration measured at different energies on both sides of the absorption edge of a particular analyte in steps of 1–2 eV. The region of interest intensity can be converted to absorption spectra and the oxidation state of the analyte can be deduced with the help of some standards. The details of such studies are available in several books and research papers [2, 3, 6].

WDXRF can also be used for oxidation state determination even with lab sources, thanks to the good resolution achieved in it, by monitoring the chemical shifts in analytical line positions. Such one application on Ni and several others is reported in the literature [7]. The TXRF, EDXRF, and WDXRF have been used in environmental, biological, industrial, forensic, art restoration, painting analysis, archaeology, forensic and several other areas for different applications.

10.7.1 Some Examples of XRF Application in Different Areas of Science and Technology

There are several such examples available in the literature [1–4]. A few shall be given here for illustration. EDXRF has been used for the investigation of surface silver enrichment in ancient high silver alloys, analysis of urinary stones, and reactor fuel using solid or solution samples [8–10]. WDXRF has been used for the analysis of corrosion products, oxidation state determination, and chemical quality control of nuclear fuel materials [11–13]. Micro-XRF has been used for the analysis of pigeon skins and geological materials, and for assessing the quality of nuclear fuel produced by different routes [14–16].

TXRF is an advanced XRF technique and finds its main application in the quality control of Si wafers [17]. However, it is being progressively used for several non-conventional applications [18–20]. Some other diverse but useful applications are for multi-element characterization of micro-greens, elemental impurities in graphemes, clays, uranium and plutonium oxides, and biological samples. In addition, it has been used for oxidation state determination of uranium in its oxides using TXRF-XANES [20–25].

10.8 Advancements in XRF Analysis

The new developments in XRF using advanced techniques and powerful sources of sample excitation, e.g. SR sources have made this technique a powerful technique to solve complex research problems of sample analysis, speciation, etc. The requirement of a very small sample amount in TXRF and the use of synchrotron sources for such applications especially for toxic, radioactive, precious, and biological materials are very much useful for trace element determinations in such materials. The oxidation state determination possibility by WDXRF using chemical shift and full width at half maximum (FWHM) intensity of different X-ray lines has a very good prospect of using this technique for oxidation state determination in the laboratory. The use of X-ray emission spectra of low atomic number elements has shown that it is having an oscillating fine structure, very much similar to that of the Extended X-ray Absorption Fine Structure (EXAFS), and can be used for structural analysis by WDXRF spectrometer in the laboratory without the requirement of a SR source [26]. Several other developments in instrumentation, sample preparation, and conjugation of XRF with other analytical techniques shall come in the next few years which will make this non-destructive technique useful not only for elemental analysis but for oxidation state determination, speciation as well as structural analysis in a simple manner.

References

1. Cullity, B. D., *Elements of X-ray Diffraction* (Addison-Wesley Publishing, Reading, Massachusetts and Menlo Park, California), 1956.
2. Bertin, E.A., *Principle and Practice of X-ray Spectrometric Analysis*, 2nd Ed. (Plenum Press, New York), 1984.
3. Van Grieken, R. E. and Markowicz, A., *Handbook of X-ray Spectrometry*, 2nd Ed., (Marcel Dekker, New York), 1993.
4. Klockenkämper, R. and Von Bohlen A., *Total Reflection X-ray Fluorescence Analysis and Related Methods*, 2nd Ed. (John Wiley, Hoboken, New Jersey), 2015.
5. Dhara, S., Parimal, P. and Misra, N. L., Direct compositional characterization of (U, Th)O_2 powders, microspheres, and pellets using TXRF, *Anal. Chem.*, 87 (2015), 120262–120267.
6. Sanyal, K., Khooha, A., Das, G., Tiwari, M. K. and Misra, N. L., Direct determination of oxidation states of uranium in mixed-valent uranium oxides using total reflection X-ray fluorescence X-ray absorption near-edge spectroscopy, *Anal. Chem.*, 89 (2017), 871–876.
7. Konishi, T., Kawai, J., Fujiwara, M., Kurisaki, T., Wakita, H. and Gohshi, Y., Chemical shift and line shape of high-resolution Ni Kα X-ray fluorescence spectra, *X-ray Spectrom.*, 28 (1999), 470–477.
8. Borges, R., Alves, L., Silva, R. J. C., Araújo, M. F., Candeias, A., Corregidor, V., Valério, P. and Barrulas, P., Investigation of surface silver enrichment in ancient high silver alloys, *Microchem. J.*, 131 (2017), 103–111.
9. Shaltout, A. A., Da, M. M., Applicability of low-cost binders for the quantitative elemental analysis of urinary stones using EDXRF based on fundamental parameter approach, *Biol. Trace Elem. Res.*, 195 (2020), 417–426.
10. Dhara, S., Kumar, S., Misra, N. L. and Aggarwal S. K., An EDXRF method for determination of uranium and thorium in AHWR fuel after dissolution, *X-ray Spectrom.*, 38 (2009), 112–116.
11. Gazulla, M. F., Ventura, M. J., Orduña, M., Rodrigo, M., and Gómez, M. P., Analysis of corrosion residues by WDXRF, *X-ray Spectrom.*, 46 (2017), 271–276.

12. Sato, K., Nishimura, A., Kaino, M. and Adachi, S., Polychromatic simultaneous WDXRF for chemical state analysis using laboratory X-ray source, *X-ray Spectrom.*, 46 (2017), 330–335.

13. Pandey, A., Dhara, S., Khan, F. A., Kelkar, A., Kumar, P., Bhatt, R. B. and Behere, P. G., Analysis of Th and U in thorium-based mixed-oxide fuel using wavelength dispersive X-ray fluorescence spectrometer, *J. Radioanal. Nucl. Chem.*, 319 (2019), 775–781.

14. Falkenberg, G., Fleissner, G., Alraun, P., Boesenberg, U. and Spiers, K., Large-scale high-resolution micro-XRF analysis of histological structures in the skin of the pigeon beak, *X-ray Spectrom*, 46 (2017), 467–473.

15. Flude, S., Haschke, M. and Storey, M., Application of benchtop micro-XRF to geological materials, *Mineral Mag.*, 81 (2017), 923–948.

16. Misra, N. L., Tiwari, M. K., Vats, B. G., Kumar, S., Singh, A. K., Lodha, G. S., Deb, S. K., Gupta, P. D. and Aggarwal, S. K., Synchrotron μ-XRF study on compositional uniformity of uranium–thorium oxide pellets prepared by different processes, *X-ray Spectrom.*, 43 (2014), 152–156.

17. Hockett, R. S., TXRF semiconductor applications, *Adv. X-ray Anal.*, 37 (1993), 565–575.

18. Dhara, S. and Misra, N. L., Elemental characterization of nuclear materials using total reflection X-ray fluorescence spectrometry, *Trend. Anal. Chem.*, 116 (2019), 31–43.

19. Misra, N. L., Advanced X-ray spectrometric techniques for characterization of nuclear materials: An overview of recent laboratory activities, *Spectrochim. Acta B*, 101 (2014), 134–139.

20. Allegretta, I., Gattullo, C. E., Renna, M., Paradiso, V. M. and Terzano, R., Rapid multi-element characterization of microgreens via total-reflection X-ray fluorescence (TXRF) spectrometry, *Food Chem.*, 296 (2019), 86–93.

21. Zhang, A., Wang, H., Zha, P., Wang, M., Ang, H., Fan, B., Ren, D., Han, Y. and Gao, S., Microwave-induced combustion of graphene for further determination of elemental impurities using ICP-OES and TXRF, *J. Anal. At. Spectrom.*, 33 (2018), 1910–1916.

22. Allegretta, I., Ciasca, B., Pizzigallo, M. D. R., Lattanzio, V. M. T., Terzano, R., A fast method for the chemical analysis of clays by total-reflection X-ray fluorescence spectroscopy (TXRF), *Appl. Clay Sci.*, 180 (2019), 105201.

23. Sanyal, K., Dhara S. and Misra, N. L., Direct multielemental trace determinations in plutonium samples by total reflection X-ray fluorescence spectrometry using a very small sample amount, *Anal. Chem.*, 90 (2018), 11070–11077.

24. Sanyal, K., Chappa, S., Pathak, N., Pandey A. K., and Misra, N. L., Trace element determinations in uranium by total reflection X-ray fluorescence spectrometry using a newly developed polymer resin for major matrix separation, *Spectrochim. Acta Part B*, 150 (2018), 18–25.

25. Gruber A., Müller, R., Wagner, A., Colucci, S., Spasić, M. V. and Leopold, K., Total reflection X-ray fluorescence spectrometry for trace determination of iron and some additional elements in biological samples, *Anal. Bioanal. Chem.*, 412 (2020), 6419–6429.

26. Kawai, J., Hayashi, K. and Tanuma, S., Extended X-ray emission fine structure of sodium, *Analyst*, 123 (1998), 617–619.

Index

absorbance 254, 256, 285
absorption 11–12, 31, 33, 38–39, 157, 182, 189, 245–246, 248–249, 254–255, 280, 284, 286, 294, 296–297
  carbonyl 189
  deuterium 99
  high-intensity 254
  low-intensity 254
  X-ray 311
absorption line 17, 130
absorption spectroscopy 12, 123
acetone 98, 183–184, 199, 259
acetophenone 183–184, 189, 249
acetylene 71, 106, 215
acid 74–75, 103, 271–272, 293–295, 314
  2-cycloheptenoic 272
  acetic 98, 117, 189, 200
  carboxylic 49, 75, 112, 115, 187–189, 193, 200
  cutin 119
  fatty 119
  glycolic 138
  humic 118
  nucleic 86
  o-anisic 295
  organic 295
  salicylic 295, 303
  thioacetic 198
α-effect 106–107, 110
alcohol 73–74, 119, 187–188, 193, 198, 248
  aliphatic 188
  dry 73
  teriary 193
aldehyde 46, 103, 112, 115, 185, 199, 271–272
alkane 103, 114, 187, 192, 247–248
alkene 43–45, 47, 60, 71, 103, 187, 190, 192, 247, 249, 269
alkyne 43, 45, 110, 187, 192, 247, 249
allene 104, 112, 192
amide 103, 112, 115, 119, 186, 201
amine 50, 74–75, 195, 248
analysis
  chemical 307, 319
  element 319
  microwave 228
  painting 320
  pesticide 303
  qualitative 95
  quantitative 285
  solid 319
  spectral 183
  structural 191, 197, 307, 321–322
  trace 319
  XRF 309, 311, 313–315, 321
angular momentum 15–16, 24, 27, 124–125, 217–219
angular velocity 3, 217
anharmonic oscillation 159, 162–163, 169, 171, 173, 239
anharmonic oscillator 159, 161–163, 169–170
anisotropy 40–41, 43, 103, 128, 141
anthracene 127, 286–287
antibonding orbital 247, 260
anti-Stokes lines 231–233, 237, 239
anti-Stokes Raman lines 237, 240

AO *see* atomic orbital
application 91, 95, 116–119, 124, 280, 285–288, 292, 299–300, 304, 307, 310–311, 319–321
  agriculture 116
  analytical 304
  medical 310
  sensor 304
  technological 7
atomic orbital (AO) 260–262
auxochrome 255, 262–264, 267
Avagadro's number 11, 182
azide 196, 276

band 157, 162–165, 169, 171–172, 177, 179, 188, 191, 193, 249–250, 252, 258, 273–274
  absorption 165, 174, 187, 191, 234, 250, 255–256, 258, 269, 272–273
  broad-spectrum 179
  carbonyl 189
  fundamental 163–165, 171
  harmonic/overtone 172
  low-intensity 165
  primary 273
  secondary 273
bathochromic shift 256, 260, 262, 264–265, 267
beam splitter 208, 210
Beer–Lambert law 252–254
bending vibration 153, 175–176
benzene 46, 48, 85, 98, 189, 249, 255, 272–274, 290
BO approximation *see* Born–Oppenheimer approximation
Bohr's magneton 124
Boltzmann constant 33
Boltzmann distribution 222
bond 67–68, 71, 75, 83–84, 153–154, 159, 181–188, 224, 234, 247–248, 255, 260, 265
  fingerprint 191
  polar 242
  terminal 185
bonding 48, 181, 242, 247–249, 260–261, 264
bond length 154, 179, 183–184, 187, 216–217
bond order 183–184
bond strength 1, 181–182, 185
Born–Oppenheimer approximation 8, 14
Born–Oppenheimer approximation (BO approximation) 8–9, 170
Bragg's law 315

carbon 37, 48, 72, 82, 95–98, 101–107, 110, 112, 117, 119, 152
  carboxyl 117–118
  natural 72
  olefinic 119
  organic 119
  saturated 71, 110
  terminal 114
  triple bonded 110
carbon atom 44, 67–68, 82, 96–98, 101–102, 116–118, 182, 265
carbonyl group 45, 104, 112–113, 184, 186, 189, 267, 269
charged particle 124, 308–309
chemical shift 35, 37, 39–43, 47–48, 59–62, 78, 82–84, 86, 88, 95, 97–99, 101–117, 119, 320–321
chemiluminescence 279, 300
chloroform 48, 98
chromophore 255–256, 269, 272
coefficient 57, 131–132
  absorption 253–254
  molar 253
  molar extinction 254
coils 87–88, 100
  radiofrequency receiver 89
  solenoid 99
collision 18, 214, 231–232

compound 57–58, 61, 71, 74–75, 77–78, 81–82, 84, 94–95, 106, 109, 113, 116, 127, 187, 191
  allenic 71
  aromatic 46–47, 71, 246, 272–273, 301
  azo 247
  carbonyl 45–46, 110, 113–115, 246–247, 249–250, 259, 267
  deuterated 77
  dicarbonyl 49
  fluorescent 255, 280
  fluorinated 84
  halogen 247
  heterocyclic 95
  humic 119
  hydrogen-bonded 188
  inorganic 86, 148, 152
  liquid 300
  nitrogen-containing 195
  non-hydrogen bonded 190
  olefinic 68
  organic 23, 43, 49, 82, 101, 116, 151–152, 186, 293
  saturated 68
  spectrum of 248, 258
  sulfur 248
  transition metal 128
  vinyl 68
computer 83, 89–90, 97, 208, 283
Conne's advantage 210
Coulomb field 308
coupling 64, 67, 71–75, 79, 82–84, 90, 97, 101–102, 117–118
  geminal 67–68
  germinal 68
  heteronuclear 71
  meta 71
  nuclear 75
  residual 101
  spin-orbit 128
  vibronic 252
  vicinal 68, 84
coupling constant 52, 59–73, 78–79, 84, 89, 137
  anisotropic 141
  geminal 67
  hyperfine 132, 138–139
cyclohexane 48, 68, 98, 109
cyclopropane 68, 109, 187

degeneracy 134–135, 137, 222
deshielding 36, 40, 44, 46–48, 104
detector 12, 32, 146–147, 204–205, 208, 210, 226–227, 231, 313–314, 316, 318
  insensitive 207
  multichannel 241
  phase-sensitive 146
  sensitive 205
  thermal 205
  thermocouple 207
  thermophile 205
deuterium 28, 72–73, 77, 133, 137, 182–183
diatomic molecule 16, 143, 153–163, 165–173, 177, 215–216, 218, 234, 236, 239–240
dipole 2, 7, 12–14, 125, 152–153, 158, 174–177, 234, 242, 286, 293, 297
  electrical 215, 233
  fluctuating 175
  induced 234
  oscillating 234
  permanent 7, 165, 213, 221
downfield shift 44, 86, 109–110, 113
drug 300, 302–303

EDXRF spectrometer 315–320
effect
  anisotropic 44
  α-substituent 105–106
  β-substituent 107
  deshielding 47

diamagnetic 103
δ-substituent 109
gauche 107
γ-substituent 107
heavy-atom 106, 110
hyperchromic 256, 265, 269, 274
hypochromic 256, 265
inductive 41, 185–186, 268
isotopic 98
mesomeric 40, 183, 185
electric field 2, 214, 233–234, 309
electronegativity 37, 41–42, 68–69, 106–107, 186
electronic transition 15, 245–252, 254, 257, 259–264, 273–274, 276, 281–282
electron magnetic resonance (EMR) 7, 123
electron spin resonance (ESR) 7, 123–124, 141, 148
electron transfer 276–277, 284, 299
emission 11–12, 14, 16–18, 31, 33, 157, 279–286, 288–289, 291, 295–296, 298, 303, 312–313
emission spectroscopy 12
EMR *see* electron magnetic resonance
energy level 7–8, 10–12, 27, 29–31, 126, 130, 132–133, 137–139, 155, 158–159, 171–173, 218–219, 221–222, 245–246, 277
energy state 10–11, 13, 17, 34, 125–126, 130, 132, 135, 158, 222, 273
enone 267, 270, 272
ESR *see* electron spin resonance
ESR spectroscopy 124, 127, 130, 132, 144, 148
ester 46, 103, 112, 185–187, 193, 271–272
ethanol 37–38, 71, 73–74, 188, 257–258, 260, 268
ether 98, 193, 200, 248
ethylene 68, 260–264
EXAFS *see* extended X-ray absorption fine structure
excitation 8, 12, 279, 282, 286–287, 289–290, 292–293, 295, 297–298, 301–302, 314–315
excited state 12, 18, 33–34, 104, 246, 251, 255, 259–260, 279–280, 285–288, 294–296, 298
extended X-ray absorption fine structure (EXAFS) 311, 321

Fellgett's advantage 210
FID *see* free induction decay
fluorescence 279–281, 283–285, 287–288, 290–296, 299–300, 302–304, 313
fluorescence quenching 298–299
fluorescence spectroscopy 279–280, 289, 298, 304
fluorine 42, 83–86
fluorophore 280, 287–288, 291–293, 295, 297–299
force constant 154, 181, 191
Forster cycle method 294, 303
Fourier transform 19, 83, 90–91, 204, 207–210
Fourier transform infrared spectroscopy (FTIR spectoscopy) 204, 207, 210
Frank–Condon principle 282
free induction decay (FID) 83, 102
frequency shift 109, 184, 190, 231
FTIR spectoscopy *see* Fourier transform infrared spectroscopy
FT-NMR spectroscopy 90–91, 99

grating monochromator 204, 210, 282

ground state 104, 161, 221, 246, 255, 259–260, 280–281, 286, 295–296, 299
gyromagnetic ratio 31, 35, 86

halide 142–143, 193, 276
halogen 38, 248, 303
harmonic oscillator 155, 157–158, 160–161, 166–167, 169
highest occupied molecular orbital (HOMO) 246, 260, 262
HOMO *see* highest occupied molecular orbital
Hooke's law 154, 159
hybridization 72, 103–104
hydrocarbon 198, 248, 301
hydrogen bond 47, 113, 188, 190–191, 258–259
hyperfine splitting 131–134, 140–141, 143
hypsochromic effect 267

instrument 38, 40, 83, 88, 95, 204, 207, 283, 285, 288, 292
  conventional 207
  dispersive 210
  dispersive-type 205
  fabricated 282
  FTIR 205, 210
  lab-based 308
  low field 78
  portable 309
  sensitive 232
  spectroscopic 19
interaction 1–2, 12–13, 17, 19, 23, 32, 125, 130, 132–133, 140–141, 143, 263, 265, 297–299, 311
  electronic 67
  hyperfine 130–131, 133, 137
  intermolecular 214
  nuclei 132
interferometer 207–208, 210
ion 34, 124, 129, 140, 142–143
  chloride 300
  indazole ammonium 296
  krypton 241
  lanthanides 81
  manganese 148
  metal 81, 148, 275–277, 299–300, 303
  vanadium 148
irradiation 79, 89–90, 102, 142, 255
isolator 146–147
isomer 75–76, 228
isotope 72, 95, 97, 101, 182, 224, 228

Jacquinot advantage 210

Kasha's rule 281, 284
ketone 103, 112, 115, 185–187, 194, 199, 272
Kramer's degeneracy 134–136

Lippert equation 297
lowest unoccupied molecular orbital (LUMO) 246, 260, 262
LUMO *see* lowest unoccupied molecular orbital

magnetic field 2–3, 12–13, 23–27, 29–30, 32–37, 40–47, 52, 87–90, 93, 96–100, 102–104, 124–126, 128–129, 136–137, 142–144, 146–147
material 17, 99, 119, 205, 246, 307–308, 319–321
  amorphous 307
  biological 321
  ferrite 146
  geological 321
  nuclear fuel 321
  photoresist 205
  pyroelectric 205
  resistive 145
  solid 118

toxic 320
transparent 208
microwave 2, 7–8, 145, 214
MO *see* molecular orbital
molar absorptivity 249–250, 252, 254–256, 264, 268
molecular orbital (MO) 41, 260–262, 264
moment of inertia 214–215, 217–218, 224, 228
monochromator 158, 204, 207, 226, 241, 283
Morse function 159
Moseley's law 312
motion 8–9, 13, 25, 153, 156, 233
anharmonic 177
circular 3, 44
circulating electron 35
continuous 10
internal 234
molecular 172, 233
nuclear 9, 14, 282
oscillation 154
rotational harmonic 172
simple harmonic 159, 177–178
thermal 18
vertical 25
multiplicative splitting 58–61

neutron 24, 27–28, 101
NMR *see* nuclear magnetic resonance
NMR signal 38, 48–49, 62, 65, 81, 87, 89, 97
NMR spectrometer 84, 86–87, 89, 99
NMR spectroscopy 23, 26, 38, 72, 75, 77, 82–83, 86–88, 91, 93, 95–98, 116–118, 124–125
NOE *see* nuclear Overhauser effect
nuclear fuel 319, 321
nuclear magnetic resonance (NMR) 7, 27, 29, 31, 33, 77, 81–86, 90–91, 93–99, 103–104, 107, 116–119, 245
nuclear Overhauser effect (NOE) 89–90, 102
nuclear spin 27, 30–31, 33, 54, 67, 82–83, 97, 101, 132, 137, 213

Overhauser effect 89
oxidation state 148, 275–277, 320

PET *see* photoinduced electron transfer
phosphorescence 279–281, 284–285, 290, 302–303
photoinduced electron transfer (PET) 299
photon 12–13, 16–17, 230–232, 279, 310–311
Planck's constant 10–11, 218, 231
polarizability 8, 233–235
polyatomic molecule 153, 173–181, 234
probability 12, 33, 116, 221–222, 254, 313
process
dynamic quenching 299
environmental 292
excitation/emission scan 289
industrial 228
nonradiative 280–281, 288
pollutant binding 118
reverse 12
scattering 230
spectroscopic 12
synchronous scan 289
protein 119, 291, 299
proton coupling 62, 67, 71, 78, 97

QP *see* quantum photometer
quality control 91, 319, 321
quantum photometer (QP) 283
quenching 299–300

radiation 1–2, 4–5, 7, 9–13, 18–19, 126–127, 152, 157–158, 166, 207–208, 218–219, 221, 226, 230–234, 241–242, 252–253

anti-Stokes Raman 232
electromagnetic 1–3, 5, 11–13, 19, 27, 31–33, 93, 123–124, 246, 255, 307–309
infrared 207
intensity of 12, 252–253
interaction of 2
microwave 144–145, 147, 214, 228
pulse 90
radiofrequency 27, 34, 90, 93
stoke 232
ultraviolet 259
wavelengths of 246, 255
Raman effect 229–233, 239
Raman lines 231, 234, 241
Raman shift 231, 235
Raman spectrometer 240–241
Raman spectroscopy 8, 229–230, 232–233, 235, 240–242
Rayleigh scattering 230, 234, 241
redshift 256, 259–260, 262, 274, 285
relaxation 33–35, 90, 130, 302
resonance 13, 32, 35–37, 39, 43–47, 52, 55–56, 60–61, 72, 74, 95, 97–99, 117, 127, 147
resonance effect 33, 183–184, 186
resonance signal 33–34, 50, 84, 86, 89, 129–130
rhodamine 283, 290–291
ring residue 265–267, 270
room temperature phosphorescence (RTP) 303–304
rotation 7, 10, 12, 29, 75–77, 152, 165, 170, 173–174, 176–177, 214–216, 316
rotational constant 170, 177, 179, 213, 218
rotational energy level 11, 167, 213, 220, 222–224, 232, 235–236
rotational level 7, 9–10, 166–167, 170, 214, 218–220, 222, 224, 230
rotational motion 9, 153, 166, 213–214
rotational spectroscopy 7, 214, 221, 230
rotational transition 176, 178, 214, 221, 238
RTP *see* room temperature phosphorescence

selection rule 12–13, 15–16, 132–133, 135, 137, 139, 157, 161, 166, 168–169, 176–180, 219–221, 235–239, 250–252, 312
sensitivity 86, 95, 97, 99–100, 205, 207, 210, 280, 283, 292, 303
SFS *see* synchronous fluorescence spectroscopy
shielding 35–36, 40–47, 98, 103
shielding effect 36, 40–42
signal 32, 39–40, 48–51, 60, 62, 65, 72, 89–90, 96–98, 100–101, 118–119, 129–130, 163, 207–208, 227
decay 90
electrical 227, 241
high-frequency 208
output 283
spectral 102
synchronous 289
unique 32
soil 118–119
solvatochromic shift 297
solvent 40, 48–49, 98–99, 189–190, 248, 257–259, 264, 268
high-polar 298
non-polar 258, 290
organic 39
protic 113

spectrometer 20, 84, 86, 88, 158, 166, 204, 207–208, 241
dispersive 207, 210
dispersive infrared 207
dispersive type 204
spectrum 44–48, 51–52, 54–56, 58–60, 62, 64–65, 77–79, 81, 83–86, 89, 101–102, 158, 165–166, 168–169, 171, 177–179, 210, 273–274, 283, 289–291
absorption 12, 151, 157, 220, 227, 246, 258, 285, 287, 292, 320
continuous 17, 207
dispersive 210
electromagnetic 7, 17, 151, 276
emission 12, 157, 283, 286, 289, 292, 297, 304
EPR 138–140
ESR 124–125, 129–130, 132–136, 139–143, 146–148
excitation 283, 286, 288, 304
fluorescence 280, 286, 292–293
infrared 158, 163, 169, 191, 198
microwave 221, 227
NMR 32, 42, 45–46, 48–52, 61–62, 65, 67, 71–74, 76–80, 82–84, 89–90, 94, 97, 101–102, 118–119
non-first-order 62
Raman 237–242
rotational 166, 213, 220–221, 224
synchronous 301
ultraviolet 257, 273
vibrational 157–158
spin coupling 50–53, 55, 58, 65, 72–73, 75, 82–84, 95, 117
spin Hamiltonian 141, 143–144
spinning 25, 28, 95, 118, 132
spinning nucleus 24, 27–29, 32
spin splitting 51, 56–58, 67, 71
spin states 32–34, 73, 79, 93, 96, 102, 125, 134–136
splitting 30, 52–56, 59–61, 71–72, 74–75, 82–83, 85, 97, 117, 123–124, 130, 133–137, 139–140
first-order 62
zero-field 128, 134–135
splitting pattern 56, 58, 60–62, 67, 136, 140
SR *see* synchrotron radiation
Stokes lines 232, 236–237
Stokes Raman lines 235, 237, 239
Stokes shift 286, 297
stretching 175, 188–189, 192–197
antisymmetric 194
asymmetric 175, 193
symmetric 152, 175, 193–195
symmetry 16, 137, 141, 143, 173–174, 234, 252, 274
axial 128
cubic 142
cylindrical 88
mirror image 286–287, 296
molecular 1
tetragonal 142
synchronous fluorescence spectroscopy (SFS) 288–289, 301–302
synchrotron radiation (SR) 309–310
system 17–18, 32–34, 110, 134–139, 143–144, 152, 154–155, 165–166, 215, 217, 261–263, 290, 294
absorbing 254
acyclic 107
aromatic 41, 272
biological 148, 299
conjugated 111, 260, 262
diene 265–266
detection 144
electron 132, 135, 137, 290
enone 269–270
molecular 191, 246–247, 249, 265

nonconjugated 246
optic 240
recording 207

technique 19–20, 23, 78–79, 83, 116–118, 148, 152, 204–205, 226, 228, 311, 318–319, 321
analytical 83, 285, 319, 321
depression 204
fluorimetric 280
fluorometric 300
microwave 227
non-destructive 322
sensitive 292
spectroscopic 1, 19–20, 242
synchronous 290
tetramethyl silane (TMS) 38–39, 47, 82, 97–98, 101, 103–104, 119
TMS *see* tetramethyl silane
total reflection XRF (TXRF) 318, 320–321
TXRF *see* total reflection XRF

ultraviolet region 5, 8, 273, 276
unpaired electrons 89, 124, 127–128, 131, 133–138, 142–143, 147–148

vacancy 142, 308, 310
van der Waals interactions 41
vibration 7, 12, 16, 146, 152–153, 155, 157–158, 165, 170–177, 179–181, 191, 233–234, 242
active 235
anharmonic 170, 179
fundamental 173–174, 239
modes of 153, 174–176
molecular 158, 181, 233
simple harmonic 154
skeletal 191
vibrational level 8–11, 157, 159–161, 163–164, 166–167, 179, 237–238
vibrational motion 9, 34, 153, 156, 158, 174, 176, 216, 234
vibrational states 11, 16, 156, 165, 167, 170, 179, 280–281, 284–285

wave function 9, 13–15, 260
wavelength 4–8, 204–205, 207, 230, 245–248, 250, 254–257, 259, 267, 269, 282–283, 289–290, 292, 312–313, 315
absorption 262, 264, 266, 271–272, 274
emission 286, 289–290, 292
excitation 280–281, 284, 287, 292
wavelength dispersive XRF spectrometry (WDXRF spectrometry) 315–316, 318–321
wave number 5–6, 155, 157, 160, 166–167, 170, 172, 182, 218–220, 236–239, 294
WDXRF spectrometry *see* wavelength dispersive XRF spectrometry
Woodward rule 269–272

XANES *see* X-ray absorption near edge spectroscopy
X-ray absorption near edge spectroscopy (XANES) 311, 320
X-ray beam 310–311, 313–315, 320
X-ray diffraction (XRD) 307, 310–311
X-ray fluorescence (XRF) 307–308, 310–322
X-ray photon 311–313
XRD *see* X-ray diffraction
XRF *see* X-ray fluorescence

Zeemann splitting factor 127